T0323663

Fundamentals of Wind Farm
Aerodynamic Layout Design

Wind Energy Engineering

Fundamentals of Wind Farm Aerodynamic Layout Design

Farschad Torabi

ELSEVIER

ACADEMIC PRESS
An imprint of Elsevier

Academic Press is an imprint of Elsevier
125 London Wall, London EC2Y 5AS, United Kingdom
525 B Street, Suite 1650, San Diego, CA 92101, United States
50 Hampshire Street, 5th Floor, Cambridge, MA 02139, United States
The Boulevard, Langford Lane, Kidlington, Oxford OX5 1GB, United Kingdom

Notices

Knowledge and best practice in this field are constantly changing. As new research and
experience broaden our understanding, changes in research methods, professional practices, or
medical treatment may become necessary.

Practitioners and researchers must always rely on their own experience and knowledge in
evaluating and using any information, methods, compounds, or experiments described herein. In
using such information or methods they should be mindful of their own safety and the safety of
others, including parties for whom they have a professional responsibility.

To the fullest extent of the law, neither the Publisher nor the authors, contributors, or editors,
assume any liability for any injury and/or damage to persons or property as a matter of products
liability, negligence or otherwise, or from any use or operation of any methods, products,
instructions, or ideas contained in the material herein.

Library of Congress Cataloging-in-Publication Data
A catalog record for this book is available from the Library of Congress

British Library Cataloguing-in-Publication Data
A catalogue record for this book is available from the British Library

ISBN: 978-0-12-823016-9

For information on all Academic Press publications
visit our website at https://www.elsevier.com/books-and-journals

Publisher: Joseph P. Hayton
Acquisitions Editor: Lisa Reading
Editorial Project Manager: Aleksandra Packowska
Production Project Manager: Nirmala Arumugam
Designer: Victoria Pearson

Typeset by VTeX

Working together
to grow libraries in
developing countries

www.elsevier.com • www.bookaid.org

To my wife Sheeva and my son Arya
who provided a lovely environment for me during the
pandemic and turned it into a great opportunity
and success!

Contents

5. Analytical model based on similarity solution

6. Numerical simulation of a wind turbine

D.   Sample wind farms

E.   Optimization methods
Mehrzad Alizadeh

F.   Implementing optimization methods in C++

Preface

Humans require energy for their lives. Without sustainable energy, the life of humans is not guaranteed. The new generations and population growth add more importance to energy generation. However, any development in energy generation must meet the standards of sustainability. Energy generation from fossil fuels causes global warming and will gradually endanger human lives if it is not stopped. For this reason, people are trying hard to find a carbon-free solution for energy generation.

Wind energy is one of the oldest sources of energy harnessed by human beings. Ancient people have used the power of wind for driving their sailing ships and boats. In addition to human beings, wind plays a vital role in the life of all living species. The fact that wind does not contaminate the air and is inexhaustible makes wind energy quite attractive for energy generation. In addition to the above characteristics, wind energy can be found almost anywhere. These characteristics have made wind energy very attractive for the scientists and industrial sectors.

Installing wind turbines in a specific land requires lots of initial planning. The site selection, installing process, power regulation and transmission, civil works, and many other issues must be checked beforehand. Each of the mentioned problems requires special attention and calculations. One of the involved issues is the layout design of the farm. Since wind turbines affect the wind speed, they have an aerodynamic effect on their downstream neighbors. According to this effect, the downstream turbines operate with a lower power. Thus, if a proper design is not chosen, the wind turbines will not work in their optimum state. The result is a substantial reduction in net annual energy production. The energy reduction may make the wind farm design economically impractical. Therefore, we have to obtain a practically economic layout for the farm.

The present book focuses on the aerodynamic optimization of wind farms. Although this is a vast topic and the physical phenomena are quite complex, we tried to lead the reader step by step through the field's concepts. It starts with simple issues and gradually explains the details of the layout design process. The present book assumes that the readers may be new to the field. Therefore, the basics of each topic are explained by different examples. Thus, the book is suitable for all people if they are either new to the topic or experienced. Different practical examples support all the chapters, and also some codes are given as appendixes to the book. Therefore, industrial wind experts can benefit from the prepared material.

This book consists of eight chapters. Chapter 1 talks about wind energy itself, the role of wind energy in human life, different types of wind turbines, and the wind turbine components. Chapter 2 deals with the wind, what the origin of wind is, and its statistical studies. Chapter 3 focuses on the basics of aerodynamics. This chapter is fundamental, and I tried to collect only the basics of aerodynamics necessary for supporting the next chapters. Obviously, many different topics should be studied to have a general overview of the aerodynamics. In this chapter, the blade element momentum method is introduced, which can be used for the simulation of a wind turbine. Chapter 4 summarizes some conventional methods that are used for the simulation of the wake of wind turbines. These models are widely used by researchers all over the world, and some of them are implemented in applications. Chapter 5 explains the wake model developed by the author. This model has been used and verified several times. The results show a good agreement with the numerical models. Chapter 6 gives the guidelines that are necessary for the simulation of a single wind turbine using CFD codes. Different scenarios that lead to the improvement of the model are discussed in this chapter. Chapter 7 extends the model developed in Chapter 6 to the simulation of a wind farm. It is shown that the simulation of a wind farm requires some considerations. The main points are discussed in detail and explained through examples. Finally, Chapter 8 deals with the optimization of wind farm, using different optimization algorithms and the contents developed in all the previous chapters. In this view, Chapter 8 is the main goal of the present book, but the knowledge of the previous chapters are necessary. It is tried to make the chapters be self-contained. In fact, all the materials that are required for the last chapter, are discussed in other ones. However, since wind turbine aerodynamic is a complex phenomenon, readers are encouraged to refer to other references.

In addition to the chapters, the book consists of 7 appendixes. Appendix A gives an introduction to the ancient Persian windmills that are considered the oldest operational windmills in the world. Appendix B collects some practical airfoils that are used in the wind turbine industry. The properties of the airfoils are given in figures and equations. Appendix C represents some operational wind turbines and gives their properties as figures and tables. To support the wind farm simulations, Appendix D presents some wind farms and their characteristics. Appendix E, written by my student Mr. Mehrzad Alizadeh, talks about different optimization methods since they are frequently used in wind farm optimization processes. To support the context of the book, the genetic algorithm, and the particle swarm optimization methods are implemented in C++ language and are presented in Appendix F. Finally, the implementation of the blade element momentum method is given in Appendix G. The data collected in these appendixes are used in the examples, problems, and explanations of different chapters.

I hope that the present book is helpful for all the people looking for a cleaner environment, and I hope that these steps will make the world a brighter place for all human beings and other living species.

Farschad Torabi
Tehran, Iran
June 2021

Chapter 1

Wind energy

1.1 History of wind turbines

Human has harnessed and used wind energy for a very long time. It is one of the most ancient energy sources for the human being. The first known usage of wind energy dates back to when Egyptians used sailboats and ships on the Nile river over 5000 years ago. Other similar boats and vessels can be seen in history all over the world, including European, Persian, Chinese, and other ancient historically great nations.

Other than sailing purposes, the first known windmills were incorporated in Persia (Iran), where people used paddle-type windmills for grinding flour from cereals such as wheat and barley (Manwell et al., 2010). The windmill construction dates back to 1000 years ago, and the whole structure was made of clay and wood. The city where the vertical windmills are installed is called Nashtifan, a place near the border of Iran and Afghanistan with a good potential for wind. Details of the ancient windmills are explained in more detail in Appendix A.

Centuries later, windmills began to appear in Europe but in different shapes. Merchants and crusaders brought the technology from Persia to Europe, and the Dutch developed the first windpumps to drain the lakes and marshes in the Rhine River Delta. Although the concept of windmills was obtained from Persia, in contrast to the vertical-axis designs, the early European windmills were horizontal-axis type where the power shaft is placed horizontally and the obtained power is transmitted by mechanical mechanisms such as gears.

During the 18th and 19th centuries, the technology of windmills reached a high level. Many windmills and windpumps were installed all over Europe, especially in the northern European countries. In the Netherlands, about 6000 to 8000 windmills and windpumps were installed, and it is estimated that European countries installed more than 23000 windmills. During these centuries, technology was developed and enhanced through trial and error, and manufacturing was state-of-the-art. During the same period and later, American colonists made thousands of windmills, windpumps, and wind sawmills all over the United States. In the late 1800s and early 1900s, the first wind turbines were introduced by coupling an electrical generator to the same windmills.

With the appearance of thermal power plants and central grids, the role of wind turbines declined, and the world started using grid power. However, after the oil crisis in the 1970s, the interest in developing alternative sources of energy, including wind energy, started. From then on, the research on enhancing

Fundamentals of Wind Farm Aerodynamic Layout Design. https://doi.org/10.1016/B978-0-12-823016-9.00007-3

wind turbines increased. In many countries, such as the United States, Germany, Denmark, and many others, universities and industries started producing more efficient wind turbines.

1.2 Pros and cons of wind energy

Like any system or source of energy, wind energy has its own pros and cons. The main advantages of wind energy are:

Clean or Green Energy It is pretty evident that wind is a part of nature, and it blows due to the thermal effects on the earth. Hence, its produces no emissions, no heat, and no other source of pollution. Consequently, wind energy is categorized as a clean or green energy source. In comparison with common thermal power plants, wind turbines cause no acid rain, smog, or greenhouse gasses.

The Energy Amount There is much more wind power than humans need even if we consider the energy consumption growth. Wind power is abundant and inexhaustible. If we harness only a small percentage of the global wind energy, it would be more than enough for all our needs.

Renewability and Sustainability Wind blows due to the thermal energy obtained from the sun. This means that as long as the sun and earth exist, wind will blow. For this reason, wind energy is considered as one of the renewable sources which we can trust.

A Domestic Source Wind energy is a domestic source, and we can install wind turbines almost in any country. We can use local grids to distribute the harnessed energy with minimal loss.

Requires Small Land Installing a wind turbine requires a minimal area for constructing the foundation. In contrast to solar energy, wind turbines can be installed in currently active fields without changing the land into a power plant. For installing solar panels, the land should be dedicated to the plant, and you cannot use the land for agriculture anymore since the panels cover the ground. However, wind turbines require only a small foundation, and the required space between the turbines should be about hundreds of meters. The difference between solar and wind energy is that solar energy is harnessed over a specific area while the wind energy is harnessed in the vertical space.

The farmers may also welcome the installation of wind turbines on their lands because they rent small pieces of their land without significantly losing their usual farming productions.

Creating Job Installing a wind turbine and harnessing wind energy creates a lot of job opportunities. Since wind energy is a multidisciplinary field, many different experts should accompany installing and maintaining a wind turbine. The installation requires civil engineers to construct the foundation, mechanical engineers for making the turbines, electrical engineers for connecting the power to the grid. Also, it requires lots of technicians for wiring, supporting, maintaining, monitoring, and other related activities.

Low Maintenance Comparing with other power plants such as nuclear, thermal, hydropower, etc., wind turbines require less maintenance. A wind turbine may work without failure for a very long period of time. The installed windmills and windpumps in Europe still work without failure after 50 years or even more. Also, the ancient Persian windmills are operational today after hundreds of years.

Cost Effective One of the most important factors or benefits of wind energy is that it is quite cost-effective (Muyeen et al., 2008). In particular, in places where wind always blows, the cost of wind power can range between two and six cents per kilowatt-hour. It should be noted that a wind turbine is designed to operate for over 20 years. In this period, it does not consume any fuel, and, as it was stated before, it is a low-maintenance device. These factors contribute to the low price of the produced energy.

So far, the main advantages of wind energy and wind turbines are discussed. But, it has its own disadvantages, too. The following are some of the main disadvantages of wind energy:

Noise Wind turbines generate noise that in some cases causes problems for people. However, in some designs, the blade design is such that the generated noise is not in the hearing range. These turbines, although apparently silent, produce noise for other living animals. In general, noise is one of the pollution sources of wind turbines.

Aesthetic Pollution Wind turbines are made big. Therefore, they make aesthetic pollution when they are installed and made operational. Especially, offshore wind turbines may make the beaches uncomfortable due to the lack of aesthetics.

Damaging Local Wildlife The rotation of turbine blades may harm the flying birds. Moreover, the noise of the turbine may have some undesirable influence on local wildlife.

Recycling The wind turbine components are made of metals, composites, nonmetallic parts, etc. Therefore, as any other object, they have many parts that, if they cannot be recycled, will produce waste which is harmful for the environment.

Generation in Remote Lands As we mentioned before, using wind turbines as local generators may reduce the wiring and power loss compared to the grid loss. However, in many cases, the local wind is not very strong. In other words, in many cases, the wind is powerful in places where electricity is not required. For example, offshore wind has more power than onshore wind (Ng and Ran, 2016). Hence, wind power plants should be installed where electricity is not needed.

Being Discontinuous Like any other natural source of energy, the generated power is discontinuous. For example, wind may stop blowing, or its velocity and direction may vary. Hence, a sophisticated design should be done

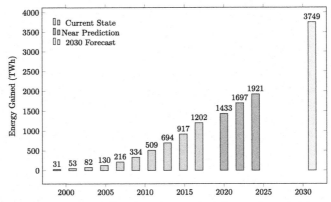

FIGURE 1.1   Current state and forecasting of energy gained from onshore wind farms (data taken from (IEA, 2019a)).

to obtain sustainable energy. In many cases, the generated power should be stored in large-scale storage devices. Or it should be coupled with other sources to obtain a reliable and sustainable source of energy.

Variation of Power   The wind power is proportional to the cube of its velocity (Burton et al., 2001). Thus, if the speed of wind becomes two times faster, the power becomes eight times larger. Consequently, the turbine design should be such that it operates in low-speed winds and can withstand high-speed wind power. This creates many design and mechanical problems, which in turn adds more cost to the project.

1.3   Trend of wind energy in the world

The energy obtained from wind has a growing trend both for onshore and offshore farms. It means that the energy cost will decrease in the future, and wind energy will become more economical.

Most installed power plants are onshore since their construction is more economical, although the available onshore power is less than offshore. The data in Fig. 1.1 shows that the energy gained from onshore farms has an exponential trend since 2000. However, to reach the goal in 2030, more effort is needed.

Constructing an offshore wind farm is much more challenging than an onshore one, despite the fact that offshore wind energy is more powerful. According to IEA report (IEA, 2019a), new offshore wind projects are achieving 40–50% capacity factor. This large factor becomes possible by inventing large wind turbines that can capture the most available resources. Such capacity factor is now available for gas-fired power plants and is more than for onshore wind turbines, being about double those of solar panels. These advantages make offshore wind turbines quite attractive.

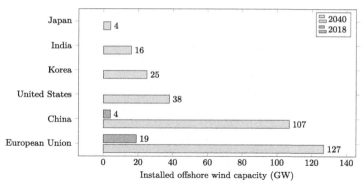

FIGURE 1.2 Current and forecast capacity of offshore wind capacity (data taken from (IEA, 2019b)).

Europe is the leader in installing offshore plants. As the data in Fig. 1.2 indicates, offshore wind farms are of great interest in many countries and regions. Until 2040, Europe will still be the leader in harnessing offshore energy, followed by China. Meanwhile, other countries, such as the United States, Korea, India, and Japan, have some plans for accessing offshore wind energy.

1.4 Wind turbine types

In general, wind turbines are categorized as *horizontal-axis wind turbines,* or HAWTs, and *vertical-axis wind turbines,* or VAWTs (Johnson, 1985). The difference between the two is related to the position of the rotating shaft and the direction of the wind. If the rotating shaft of the turbine is aligned with wind direction, it is called an HAWT, and if the rotating shaft of the turbine is perpendicular to the wind direction, it is called a VAWT. By this definition, for being a VAWT, the rotating shaft should not necessarily be installed perpendicular to the ground. The rotating shaft may be installed parallel to the ground, but it is still perpendicular to the wind direction.

1.4.1 Horizontal-axis wind turbine

Such turbines are the most commonly used turbines. They may have one, two, or three blades that convert wind energy into mechanical energy in the shape of a rotating shaft, know as a *low-speed shaft*. The mechanical energy of the low-speed shaft is transmitted to a generator via a gearbox to increase the shaft velocity. The gearbox increases the shaft angular velocity to make it suitable for generating power by the generator. In some cases, the generator is designed so that the low-speed shaft's rotational speed would be enough for making electricity. Such wind generators, made using a permanent magnet, are called direct-drive generators. All the components, including the blades, low-speed

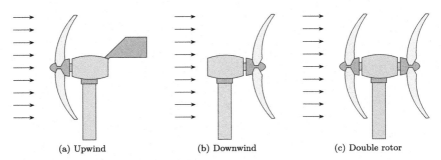

(a) Upwind　　　　　　　(b) Downwind　　　　　　(c) Double rotor

FIGURE 1.3 Different configuration of HAWT.

shaft, gearbox, high-speed shaft, and the generator are installed on top of a tower in a case that is called *nacelle*.

Current commercial wind turbines used for the generation of electrical power are three-bladed in which the blade size may be up to 100 meters and more. The technology prefers to increase the size of the blades as large as possible so that more power can be obtained with a single turbine. Wind turbines with 100-meter blades can produce up to 10 MW power and have been used in offshore farms. However, the manufacturers have plans to build giant turbines with blades up to 120 meters to produce over 15 MW power. The three-bladed turbines also have lower torque ripples which makes them superior over one- and two-bladed designs.

In order to get the most benefit from wind energy, HAWT should point toward the wind. Hence, proper mechanisms should be installed on HAWTs to detect and rotate the turbine. In small turbines, a simple vane is enough, but in larger ones necessary sensors and yaw systems should be included. Based on this, HAWTs are divided into upwind, downwind, and double rotor configurations (Wagner and Mathur, 2013). Upwind type turbines rotate towards the wind while their blades hit the undisturbed wind first. Fig. 1.3a shows such a configuration, and, as it can be seen, the wind passes the nacelle after passing the blades. In downwind type turbines, wind first passes over the nacelle and only then the blades as shown in Fig. 1.3b. In a double rotor design, both sides have a rotor, as shown in Fig. 1.3c.

The upwind type turbine is more common among the above three configurations because undisturbed wind first touches the blades. It means that the wind is less turbulent and exerts less dynamic stress on the blades. In downwind turbines, the wind touches the blades after passing over the nacelle. Consequently, the flow becomes more turbulent. However, as can be seen in Fig. 1.3, downwind configurations have the advantage that the blades do not hit the tower due to their deflection. In upwind systems, this fact should be considered in designing the whole setup.

Double rotor type turbines are not very common and can be seen in some small instances. Having two counterrotating blades makes the configuration

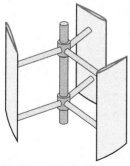

(a) The main shaft is perpendicular to the ground

(b) The main shaft is parallel to the
ground

FIGURE 1.4 Illustration of VAWT.

more complicated than upwind or downwind ones. However, the counter-
torques generated from both rotors simultaneously cancel the horizontal forces
and may lead to a more stable turbine with lower vibrations. At the same time,
having two rotors means that the turbine generates more power.

1.4.2 Vertical-axis wind turbine

In contrast to an HAWT, the rotating shaft of a VAWT is perpendicular to the
wind direction (Mathew, 2006). Being perpendicular to the wind direction does
not mean that the shaft is perpendicular to the ground. The rotating shaft can be
installed horizontally, but its rotation is still perpendicular to the wind direction.
The two different configurations are illustrated in Fig. 1.4.

The vertical arrangement means that such turbines are not sensitive to the
wind direction. Therefore, they do not require any yawing system. In addition,
VAWTs have lower rotational speed and hence produce less noise. In general,
VAWTs are less costly than HAWTs, but they are comparably less economical.
The reason is that VAWTs have lower efficiency than HAWTs. For this reason,
in the construction of large wind farms, HAWTs are used instead of VAWTs.

Since VAWTs do not require to be rotated toward the wind, they are more
efficient in gusty winds because they are already facing the gust. Another advantage of VAWTs over HAWTs is their larger surface area. In comparison to
HAWTs, the VAWTs have a greater surface area for energy capture. Thus, their
energy input can be many times greater compared to HAWTs. The above characteristics result in the fact that VAWTs can be installed in more locations, for
example, on roofs, along highways, in parking lots.

For instance, installing a VAWT on the roof of a building will cause the
turbine to experience a higher wind speed because the building pushes the wind
to pass over its roof. Thus the wind speed on a building roof is higher than the
far-field. This effect can be seen in Fig. 1.5.

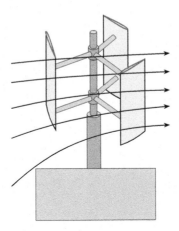

FIGURE 1.5 Effect of buildings on wind speed.

Since VAWTs require fewer components, such as yaw systems, they are always less complicated than HAWTs and thus are less expensive. In addition, they can be made to produce from milliwatts to megawatts of power. One of the advantages of VAWTs is that the generator can be installed on the ground, making it relatively simpler for maintenance, meaning that their maintenance is also less expensive.

Another advantage of VAWTs is that their blades rotate slowly, which is very important in different ways. First of all, VAWTs are commonly considered less dangerous for birds, other animals, and humans. Secondly, the fact that they produce less noise is significant from the environmental point of view since the noise of wind turbines may harm the wildlife and should be avoided as far as possible. Finally, the low-speed rotation also makes them produce less mechanical wearing.

In addition to their benefits, VAWTs have some drawbacks that have made them unsuitable for large wind farms. First of all, they are not as efficient as HAWTs. HAWTs have a capacity coefficient as large as 50%, while this value is significantly less for VAWTs. Their lower capacity factor is because at each time instant, only one blade is at its best performance, and the other blades are not facing the wind; hence are far from their optimal position. Moreover, the blade rotation causes additional drag force, which in turn reduces the output power.

Another disadvantage of VAWTs is that they usually require a prerotation or push for starting to work. From the scientific point of view, it is because their efficiency at zero rotational speed is zero. It means that, whatever the speed of the wind, the output power, which is translated into shaft rotation, is zero, or in simpler words, it does not rotate.

There are many different types of VAWTs, and every year new designs emerge. Some of the most famous are as follows:

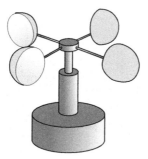

FIGURE 1.6 Cup anemometer.

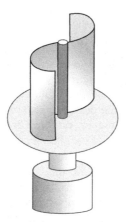

FIGURE 1.7 Savonius VAWT.

Cup anemometer The very first anemometers, such as that shown in Fig. 1.6, are considered vertical, in which the generated power is about a few milliwatts. The generated output is calibrated with wind speed; hence we can use them for speed measurements. The advantage of this type is that it is not sensitive to wind direction and rotates even at a very small wind speed.

Savonius While cup anemometer generators are not used for power generation, Savonius types are used for this purpose. The turbine is named after its Finnish engineer inventor Sigurd Johannes Savonius, who introduced it in 1922. In this turbine, shown in Fig. 1.7, some curly blades are installed on a vertical axis in such a way that when the wind blows, it exerts significant force in the concave surface and less on the convex ones. The imbalanced forces cause the turbine to rotate.

As it is clear, the wind direction does not contribute to the rotation direction, but the exerted power on the vertical shaft oscillates as the blades rotate and make different angles with the wind direction.

FIGURE 1.8 Helical Savonius VAWT.

In a Savonius type turbine, the number of blades depends on the size and design of the turbine. Blade numbers can be two for small turbines, and three or more for larger ones. One of the benefits of the Savonius turbines is that they are self-starting devices, meaning that if they are not moving and the wind blows, without any prerotation they can begin to rotate.

Helical Savonius The problem of oscillating power explained in Savonius turbines can be minimized by making helical blades as shown in Fig. 1.8. In helical Savonius turbines, the direction of the wind and the position of blades do no change as the vertical shaft rotates. The result is a more smooth rotation which, in turn, results in a more uniform power generation. This effect significantly reduces the mechanical fatigue of the main shaft.

Lift-Type VAWT The helical Savonius type turbines also use the drag force for rotation. It is obvious that the lift force can be much greater in airfoils compared to the drag force. For this reason, the lift-type VAWTs are fabricated and developed. In such turbines, instead of using cups, some airfoils are attached to the vertical axis as shown in Fig. 1.9. When the wind blows, the unbalanced lift on each airfoil makes a rotational torque on the central axis and causes the rotor to rotate.

Darrieus In a particular design, the blades of the lift-type turbine are curved and make a shape like that shown in Fig. 1.10. This type of turbine is called Darrieus after its inventor Georges Jean Marie Darrieus who was a French aeronautical engineer and presented this design in 1926.

Darrieus turbines are superior to those of Savonius type in generating power and efficiency. However, one of their major problems is that they are not self-starting devices. Even if the wind speed is quite suitable for power generation, the turbine does not rotate if it is initially stationary. Another drawback of Darrieus generators is their instability in high-speed winds.

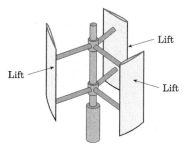

FIGURE 1.9 VAWT making use of the lift coefficient.

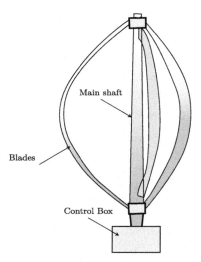

FIGURE 1.10 Darrieus VAWT.

Combined Darrieus–Savonius The self-starting problem of Darrieus turbines can be overcome by combining it with a Savonius turbine. The configuration is shown in Fig. 1.11 where a Savonius turbine is located on the main shaft of a Darrieus generator. This configuration helps the Darrieus rotor rotate because Savonius-type turbines are self-starting, and after the initial rotation starts, Darrieus blades start to generate the main torque.

1.5 Wind turbine components

A wind turbine consists of different parts, including the rotor and generator. But there are many different components for producing electricity, many of which are available in both horizontal- and vertical-axis designs. But some components are used either in horizontal- or vertical-axis turbines. Here we discuss the main components used in HAWTs, and for VAWTs, we discuss the differences.

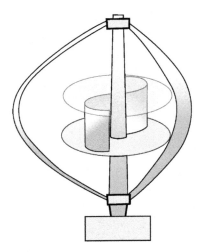

FIGURE 1.11 Combined Darrieus and Savonius turbines.

1.5.1 Blades, hub, and low-speed shaft

In small-scale HAWTs, the blades are usually made from steel, and for larger turbines, they are made of fiberglass materials. The design of blades is one of the most challenging steps of turbine design because complex aerodynamic phenomena exist during a turbine's operation.

Fig. 1.12 shows a typical blade used in HAWTs. As shown in the figure, the blade has a root that allows it to be connected to a hub. Usually, three blades are connected to a hub to construct a rotor. It should be noted that one-, two-, and more-bladed turbines are available on the market. The higher the number of blades, the more power can be generated, but it increases the aerodynamic wake interaction, resulting in vibration and mechanical fatigue. Hence the blade number should be optimized for each individual turbine. In recent years, three-bladed turbines are considered to be the best choice.

A cross-section of any blade or wing is called an *airfoil*. The construction of a blade starts by defining its cross-sections at the different radii. Usually, different airfoils are used to construct a single blade. It means that, at each section of a blade, a different airfoil may be used. As shown in Fig. 1.12, the blade is divided into three main sections, namely the root, the main span or primer, and the tip. The airfoils that are suitable for root sections must be thick since they bear the mechanical load. The main span generates the majority of the blade torque. Thus it requires airfoils that create more lift than drag. Finally, in the tip section, the airfoils should appropriately act against the vortexes known as *tip vortexes*. In general, the designer is free to use as many different airfoils as is required at each of the three sections. For small-scale turbines, usually, a single profile is chosen for each section, but for large-scale ones, many different airfoils are used.

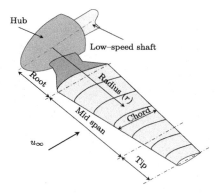

FIGURE 1.12 A typical blade.

At the beginning of the development of advanced wind turbines, the manufacturers started using NACA airfoils in making the blades. But different research labs and groups began to develop airfoils that are more suitable for wind turbines. There are many differences between the aerodynamic physical phenomena of airplane wings and wind turbine blades. In a wind turbine, the wake and tip vortex interaction and the blade rotation cause different phenomena, for which the airplane airfoils are not optimized. Nowadays, there are lots of various airfoil families developed explicitly for the wind industry. Some of the famous families are introduced in detail in Appendix B.

As mentioned before, the blades are connected to a hub in order to make the rotor. The main rotor is connected to a low-speed shaft as shown in Fig. 1.12. Therefore, the rotational speed of the shaft is exactly equal to the rotational speed of the blades. In fluid mechanics, the flow is considered to be compressible if the relative Mach number Ma exceeds 0.3. When the flow becomes compressible, a part of the input energy is dedicated to compressing the fluid. In other words, in a compressible flow, a certain amount of energy is lost. Therefore, to avoid energy losses due to compressibility, the rotational speed of blades is chosen such that the Mach number at the tip of the blades becomes less than 0.3. To have a better understanding, although the speed of sound is a function of temperature and pressure, normally, it is about $330\,\mathrm{m\,s^{-1}}$. Therefore, the rotational speed of the blades is chosen such that the tip speed of the blades becomes less than $100\,\mathrm{m\,s^{-1}}$.

Example 1. For a wind turbine with diameter $D = 70\,\mathrm{m}$, what is the maximum rotational speed?

Answer. As discussed before, the maximum rotational speed should be selected such that the tipspeed becomes less than $100\,\mathrm{m\,s^{-1}}$. For a $70\,\mathrm{m}$ rotor diameter, the tipspeed is

$$v_{\text{tip}} = r\omega = \frac{D}{2}\omega.$$

TABLE 1.1 The rotational speed of generator's shaft in rpm.

Poles	at 60 Hz	at 50 Hz
2	3600	3000
4	1800	1500
6	1200	1000
8	900	750
10	720	600
12	600	500
14	514.3	428.6
16	450	375
18	400	333.33
20	360	300

Applying the available data, we have

$$v_{\text{tip}} = \frac{70}{2}\omega = 100.$$

It results in $\omega = 2.857\,\text{rad}\,\text{s}^{-1}$, or $\omega = 27.28\,\text{rpm}$. □

1.5.2 Generator

The frequency of the local grid is standardized as either 50 Hz with 220–240 V or 50 Hz with 100–127 V. Neither of these is superior to the other. However, from a generation point of view, we have to take care of the grid frequency because the power output should be synchronized with it.

Usually, a generator is made of one, two, or more pairs of magnet poles known as the stator. Inside the stator, a rotating coil known as a rotor converts the input mechanical force to electricity according to the first law of Faraday. If the generator has only one pair of magnetic poles, then the generated electricity has the same frequency as the rotating shaft. If it contains two pairs, the output frequency will be twice the frequency of the shaft. The higher the number of poles, the lower the shaft rotational speed required. Accordingly, regarding the grid frequency, the rotational speed of the generator shaft is summarized in Table 1.1.

Large wind turbines are made such that the generated power is three-phase AC with 690 V output. The output voltage is then increased to about 10 to 30 kV for transmitting to remote places. However, in small-scale wind turbines, the output may be very different depending on the application. For example, some turbines generate even DC power, and then an inverter is required to convert the power to a suitable frequency and voltage.

In practice, the generators can produce more power than their nominal capacity if kept in good temperature conditions. For small-scale generators, air-cooled heat sinks are enough for keeping them in a reasonable range. But for large ones, water-cooled systems are designed to ensure that the generator does not exceed the temperature limits. If the generator is kept cool in a normal range, it can produce about 10% more power than its nominal capacity.

As the wind speed increases, the wind power increases, too. Hence, the shaft power increases if proper modifications are not considered. The higher shaft power results in higher generator input load, which results in more power generation, heat build-up, and consequently, damage to the generator. In normal turbines, the shaft power is controlled using different scenarios, including (a) proper design of the blades such that it does not acquire power even if the wind power is increased, and (b) reducing the shaft power by rotating the blades to capture less power. But in some turbines, the manufacturers may use two different generators in a single wind turbine for low and high winds. In this design, one generator is optimized for operating in low-speed winds, and the other is used in a higher-speed wind. Other manufacturers have used variable-pole generators in which the generators can remove or add magnetic poles. Therefore, their output power and frequency can be adjusted according to the shaft rotational speed. These designs perform better, but also cost more. Hence, their utilization requires economic calculations.

1.5.3 Gearbox and high-speed shaft

Normally, the low-speed shaft speed is not enough for generating a regulated power since, according to Table 1.1, there should be too many pole pairs. Although such a generator exists and some turbines are equipped with generators that can produce regulated power with very low input rotational speed, the majority of designs are such that the rotational speed of the hub is not enough for producing 50 or 60 Hz power. For these turbines, the low-speed shaft's rotational speed should be increased by means of a gearbox. In a gearbox, the input power is carried by the low-speed shaft, and the output is transmitted by the high-speed shaft. The design of the gearbox is such that the rotational speed of the high-speed shaft be suitable for generating a regulated power with a frequency of 50 or 60 Hz.

Single-ratio gearboxes are used to increase the shaft speed. It means that they do not shift gears to reduce or increase the gear ratio. If the required gear ratio is very high, a multistage gearbox is used to reduce mechanical failure and wearing. In practical applications, the ratio of each gear stage should be less than 6. It means that the diameter ratio of successive gears is chosen less than 6. So, using each stage can increase the rotational speed by six times at most. If a larger increase is desired, we can use a two-stage gearbox. It means that the high-speed shaft could become $6^2 = 36$ times faster. If this speed is still not enough, a three-stage gearbox can be used in which the ratio of the speed of output to input shafts would be $6^3 = 216$. In practice, this increase is much more than enough, meaning that for wind turbines, the gearbox does not have more than three stages.

Some examples of wind turbines are collected in Appendix C. As it can be seen, there are some direct-drive turbines in which the low-speed shaft is directly connected to the generator, including Enercon E-30 (Section C.8) and Enercon E-40/6.44 (Section C.13). These turbines do not have any gearbox. Also, there are one-, two-, and three-stage gearboxes. As the size of the blades increases, the gearboxes must have a greater gear ratio.

A study of the turbines reveals that the manufacturers use different types of gearboxes. Spur gears are common, but other types, including planetary, helical, and other commonly used types, are also used. Some turbines make use of different technologies. For example, they may use spur gear in one stage and planetary types for the other two stages. Many combinations can be seen in different systems.

1.5.4 Tower

In HAWTs, the whole plant is elevated over a tower. As the height of the wind turbines becomes larger, the towers become taller and taller. For different turbine sizes and types, different towers are used. The more common types are as follows:

Guyed Pole This is the most simple type which is useful for small turbines and usually is good for small-scale turbines with power up to tenths of kilowatts. The tower consists of a simple pole that is supported by guy lines for stability. The pole itself supports the weight of the turbine, but it cannot withstand the lateral forces. Hence, the guy lines support the pole to make it stable at its position.

Lattice or Truss For larger turbines, lattice-type towers are more common, where the labor cost is not too large. Indeed, for making the truss, lots of welding is required, which adds to the manufacturing cost. In contrast, the material cost is much lower than the tubular towers since less material is used for constructing the tower. According to aesthetic reasons, lattice type turbines are almost never produced for large wind turbines.

Tubular Steel The most commonly used tower has tubular steel design. In such a turbine, more material is used for constructing the tower; however, minimum labor cost is required for hinging the tower parts and the whole construction requires less time. For places where the labor cost adds too much to the project, tubular steel towers are quite common.

Concrete Nowadays, the height of turbines is becoming higher and higher. For this reason, for becoming feasible to build, and for economic reasons, the concrete towers are emerging. Making tall towers out of concrete is not a problem for modern turbines.

Hybrid Types Hybrid type towers are made by combining any of the types mentioned above. For example, one may build a lattice tower with a guyed pole on its top. Other combinations may also be possible.

1.5.5 Yaw systems

HAWTs should always face the wind; thus, a yawing system is necessary. In small turbines, incorporating a simple vane is enough since the system is not heavy and the drag force exerted on the vane has enough power to rotate the system toward the wind. In larger systems, however, a sophisticated mechanism should be designed, which includes advanced sensors to detect the wind direction and a proper motor and gear to rotate the turbine.

In designing the yawing system, it should be noticed that the turbine may be upwind or downwind, as shown in Fig. 1.3. One of the advantages of the downwind type is that by designing a proper shape for the nacelle, the turbine automatically rotates toward the wind. But in upwind designs, as explained, a vane should be mounted on top of the nacelle. The difference is obvious in the same figure.

Any yaw system should take care of the continuous rotation in one direction because the generated power of the turbine is transmitted by cables through the tower. Therefore, if a turbine continuously rotates in one direction, the cables twist, and this causes problems. Consequently, the yawing system must be aware of the rotation count, and if it reaches a predefined value, the turbine must be rotated in the other direction.

1.5.6 Anemometer

Detecting and sensing the wind speed is crucial for the secure operation of wind turbines. If the wind speed exceeds some limits, the exerted power becomes too much and may result in bursting or other failures. Therefore, we have to acquire exact information about the wind speed, and if the wind speed exceeds a specific limit (depending on the design of the turbine), the turbines must be stopped. The accuracy of the wind speed sensors is important since the power of wind is proportional to the third power of the velocity. The following example shows the importance of measurement accuracy.

Example 2. An anemometer used in a wind turbine has an accuracy of 10%. Calculate the error of power estimation.

Answer. The power of wind is proportional to the third power of the velocity, i.e.,

$$P \propto u^3.$$

The error associated with the accuracy of the device is calculated as the difference between the measured and exact values,

$$E_P = \frac{(u_m^3 - u_e^3)}{u_e^3} \times 100,$$

where u_m is the measured and u_e is the exact value, while E_P is the associated error. Thus the resulting error is

$$E_P = \frac{((1.1u_e)^3 - u_e^3)}{u_e^3} \times 100 = 33.1\%. \qquad \square$$

The above example shows that for having good accuracy in estimating the wind power, an accurate sensor is crucial.

There are many different technologies for measuring wind speed, including mechanical, hot wire, ultrasonic, and acoustic resonance. The simplest ones are mechanical that are made of a vertical-axis 3-cup sensor. The sensor was invented by the Irish astronomer and physicist Dr. John Thomas Romney Robinson in 1845. The modern 3-cup sensors can measure the wind speed up to 60 mph equal to $97 \, \mathrm{km \, h^{-1}}$. Mechanical sensors are able to measure the wind speed with an error up to 3%, which is in an acceptable range for the wind turbine industry.

Digital anemometers usually have higher accuracy and can measure wind velocity with an accuracy around 1.5% to 2%. Among the various available technologies, hot wire sensors are quite common in measuring wind velocity and direction. Fig. 1.13a shows a typical hot wire in which a current passes through a specific wire that is warmed to a temperature higher than that of the atmosphere. Since the resistance of the metals depends on their temperature, when the flow passes through the wire, its resistance changes. By measuring the current and resistance of the wire, the velocity of the wind can be estimated. Another technology is called an ultrasonic sensor, in which a transducer emits a sound with a specific property and receives it with a receiver (Fig. 1.13b). If the wind speed changes, the receiving time of the wave pulses varies, and it can be calibrated for measuring the speed of the wind. The direction of the wind can also be measured using different transducer pairs. Other digital devices use different technologies for measuring the speed and direction of the wind.

The main advantage of digital sensors is that they are not sensitive to icing, which is a problem for mechanical sensors. In cold weather, icing may cause the mechanical sensors to not operate normally. In addition to icing, solid particles such as dust have a direct effect on the accuracy of the mechanical sensors. Therefore, mechanical sensors should be controlled and maintained with a regular program to make sure that their functionality and accuracy are valid. Digital sensors are not as sensitive to icing and dirt as mechanical sensors. But they should also be regularly checked and cleaned.

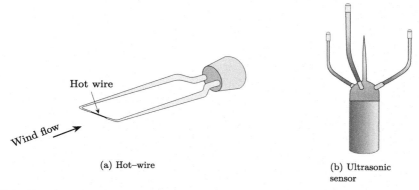

(a) Hot–wire

(b) Ultrasonic sensor

FIGURE 1.13 Different types of digital wind speed sensor.

In contrast, mechanical sensors are more tolerant of bad conditions and show longer service life. Digital sensors encounter some problems against lightning or rain. Specifically, after lightning, hot wire sensors should be monitored to make sure that they are still working. Digital sensors should also be checked for dust and dirt since their accuracy is affected if they are not clean. However, they are less sensitive to dirt compared to the mechanical sensors. Another advantage of mechanical sensors is that they have their own generator and do not require electricity for functioning. In contrast, digital sensors require electricity for functioning. So, if the grid power is off, mechanical sensors can still deliver weather data, whereas digital sensors cannot.

Example 3. Compare the associated error of power estimation between the mechanical and digital sensors.

Answer. For a mechanical sensor, the measurement error is about 3%. Therefore, the associated error for power estimation is

$$E_P = \frac{((1.03u_e)^3 - u_e^3)}{u_e^3} \times 100 = 9.27\%.$$

For a typical digital sensor with an accuracy of 98%, the error is

$$E_P = \frac{((1.02u_e)^3 - u_e^3)}{u_e^3} \times 100 = 6.09\%.$$

It shows that increasing the accuracy by only 1% will result in an accuracy of about 3% for power estimation. In many practical cases, this increase can be negligible; hence a cheaper sensor is a better choice, and in some practical cases, the more accurate sensor is used. ☐

In addition to wind speed for HAWTs (but not for VAWTs), the direction of the wind is really important because the turbine should face the wind. As men-

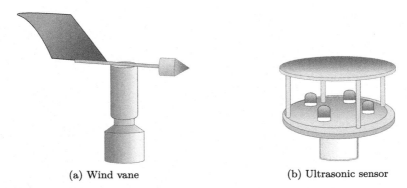

(a) Wind vane   (b) Ultrasonic sensor

FIGURE 1.14 Different types of wind direction sensors.

tioned above, a simple vane is enough for small turbines; however, we have to have exact information about the wind and its direction for large ones. Hence, an anemometer plays an important role in the determination of the wind direction.

A wind vane is one of the most common solutions of direction measurement. As shown in Fig. 1.14a, it looks like a small wind turbine itself. The tail of the sensor is always positioned along the wind direction. Since it is a mechanical instrument, a small wind speed can overcome the friction and make the sensor rotate. Usually, the industrial sensors start rotating at wind speeds above $1.5 \, \mathrm{m \, s^{-1}}$. According to standards, the accuracy of the direction sensing should be less than 5 degrees. The mechanical sensors are able to fulfill such criteria.

Other digital sensors are also available, including ultrasonic sensors that can measure wind speed and direction simultaneously. An example of such a sensor is shown in Fig. 1.14b. The device is able to measure the wind direction with less than 3° of accuracy. Also, the speed measurement can be done with an accuracy of 3%.

Finally, it should be remarked that the accuracy of the sensors is measured under specific conditions. For example, if a sensor has an accuracy of 3% in speed sensing, you need to check the technical data of that sensor to make sure in what range of wind speed you can trust that accuracy. For some sensors, the technical data are given in different ranges. For example, you may see 3% accuracy for wind speed up to $35 \, \mathrm{m \, s^{-1}}$ and 5% for higher wind speeds. Consequently, checking the provided technical data is crucial when selecting a sensor for your wind turbine.

1.5.7 Brakes

For many reasons, the rotor must be stopped; therefore, we need to have some brakes. If the wind speed exceeds some limits, the rotational speed of the rotor becomes too large, and the generator may burst. Moreover, we have to stop the rotor; otherwise, it is not possible to check the inside of the nacelle. Usually, there are three types of brakes used in wind turbines, namely (a) mechanical, (b) electrical, and (c) aerodynamic brakes. Each of these break types has its own characteristics and is used according to the needs of the turbine. Among the three, mechanical brakes should always be present, but using an electrical or aerodynamic system depends on the design. Now we present the main descriptions of these systems:

Electrical For moderate wind turbines, electrical brakes are used to slow down the rotor speed. When the turbine is operating in high-speed winds, the brakes can control the rotation of the shaft by converting the mechanical energy into heat. In addition to this operation, electrical brakes are used to slow down the rotor when it is going to be stopped for maintenance.

Aerodynamic For large turbines, electrical brakes are not suitable because the load on the shaft is too much. In these cases, aerodynamic brakes are used instead. These brakes usually are located at the tip of the blades, and by rotating on top of the blades, the aerodynamic force is exerted in the opposite direction to the rotation, causing the rotor to slow down.

Mechanical Whether the turbine uses electrical or aerodynamic brakes, mechanical brakes must be installed in any wind turbine. The previously described brakes can only slow down the rotor speed, and none of them are able to lock the rotor and shaft. When the turbine is going to be repaired, the rotor should be locked. In this case, mechanical brakes should lock the rotor to ensure that no rotation occurs.

Mechanical brakes are the same as those used in vehicles and may be in the shape of drum or disk brakes. Therefore, they should be used after the rotor speed is really low. If the mechanical brakes are used at high speeds, they may be damaged.

In all the medium- to large-scale turbines, two types of brakes are used. Using mechanical brakes is a must, and depending on the size and design, electrical or aerodynamic brakes should also be used. For small-scale turbines, i.e., in the order of one kilowatt, mechanical brakes are enough.

1.5.8 Fluctuation monitoring systems

When the fluctuations of the wind turbine are too much, it becomes a failure mode, and the turbine should be stopped. In turbulent winds, the frequency of fluctuations may become the same as the natural frequency of the turbine. This fact causes the turbine fluctuations to become larger and larger, which gradually leads to turbine crash and breaking down. Consequently, a fluctuation monitoring system is crucial in wind turbines and when the level of fluctuations exceeds the acceptable limits, the braking systems should be activated first to slow down, and, if it is not enough, stop the system.

1.5.9 Lubrication system

Mechanical devices require lubrication for reducing friction and increasing life. Gearbox, generator, yawing system, and other moving parts require lubrication. For the lubrication system, either central or local systems can be used. In central systems, an oil reservoir exists, and oil is pumped to the places where needed. Cooling systems and filtrations are also available for regulating the oil quality and temperature. In local methods, for each component, a different oil reservoir is placed near that component, and oil regulation takes place locally.

Each system has its own pros and cons and utilizing any of the systems depends on the design and the experiences of the manufacturer. But in general, it should be noted that the lubrication characteristic of any synthetic oil strongly depends on temperature and purity. For hot climates, the oil should be cooled down to an acceptable level, while in cold regions, the oil should be warmed. Proper filtration should also be mounted to remove solid particles from oil to make sure that the oil is within a good range of purity.

1.5.10 Foundation

For onshore turbines, a standard concrete foundation is suitable for large turbines, and for small-scale ones, any platform that can withstand the weight and vibration of the turbine is enough. For offshore plants, however, the foundation is a great challenge (Wagner and Mathur, 2013). Depending on the composition of the seabed, depth of the sea, and other conditions, different technologies have been developed and used. Some of the conventional foundations are as follows:

Gravity foundation The simplest is the gravity foundation, where a very heavy platform is constructed on the seabed. The weight of the foundation makes the plant stable. Fig. 1.15a shows such a foundation. This type of platform is usually used in shallow waters where the maximum depth is about 30 meters. Gravity foundations are made of precast concrete.

Monopile foundation Another solution is called monopile, in which a single pile is inserted into the seabed by a hammer and makes a stable base for installing the turbine. Monopiles are used in many other sea structures and are well developed. Such a system for establishing an offshore wind turbine is shown in Fig. 1.15b.

Monopiles are good choices for waters with less than 25 meters depth. The pile itself is about 30 meters and is driven into the sea bed by force. Hence, it makes a good firm basement for a single turbine.

Tripod For deeper waters, for example, up to 35 meters, a tripod as shown in Fig. 1.15c is used. A tripod is installed on three steel piles that are inserted in the seabed. The tripod itself is made of steel which is welded together and makes a good basement for the plant.

Jacket For deeper waters, usually more than 40 meters, a jacket foundation as shown in Fig. 1.15d is used. A jacket is made of many different pieces of

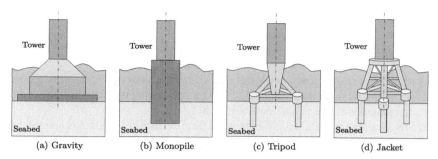

(a) Gravity   (b) Monopile   (c) Tripod   (d) Jacket

FIGURE 1.15 Different common foundations for offshore plants.

steel beams that are welded together and make a truss shape foundation. The weight of each jacket may reach 500 tons!

Many other different systems are developed and used by many manufacturers for different situations. For example, guyed monopiles are used for some turbines in which a monopile is supported by guy-ropes to make a firmer foundation. Other types can also be found in the literature and industrial works.

1.5.11 Other components

A wind turbine is a complex mechanism, in which different fields of engineering are involved. Each part of the turbine requires many different subsystems and parts that are not listed in this text. Controlling devices, sensors, different accessories, voltage regulators, and other facilities are required for a safe, secure, and smooth operation. Different parts can be found in different handbooks and related articles and textbooks.

1.6 Summary

In this chapter, wind energy as well as its advantages and disadvantages are explained. Different wind turbines are introduced, and for an HAWT, different parts are discussed. It is good to mention that VAWTs also have the same components, and only the design differs. For example, the gearbox, generator, wind speed sensors, and many other components also exist in VAWTs. However, some parts may not be needed. For example, wind direction is not important for VAWTs. As another example, VAWTs usually are not mounted on a tower.

1.7 Problems

1. Describe the main difference between an HAWT and a VAWT.
2. The low-speed shaft of a turbine rotates at 30 rpm. How many gear stages are required for driving a 6-pole generator for generating 60 Hz electricity?

3. If the above generator produces 50 Hz output, calculate the number of stages again.
4. If a turbine rotates at 52 rpm, calculate the number of gear stages for driving a 20-pole generator if the generator produces 60 Hz output.
5. Repeat the above example for the case of 50 Hz frequency.
6. An anemometer has an accuracy of 5% for speed sensing. Calculate the error associated with the calculation of power.
7. We are going to build an offshore wind farm. Which foundation is suitable if the depth of water is 15 meters?

References

Burton, Tony, Sharpe, David, Jenkins, Nick, Bossanyi, Ervin, 2001. Wind Energy Handbook, vol. 2. Wiley Online Library.

IEA, 2019a. Offshore wind outlook 2019. https://www.iea.org/reports/offshore-wind-outlook-2019. (Accessed 20 January 2020).

IEA, 2019b. Installed offshore wind capacity, 2018 and 2040, stated policies scenario. https://www.iea.org/data-and-statistics/charts/installed-offshore-wind-capacity-2018-and-2040-stated-policies-scenario. (Accessed 20 January 2020).

Johnson, Gary L., 1985. Wind Energy Systems. Citeseer.

Manwell, James F., McGowan, Jon G., Rogers, Anthony L., 2010. Wind Energy Explained: Theory, Design and Application. John Wiley & Sons.

Mathew, Sathyajith, 2006. Wind Energy: Fundamentals, Resource Analysis and Economics. Springer.

Muyeen, S.M., Tamura, Junji, Murata, Toshiaki, 2008. Stability Augmentation of a Grid-Connected Wind Farm. Springer Science & Business Media.

Ng, Chong, Ran, Li, 2016. Offshore Wind Farms: Technologies, Design and Operation. Woodhead Publishing.

Wagner, H.J., Mathur, J., 2013. Introduction to Wind Energy Systems, Green Energy and Technology.

Chapter 2

Wind properties and power generation

From the thermodynamic point of view, the earth's atmosphere can be considered a thermal engine between two heat sources. The hot source is the heat of the sun, and the cold source is the infinite space. One of the outputs of such an engine is the flowing of air, best known as wind. Although thermal effects play the most important role in the generation of wind, it is not the only involved parameter. Earth's rotation, Coriolis effect, atmospheric composition and gradients of parameters, geographical barriers and obstacles, and many other factors directly affect the wind speed, power, and direction.

It is estimated that about 1–2% of solar heat is converted into wind. This portion is very important for the existence of all the species in the world. Wind plays an important role in the pollination of plants, creation of food, and movement of clouds, resulting in rain and snow, as well as many other vital global activities.

The composition of air differs from place to place due to factors such as humidity and atmospheric pressure (Ng and Ran, 2016). This composition affects some parameters that are very important in power generation issues. As an example of these important factors, one can name the density of the air, which has a direct effect on wind power. Therefore, in any place, we need to gain enough information before constructing any wind power plant.

In addition to wind power, wind direction is also an important parameter when dealing with wind farms. Unfortunately, both the wind power and direction are random parameters thus cannot be estimated or predicted for long periods. Consequently, for designing and estimating wind power generation, we need to work with statistical data.

In the present chapter, these factors are discussed in more detail. To cover the topic, firstly, we describe the wind, its origin, and the way it is studied. Secondly, the statistical investigation of wind is presented. Thirdly, the relation between the wind statistical data and its power is explained. Finally, we discuss the portion of power that can be extracted from wind with respect to its velocity and power.

2.1 Atmospheric properties

The earth's atmosphere is mainly composed of N_2 and O_2 but it also contains some other gases (Johnson, 1985). In addition to these gases, argon, neon, krypton, xenon, helium, and CO_2 exist in the atmosphere. Moreover, different amounts of water vapor also exist in different places. It is more humid near the seas and less humid in arid regions.

For wind industrial engineering purposes, the best model for atmosphere is the famous model of an ideal gas in which the pressure, p, temperature, T, and volume, V, of n kilomoles of a specific gas are related by

$$pV = nRT, \tag{2.1}$$

where the units of pressure, volume, and temperature are respectively expressed in Pa, m^3, and K. In this equation, $R = 8.3145\,\text{kJ}\,\text{kmol}^{-1}\,\text{K}^{-1}$ is the universal gas constant.

From Eq. (2.1) and the definition of density, we can obtain the relation of the density of air, its pressure, and temperature. We know that density is defined as

$$\rho = \frac{M}{V}, \tag{2.2}$$

where M is the mass of one kilomole of gas; thus substituting Eq. (2.2) into Eq. (2.1), we have

$$p = \rho \frac{R}{M} T. \tag{2.3}$$

Although the composition of air differs from place to place, the average molecular weight of atmosphere is assumed to be $M = 28.97$ kilogram per kilomole.

The atmospheric pressure is a function of altitude and can be approximated by the following equation, as long as we are in the troposphere which is no more than 18 km above the sea level:

$$p = p_0 \left(1 - \frac{Lh}{T_0}\right)^{\frac{gM}{RL}}, \tag{2.4}$$

where $p_0 = 101325\,\text{Pa}$ and $T_0 = 288.15\,\text{K}$ are respectively the sea-level standard atmospheric pressure and temperature. Also $g = 9.80665\,\text{m}\,\text{s}^{-1}$ is the earth's surface gravitational acceleration, and finally, $L = 0.0065\,\text{K}\,\text{m}^{-1}$ is the temperature lapse rate.

Example 2.1. For a region near an ocean with temperature about 30°C, calculate the air density.

Answer. For the regions at the free sea level, the atmospheric pressure is 101325 Pa. Thus from Eq. (2.3) we have

$$\rho = \frac{pM}{RT} = \frac{101325 \times 0.02897}{8.3145 \times (273.15 + 30)} = 1.1646 \, \text{kg m}^{-3}.$$

Note that all the units are converted to standard units. □

Example 2.2. For a region with altitude of 1000 m and temperature of about 30°C, recalculate the air density.

Answer. The difference between the present example and Example 2.1 is that the altitude is not the same. Therefore we have to first calculate the pressure at the given altitude from Eq. (2.4):

$$p = p_0 \left(1 - \frac{Lh}{T_0}\right)^{\frac{gM}{RL}} = 101325 \left(1 - \frac{0.0065 \times 1000}{288.15}\right)^{\frac{9.8 \times 0.02897}{8.3145 \times 0.0065}}$$
$$= 89880.0 \, \text{Pa}.$$

Then from Eq. (2.3) we have:

$$\rho = \frac{pM}{RT} = \frac{89880.0 \times 0.02897}{8.3145 \times (273.15 + 30)} = 1.033 \, \text{kg m}^{-3}.$$

Comparing the results of these examples, we see that, due to the altitude change, the density drops by about 12%. □

2.1.1 Global wind direction

In gaseous planets, such as Jupiter or Neptune, the atmospheric wind is mainly based on thermal effects, gravity, and Coriolis effect. On solid planets, such as the Earth, in addition to the above mentioned factors, the atmospheric winds strongly depend on surface geographical features such as mountains, hills, lakes, and rivers.

When the Sun shines on the equator, the air near the Earth becomes warmer; hence its density drops. The low-density air makes a low-pressure region around the equator and rises vertically to the higher altitudes. In contrast to the equator, in polar regions, a reverse condition exists, and the air moves downward. In this region, the cold and thus heavier air makes a high-pressure region. Hence, in these regions (i.e., the high- and low-pressure zones), the wind is not strong, and we may have calm weather for days or weeks. However, the rise and descend of air in certain latitudes cause the wind to form in some latitudes in-between. Between the high- and low-pressure regions, the wind blows from the high- to the low-pressure zones. Since the Earth's radius is large compared to the thickness of its atmosphere, there are different high- and low-pressure zones between the equator and the poles.

In Fig. 2.1 a global pattern of winds is shown. As it is illustrated by the global map, the wind direction is not from the north to the south (or vice versa). The rotation of the Earth, hence the Coriolis effect, causes the slopes in the wind

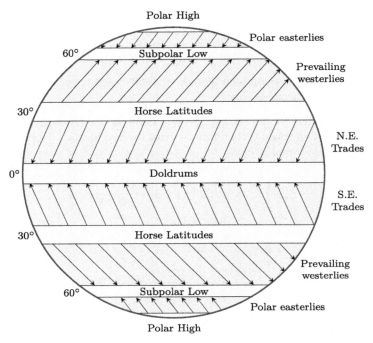

FIGURE 2.1 Global wind direction.

direction. The bands over which no direction is shown are not windy, and most of the time, the flow pattern is upward or downward. Some of the bands have their own names according to historical sailing documents. For example, the *horse latitudes* are the area where the ships used to stop moving because there was no wind to move them. Then if the still conditions were continuing for a long time, the sailors had to drop horses into the sea in order to save fresh water and food.

In contrast to these regions, the bands on which the wind directions are shown usually have strong winds. For instance, the term *trade winds* refers to the area in which wind normally blows in the direction shown in the figure. The wind is also strong enough and provides good conditions for sailing trade ships!

From the physical point of view, there are maps of global winds available in the open literature; however, the effects of geographical features are so important that, in general, we cannot trust global maps. For wind industry purposes, we have to obtain local data for any specific region in which a wind turbine or wind farm is to be installed.

The data for site selection should be available either from ground weather stations or may be provided by remote sensing satellites. Both data are useful for evaluation and processing. The only point is that the global map is not enough.

2.1.2 Turbulence

Turbulence can be described as fluctuations of any fluid flow, which is the result of interaction between inertial forces and viscosity (Manwell et al., 2010). Inertial forces tend to increase the fluctuations while the viscosity dissipates or damps them by converting the fluid energy into heat. Turbulent fluctuations have a relatively fast time-scale compared to the mean flow. In that view, atmospheric wind can be considered as a mean flow with a time-scale of about one to several hours, combined by fluctuations with a time-scale of around 10 minutes. Turbulent flow is one of the most complex phenomena in science, and its behavior is still under debate. However, it can be described by certain physical conservation laws such as continuity and conservation of energy and momentum.

Turbulence fluctuations vary in time and may become stronger or weaker depending on whether the inertial or viscous terms become dominant. Since the time-scale of turbulence is relatively smaller than the time-scale of the wind, we can define the strength of turbulence by means of *turbulence intensity* as follows:

$$I = \frac{s}{\bar{u}}, \tag{2.5}$$

where s is the standard deviation of wind velocity, and $\bar{u}$ is the mean wind velocity. As mentioned above, the nature of turbulence intensity is temporal; hence, turbulence intensity is valid for a small period of time, e.g., from 10 min up to 1 hour.

Turbulence is a property of atmospheric wind that is caused by many reasons, including surface roughness, Coriolis effect, thermal effect, and other atmospheric properties. At lower heights, surface roughness has a great role in making turbulence in addition to the wind velocity. This layer, which is the most important layer for wind turbines, is called the atmospheric boundary layer (ABL). The reason for ABL to be important for wind turbines is because the height of wind turbines is such that they are located inside the ABL. Therefore, any accurate estimation or study of a wind turbine strongly depends on the accurate information about ABL.

2.1.3 Variation of wind with height

When the wind blows, a boundary layer takes place on the Earth's surface. Consequently, the higher we move from the surface, the faster the wind becomes. The variation of wind with height is discussed in many different references and is modeled by different functions (Mathew, 2006). Since the velocity is not uniform with respect to the height, for wind speed measurement purposes, a reference height should be considered so that the speed of wind in all the global locations becomes comparable. According to the standard, the wind speed should be measured at the height of 10 m above the surface. The meteorological data that is available from any weather station are taken at the height of 10 m. Therefore, for having a good estimate of the variation of wind velocity with respect to height, either we need to use the models or install measuring facilities higher. A typical example of such measured data is shown in Table 2.1. It is clear that the sampling rate was 10 min, and the measurements are reported at the heights of 10, 30 and 40 m.

There are many different models for the estimation of wind velocity in different heights other than 10 m (Manwell et al., 2010). The most simple model is called the exponential model, in which it is assumed that the relation between wind speed at any height z is related to a reference point like z_A by the following equation:

$$\frac{u(z)}{u(z_A)} = \left(\frac{z}{z_A}\right)^\alpha. \tag{2.6}$$

In this equation, $u(z)$ is the velocity at height z that is to be calculated and $u(z_A)$ is the velocity recorded or measured at z_A. The exponent α is called Helmann exponent and is considered to be $\alpha = 0.14$ if no data is available. It is quite obvious that α depends on the terrain and other characteristics of each region and is not a constant value. But since in weather stations the velocity of wind is given at only one reference point, we have no choice but to use the same value. It is worth mentioning that this model sometimes is called the 1/7-exponent model because $1/7 = 0.1428$, which is very close to 0.14 or α.

For stations where the wind speed is tabulated in more than one height, the exponent α can be obtained using the values at different heights. Assuming that the wind velocity is given in two different heights, namely $u(z_1)$ at z_1 and $u(z_2)$ at z_2 (such as those given in Table 2.1), by taking the logarithm of both sides of Eq. (2.6), we can write

$$\log\left(\frac{u(z)}{u(z_A)}\right) = \log\left((\frac{z}{z_A})^\alpha\right) = \alpha \log\left(\frac{z}{z_A}\right). \tag{2.7}$$

This equation yields

$$\alpha = \frac{\log\left(\frac{u(z)}{u(z_A)}\right)}{\log\left(\frac{z}{z_A}\right)}. \tag{2.8}$$

Note that α is a unique value for a region; hence we need to calculate it for different rows of the datasheet and finally compute an average value.

Example 2.3. For data given in Table 2.1 at 3:30 AM, calculate α and compare the result with 0.14.

Answer. To calculate α, we need wind speed at two different heights. At 3:30 AM, the following data is available:

$$u(10) = 5.5, \qquad u(30) = 6.2, \qquad u(40) = 6.6.$$

TABLE 2.1 A typical wind speed data at 10, 30 and 40 meters heights. The unit of data is m s^{-1} (SATBA, 2020).

Time	40 m	30 m	10 m	Time	40 m	30 m	10 m	Time	40 m	30 m	10 m	Time	40 m	30 m	10 m
0:00	6.8	6.5	5.8	6:00	10.3	9.5	8.1	12:00	13.3	12.1	11.5	18:00	12.7	11.2	10
0:10	7.1	6.7	6.2	6:10	10.1	9.6	8.3	12:10	11.5	10.2	9.3	18:10	13.4	12	10.9
0:20	8.2	7.8	7.1	6:20	11.5	10.8	9.1	12:20	14	12.5	11.7	18:20	11.4	10	8.8
0:30	9.1	8.7	7.9	6:30	10.9	10.2	9	12:30	12.7	11.6	10.8	18:30	13.7	11.9	10.7
0:40	9.6	9.2	8.4	6:40	10.7	10.2	9	12:40	13.5	12.1	11.4	18:40	12.8	11.3	9.9
0:50	8.2	7.9	7.1	6:50	11.2	10.6	9.5	12:50	13.1	12.1	11.4	18:50	13.8	12.2	10.9
1:00	7.5	7.1	5.9	7:00	11.3	11	9.7	13:00	14	12.4	11.6	19:00	14.1	12.4	11.3
1:10	7	6.6	5.5	7:10	12.7	12.2	10.9	13:10	13	11.6	10.8	19:10	13.4	12.1	10.4
1:20	7.2	7	6.1	7:20	9.4	9	8.1	13:20	13.8	12.2	11.1	19:20	13	11.8	10.2
1:30	8.2	8	7	7:30	9.9	9.4	8.2	13:30	13.3	11.8	10.9	19:30	13.1	11.9	10.5
1:40	6	5.7	4.8	7:40	9	8.6	7.7	13:40	14.5	13.1	12.1	19:40	12.6	11.6	10.1
1:50	7.7	7.7	6.8	7:50	10	9.2	8.5	13:50	14.5	13.2	11.9	19:50	12.4	11.6	10.1
2:00	7.3	7.1	6.4	8:00	9.2	8.7	8	14:00	14.6	13	12.3	20:00	11.2	10.7	9.5
2:10	8	7.9	6.7	8:10	8.5	7.9	7.1	14:10	14.3	12.8	11.8	20:10	11.9	11.1	9.9
2:20	8.2	8.3	7.7	8:20	9.2	8.5	7.5	14:20	13.2	12.3	11.4	20:20	10.4	10	9
2:30	6.2	6.1	5.5	8:30	6.9	6.4	5.7	14:30	13.6	12.1	11.3	20:30	10.7	10.2	9.1
2:40	5.7	5.5	4.7	8:40	7.8	7.2	6.4	14:40	13.4	12	11.2	20:40	11.6	10.5	9.4
2:50	6.6	6.5	5.7	8:50	6	5.5	4.7	14:50	16.5	14.9	13.7	20:50	12.2	11.2	9.5

continued on next page

TABLE 2.1 (continued)

Time	40 m	30 m	10 m	Time	40 m	30 m	10 m	Time	40 m	30 m	10 m	Time	40 m	30 m	10 m
3:00	5.5	5.3	4.8	9:00	7.2	6.9	6.2	15:00	15.8	13.9	13.1	21:00	15.8	14	12.3
3:10	6	5.8	5.5	9:10	7.3	7	6.5	15:10	15.2	13.6	12.6	21:10	17.5	15.5	14
3:20	6.4	6.3	5.7	9:20	7.6	6.9	6.5	15:20	14.6	13.1	12.3	21:20	14.9	13.4	11.7
3:30	6.6	6.2	5.5	9:30	8.6	8.2	7.5	15:30	13.8	12.5	11.7	21:30	13.5	12.4	11
3:40	8	7.6	6.9	9:40	8.4	8.1	7.6	15:40	16.3	14.8	13.6	21:40	12.8	11.9	10.6
3:50	7.9	7.6	6.7	9:50	7.9	7.6	7	15:50	15.3	13.6	12.6	21:50	11.9	11	9.3
4:00	7.6	7.2	6.6	10:00	7.5	7.2	6.9	16:00	16.4	14.8	13.7	22:00	12.9	11.4	9.9
4:10	7.7	7.4	6.8	10:10	9	8.5	7.9	16:10	15.8	14.2	12.9	22:10	11.8	10.9	9.4
4:20	8.2	7.9	7	10:20	8.7	8.2	7.6	16:20	15.7	13.9	12.8	22:20	11.9	10.8	9.2
4:30	8.1	7.3	6.5	10:30	10.7	9.6	8.7	16:30	14.9	12.7	12.2	22:30	14.4	12.8	11.3
4:40	8	7.5	6.4	10:40	11.3	10.4	9.7	16:40	15.9	13.9	12.7	22:40	16.8	14.7	12.9
4:50	7.8	7.4	6.6	10:50	11	9.9	9.2	16:50	15.6	13.6	12.8	22:50	17.2	14.8	13.2
5:00	7.1	6.6	5.7	11:00	13	11.3	10.6	17:00	15.1	13.5	12.4	23:00	17.2	14.7	13.2
5:10	7.4	7.1	6.2	11:10	13.3	11.8	10.6	17:10	14.7	12.9	11.6	23:10	17	14.6	12.8
5:20	6.3	5.7	5	11:20	11	10.1	9.5	17:20	15.7	14	12.5	23:20	15.1	13.1	11.6
5:30	8.6	8	7.1	11:30	12.7	11.5	10.5	17:30	15.4	13.4	12.3	23:30	15	13.1	11.7
5:40	9.3	8.8	7.8	11:40	12.3	11.2	10.1	17:40	14.8	12.9	11.3	23:40	15.6	13.6	11.8
5:50	11.3	10.9	9.5	11:50	14.5	13.1	12	17:50	15.2	13.2	12.1	23:50	13.8	12	10.4

Using data at 10 and 30 meters in Eq. (2.8) yields

$$\alpha = \frac{\log\left(\frac{6.2}{5.5}\right)}{\log\left(\frac{30}{10}\right)} \sim 0.11.$$

Now, if we use data at 10 and 40 meters, we have

$$\alpha = \frac{\log\left(\frac{6.6}{5.5}\right)}{\log\left(\frac{40}{10}\right)} \sim 0.13.$$

Using $\alpha = 0.11$ has about 21% and $\alpha = 0.13$ about 7% difference with $\alpha = 0.14$. $\square$

Although from the above example it seems that α is always around 0.14, a closer observation reveals that it is not true in all cases. For example, if we repeat the same procedure for different rows, we will find different values. Since we are obtaining a unique value for α for a specific region, we have to first calculate α for all the available data; and then we can use their average value.

Example 2.4. For data given in Table 2.1 calculate α for some selected times and obtain the averaged value.

Answer. Following the same procedure as shown in Example 2.3, we can find the following data:

Time	Velocity at 40 m	Velocity at 30 m	Velocity at 10 m	α with $z_1 = 10$ and $z_2 = 40$	α with $z_1 = 10$ and $z_2 = 30$
5	7.4	7.1	6.2	0.1276	0.1233
11	13	11.3	10.6	0.1472	0.0582
16	16.4	14.8	13.7	0.1297	0.0702
22	12.9	11.4	9.9	0.1909	0.1284
	Averaged α			0.145	0.098

The above example shows that the averaged value gives a more reasonable estimate for α. The difference between the two averaged values appears because we used few values for calculating the averaged value. Using more data will give closer values from both heights.

A more complete datasheet can be accessed from (SATBA, 2020). The reader is encouraged to get data and investigate the problem with real values. $\square$

Another model for the prediction of wind velocity at higher heights is called the logarithmic model. The model predicts the wind profile by the following equation:

$$\frac{u(z)}{u^*} = \frac{1}{\kappa}\left(\ln\frac{z}{z_0} + \Psi\right), \tag{2.9}$$

TABLE 2.2 Typical surface roughness lengths (Wikipedia, 2021).

Type of terrain	Roughness length z_0 (m)
Regular large obstacle coverage	1
Cities, forests	0.7
Suburbs, wooded countryside	0.3
Villages, countryside with occasional trees and hedges	0.1
Open flat terrain, grass, few trees and buildings	0.03
Mud flats, snow, no obstacles	0.005
Flat desert, rough sea	0.001
Open sea	0.0002

where u^* is called friction velocity which must be obtained using the surface roughness. The von Karman constant κ can be assumed to be $\kappa = 0.4$, although its exact value is under debate. The parameter Ψ is a function that depends on the stability of the atmosphere. For a neutral atmosphere where wind is strong (which is the case in the wind energy applications), it can be obtained using

$$\Psi = 34.5 f \frac{z}{u^*}. \tag{2.10}$$

In Eq. (2.10), f is the Coriolis parameter of the Earth defined as

$$f = 2\Omega \sin(|L|), \tag{2.11}$$

in which Ω is the angular velocity of the Earth and L is the latitude.

The last parameter to be discussed in Eq. (2.9) is z_0, which is called the surface roughness length. This parameter has units of length and can be obtained from the values given in Table 2.2.

Example 2.5. For windy cities of Iran, namely Zabol and Manjil, calculate the Coriolis parameter.

Answer. Manjil is a city in the Central District of Rudbar County, Gilan Province, northern Iran. The city is located at (49.4110°E, 36.7399°N). Zabol is a city and capital of Zabol County, Sistan and Baluchestan Province, eastern Iran. The city is located at (61.4902°E, 31.0324°N). Both cities are windy and good places for installing wind farms.

Every 24 hours, the Earth completes a full rotation around its own axis. Therefore, the angular velocity of the Earth is roughly

$$\Omega = \frac{2\pi}{24 \times 3600} = 7.2722 \times 10^{-5} \, \text{rad s}^{-1}.$$

A more accurate angular velocity has been measured to be

$$\Omega = 7.292124 \times 10^{-5} \, \text{rad s}^{-1}.$$

Consequently, for Manjil, the Coriolis parameter is

$$f = 2 \times 7.292124 \times 10^{-5} \sin(|36.7399\frac{\pi}{180}|) = 8.724 \times 10^{-5},$$

and for Zabol, we have

$$f = 2 \times 7.292124 \times 10^{-5} \sin(|31.0324\frac{\pi}{180}|) = 7.5185 \times 10^{-5}.$$

The results show that the Coriolis parameter is very small for different latitudes.
□

Example 2.6. For the cities of Example 2.5, and the height of 40 meters, calculate parameter Ψ and compare it with $\ln(z/z_0)$ for different values of z_0.

Answer. For a moderate value for $u^* = 1\,\mathrm{m\,s^{-1}}$, we have

$$\Psi_{\mathrm{Manjil}} = 34.5 \times 8.724 \times 10^{-5} \times \frac{40}{1} = 0.120$$

and

$$\Psi_{\mathrm{Zabol}} = 34.5 \times 7.5185 \times 10^{-5} \times \frac{40}{1} = 0.103.$$

Now, we can produce the following tabulated data:

(z_0)	1	0.7	0.3	0.1	0.03	0.005	0.001	0.0002
$\ln(z/z_0)$	3.67	4.04	4.89	5.99	7.19	8.99	10.59	12.21

The results show that parameter Ψ is very small compared to $\ln(z/z_0)$ and can be dropped out from Eq. (2.9).
□

Studying the above examples, one deduces that for wind speeds of interest, parameter Ψ has a very small value compared to $\ln(z/z_0)$. Thus it can be removed form Eq. (2.9) resulting in the following equation:

$$u(z) = \frac{u^*}{\kappa} \ln \frac{z}{z_0}. \tag{2.12}$$

Eq. (2.12) is called the logarithmic atmospheric wind profile. If we measure the wind speed at a reference point, then we can calculate the wind speed at any level, using Eq. (2.12). Suppose that we gather information from a wind station at elevation z_r, then for any elevation we can write

$$\frac{u(z)}{u(z_r)} = \frac{\ln \frac{z}{z_0}}{\ln \frac{z_r}{z_0}}. \tag{2.13}$$

Example 2.7. Referring to data from Table 2.1, assume that the reference height is 10 m. Calculate the wind speed at 30 and 40 meters from Eq. (2.13) and compare the results with tabulated data. Compare the results for some selected hours.

Answer. The tabulated measured data are:

Time z, m	0:00	3:00	6:00	9:00	12:00	15:00	18:00	21:00
10	5.8	4.8	8.1	6.2	11.5	13.1	10.0	12.3
30	6.5	5.3	9.5	6.9	12.1	13.9	11.2	14.0
40	6.8	5.5	10.3	7.6	13.3	15.8	12.7	15.8

Since the surface roughness of the site is not known, we perform the calculations for different z_0. The following table shows the results for $z = 30$ m.

Time z_0	0:00	3:00	6:00	9:00	12:00	15:00	18:00	21:00
0.7	8.196	6.78	11.4	8.76	16.25	18.511	14.13	17.381
0.3	7.617	6.30	10.6	8.14	15.10	17.204	13.13	16.153
0.1	7.183	5.94	10.0	7.67	14.24	16.225	12.38	15.234
0.03	6.896	5.70	9.63	7.37	13.67	15.577	11.89	14.626
0.001	6.491	5.37	9.06	6.93	12.87	14.662	11.19	13.767

As we can see, the calculated data are very close to the measured if $z_0 = 0.001$. Hence, it means that the site in which the data are measured was a flat desert.

Repeating the same procedure to obtain data at $z = 40$ yields

Time z_0	0:00	3:00	6:00	9:00	12:00	15:00	18:00	21:00
0.7	8.82	7.30	12.3	9.43	17.49	19.92	15.21	18.71
0.3	8.09	6.69	11.3	8.65	16.04	18.27	13.95	17.16
0.1	7.54	6.24	10.5	8.06	14.96	17.04	13.01	16.00
0.03	7.18	5.94	10.0	7.67	14.24	16.22	12.38	15.23
0.001	6.67	5.52	9.31	7.13	13.23	15.07	11.50	14.15

Again the results show that the surface roughness for which the measured and calculated date are close to each other is about $z_0 = 0.001$. □

2.2 Statistical study of wind

Wind is not predictable for long periods. Meteorological forecasts usually are accurate just for some days and are not adequate for long-term predictions. But for installing wind power plants, we have to estimate the wind speed and direction. It is worth mentioning that if we are going to install a single turbine, the wind direction is not important because the yawing system of the turbine rotates the turbine toward the wind. However, the wind direction is important for situations where we want to install a wind farm in which some turbines are located in the wake of others.

Since wind is a random phenomenon, both from the speed and direction point of view, we have to work with statistical data. The speed and direction of wind strongly depend on the geography of the field where the plant is going to be installed. Unfortunately, the weather data of a location cannot be extended and used for other locations even if they are in the same region. For instance, the existence of a lake, woods, or mountain strongly affects the strength and direction of the wind. Therefore, for any specific region, we have to access local statistical data for that exact region.

Statistical data are available from weather stations, but the problem is that the values may not be exactly used for the specific region that you like. For this reason, you have two choices: you can either install a weather station in that place and start gathering data or use remote sensing data that are available from satellite monitoring systems. The former option requires lots of time since measured data of a year is not enough for decision making. To have good information, the measured data must be available for at least three years. The main advantage of remote data is that they are available for any location for more than 30 years.

Now that the necessary data is available, it should be processed and converted to information. For wind industry purposes, the probability density function of the wind is the most important factor. In other words, the weather data should be converted into the density function both for speed and direction. As mentioned, the density function of the wind direction is only used in wind farm design and is not important if you are dealing with a single turbine.

The instruments used for data acquisition have stepwise resolution and do not give continuous values. Normally, wind speed measurements are reported with $1 \, \text{m s}^{-1}$ resolution, meaning that the speed of wind is tabulated by integer values. Hence, for example, you do not see any measured data having decimal values. In modern instruments, the data are given with only one decimal value. Therefore, whether there are old instruments or new and advanced ones, the tabulated data are stepwise and are not continuous. Table 2.1 shows sample data obtained from a real weather station. The table shows that the wind speed is measured at heights of 10, 30, and 40 meters, and are given for every 10 minutes. Moreover, the resolution of data is up to one decimal point.

For many applications, including wind turbines, a resolution of $1 \, \text{m s}^{-1}$ is enough; hence the tabulated data are integers. In statistics, such data is tabulated as *grouped data,* meaning that all the individual records are divided into some groups, also known as classes. Each class of data has a mean value that is the indicator of that group. For example, Table 2.1 shows a sample of measured wind velocity at Torbate Jaam, a city that is located in the eastern Iran. As you can see, the resolution of the measurement is $0.1 \, \text{m s}^{-1}$. Therefore, the data can be grouped into integer velocities as shown in Table 2.3, where the resolution

TABLE 2.3 Wind speed data at different heights.

Wind speed class (m s^{-1})	m_i		
	40 m	30 m	10 m
0	1313	758	949
1	4260	4670	4243
2	7299	7616	8322
3	7075	7483	8815
4	5334	5798	6971
5	4286	4952	5286
6	3875	4175	3918
7	3304	3561	3197
8	2941	3069	2930
9	2620	2633	2208
10	2392	2140	1822
11	1983	1664	1376
12	1522	1310	987
13	1146	1022	632
14	969	661	441
15	811	433	231
16	477	294	109
17	370	164	57
18	287	80	35
19	139	32	18
20	84	22	7
21	30	11	5
22	18	6	1
23	15	2	0
24	6	3	0
25	4	1	0

is $1 \, \mathrm{m\,s^{-1}}$, and note that the measured values are for a whole year and not for a specific day (The data shown in Table 2.1 is just a piece of the whole data that is available at SATBA (2020); but Table 2.3 is obtained from the whole data). Finally, note that each class of data has a lower and upper limit.

Example 2.8. For the data from Table 2.3, calculate the lower and upper limit of classes 0, 5, 10, and $25 \, \mathrm{m\,s^{-1}}$.

Answer. The actual measured data has a resolution of $0.1 \, \mathrm{m\,s^{-1}}$ and are divided into $1 \, \mathrm{m\,s^{-1}}$ classes. Hence, each class represents the frequency distribution of data around itself with a radius of $1\,\mathrm{m\,s^{-1}}$. For some selected classes, the following table indicates their lower and upper limits:

class	lower limit	upper limit
0	0	0.5
5	4.6	5.5
10	9.6	10.5
25	24.6	25.5

Since there is no negative value, class 0 has no negative lower limit. For all the other classes except 0, the class value is located at the middle of its range.   □

In making Table 2.3, the values of the table indicate the number of times that a specific velocity is recorded, also known as the frequency distribution of the data. For example, at the height of 10 m, wind speed was recorded to be $5\,\mathrm{m\,s^{-1}}$ for 5286 times, and $20\,\mathrm{m\,s^{-1}}$ for 7 times. At this location, no $25\,\mathrm{m\,s^{-1}}$ was recorded at 10 m but 1 time at 30 m and 4 times at 40 m.

At each height, 52560 data points are reported as shown in Table 2.3. Knowing that the data is for a whole year (8760 hours), it means that the temporal resolution of data acquisition is 10 minutes. Typically, data of a single year is not enough for accurate calculations, and we need more data. Therefore the total number of available data is usually very large. However, it doesn't matter how huge the input data is because we sort the data file and rearrange it in the same format as shown in Table 2.3. The wind speed rarely exceeds $30\,\mathrm{m\,s^{-1}}$; hence the total number of rows is only about 30!

2.2.1   Mean and variance

The first meaningful parameter that can be deduced from weather data is the mean value of the wind speed. The mean wind speed of a region quickly reveals whether the region is appropriate for the installation of wind turbines or not. For example, assume that the mean value for wind speed in a region is $2\,\mathrm{m\,s^{-1}}$. It is quite evident that the region does not have a good potential for utilizing wind energy. However, if the mean wind speed of another region is $6\,\mathrm{m\,s^{-1}}$, it means that the region is quite suitable.

Obtaining the mean value for measured data is quite easy and straightforward. We can use the mean value formula for its calculation:

$$\bar{u} = \frac{1}{n}\sum_{i=1}^{n} u_i, \tag{2.14}$$

where $\bar{u}$ is the mean wind speed velocity and u_i are the measured values. Fortunately, with the advanced statistical languages and packages, obtaining the mean value is quite easy. For example, modern spreadsheets like Microsoft Excel or LibreOffice Calc have built-in internal functions to calculate mean values. Hence, the mean value of massive data can be calculated just with a few clicks.

Example 2.9. For data at $z = 10$ m given in Table 2.1, calculate the mean value.

Answer. The number of data points is $n = 144$. Hence, the mean value can be easily found as

$$\bar{u} = \frac{1}{144} \sum_{i=1}^{144} u_i \approx 9.37.$$

The value shows that the region has good potential for wind engineering and wind turbine industries, but it should be noted that the data are only for one day, and you cannot trust the validity of the mean value. $\square$

Wind data may be available in the grouped format as shown in Table 2.3. In this case, Eq. (2.14) is equivalently written as

$$\bar{u} = \frac{1}{n} \sum_{i=1}^{w} m_i u_i. \tag{2.15}$$

where w is the total number of groups or classes, which is equal to the table row count and m_i is the number of repeating values or the frequency of u_i.

Example 2.10. Referring to Table 2.3, calculate the mean value for elevation $z = 10$ m.

Answer. The mean value is calculated using Eq. (2.15):

$$\bar{u} = \frac{1}{52560} \sum_{i=1}^{22} m_i u_i = \frac{1}{52560} (949 \times 0 + 4243 \times 1 + 8322 \times 2 + 8815 \times 3$$

$$+ \cdots + 7 \times 20 + 5 \times 21 + 1 \times 22) = \frac{257852}{52560} \approx 4.91.$$

The mean value in this example is calculated for a whole year; thus, the mean value is more reliable than the value calculated in Example 2.9. $\square$

The mean value is a good indicator, though it is not the only meaningful parameter. Example 2.10 shows that the mean wind speed at the selected location is $\bar{u} = 4.91 \, \mathrm{m\,s^{-1}}$, but does it mean that wind blows with the same velocity at every moment? Definitely not! Obviously, Table 2.3 shows that it has different speeds at different times. But the question is how wind speed varies about the mean value. The parameter that answers this question is called *standard deviation*. In other words, standard deviation, usually shown by s, is a measure of how spread the data are. The square of the standard deviation is called *variance*. For discrete numbers such as weather data, the variance is calculated using

$$s^2 = \frac{1}{n-1} \sum_{i=1}^{n} (u_i - \bar{u})^2. \tag{2.16}$$

Note that in Eq. (2.16), the sum is divided by $(n-1)$ instead of n to have a better consistency with actual data. This fact is well-discussed in statistical books such as (Mann, 2007; Heumann et al., 2016).

Eq. (2.16) is suitable for ungrouped or raw data such as those in Table 2.1. For grouped data such as data in Table 2.3, Eq. (2.17) should be used instead:

$$s^2 = \frac{1}{n-1} \sum_{i=1}^{w} m_i (u_i - \bar{u})^2. \tag{2.17}$$

The square root of variance is the standard deviation (SD):

$$s = \sqrt{s^2}. \tag{2.18}$$

In general, SD has the same units as its corresponding data. More specifically, in our case the SD of the wind speed has the units of $m\,s^{-1}$.

Example 2.11. Referring to Table 2.1, calculate the variance and SD for elevation $z = 10\,m$ for that specific day.

Answer. Since Table 2.1 presents ungrouped data, we have to use Eq. (2.16) for calculation of SD with $\bar{u} = 9.37$ obtained from Example 2.9:

$$s^2 = \frac{1}{n-1} \sum_{i=1}^{n} (u_i - \bar{u})^2 = \frac{1}{144-1} \left((5.8 - 9.37)^2 + (6.2 - 9.37)^2 + \cdots \right.$$

$$\left. + (11.8 - 9.37)^2 + (10.4 - 9.37)^2 \right) \approx 6.176.$$

Then the SD, s, becomes

$$s = \sqrt{6.176} = 2.48.$$

Consequently, at the specific day, the mean velocity at the height of $10\,m$ is $\bar{u} = 9.37\,m\,s^{-1}$ with $s = 2.48\,m\,s^{-1}$. □

Example 2.12. Referring to Table 2.3, calculate the variance and SD for elevation $z = 10\,m$ for that region in a whole year.

Answer. In contrast to the previous example, the data in Table 2.3 are tabulated in a grouped format. Hence, we have to use Eq. (2.17). Referring to Example 2.10, the mean value is $\bar{u} = 4.91\,m\,s^{-1}$, so

$$s^2 = \frac{1}{n-1} \sum_{i=1}^{w} m_i (u_i - \bar{u})^2$$

$$= \frac{1}{52560-1} \left(949(0 - 4.91)^2 + 4243(1 - 4.91)^2 \right.$$

$$+\cdots+5(21-4.91)^2+1(22-4.91)^2\Big) \approx 10.72.$$

Thus the standard deviation s is

$$s = \sqrt{10.72} = 3.27.$$

Consequently, at the selected city, the mean velocity at the height of 10 m is $\bar{u} = 4.91\,\mathrm{m\,s^{-1}}$ with $s = 3.27\,\mathrm{m\,s^{-1}}$. The data is valid for a whole year. □

2.2.2 Probability

Another important information that can be obtained from statistical data is the probability distribution. The probability that wind blows with the speed of u_i in a year is equal to its frequency divided by the total number of data. Mathematically, it is

$$p_i = \frac{m_i}{n}. \tag{2.19}$$

Table 2.4 shows the probability of wind speeds given in Table 2.3. This table is obtained simply by dividing each table value by $n = 52560$.

Needless to say that the sum of all the probabilities is equal to 1. In mathematical form,

$$\sum_{i=1}^{w} p_i = 1. \tag{2.20}$$

One way for analyzing probability distribution is to draw the probability table on a bar chart, which is known as a *histogram*. Fig. 2.2 shows the histogram of Table 2.4 for probability at the height of 10 m. The figure illustrates a more meaningful representation of probability.

2.2.2.1 Weibull probability density function

While the probability table, or histogram, is very informative, it would be much better to express the probability by a continuous function for wind turbine calculations. The idea is to choose a proper function and find its coefficients such that the function best fits the tabulated data. In general, we can choose any continuous function and fit its coefficients to have a minimum error with the measured data. In selecting the function, it should be kept in mind that, whatever the function, its integral over the whole range should be equal to 1 because the integral of that continuous function is exactly the same as the sum of all the probabilities, which is equal to unity. The fitted continuous function is called the *probability density function*, or PDF, and normally designated by $f(u)$. As stated before, any chosen PDF should satisfy the following relation:

$$\int_0^\infty f(u) = 1. \tag{2.21}$$

TABLE 2.4 Wind speed probability at different heights.

Wind speed class (m s^{-1})	m_i		
	40 m	30 m	10 m
0	0.02498	0.014421	0.01805
1	0.08105	0.088850	0.08072
2	0.13886	0.144901	0.15833
3	0.13460	0.142370	0.16771
4	0.10148	0.110312	0.13262
5	0.08154	0.094216	0.10057
6	0.07372	0.079433	0.07454
7	0.06286	0.067751	0.06082
8	0.05595	0.058390	0.05574
9	0.04984	0.050095	0.04200
10	0.04550	0.040715	0.03466
11	0.03772	0.031659	0.02617
12	0.02895	0.024923	0.01877
13	0.02180	0.01944	0.01202
14	0.01843	0.01257	0.00839
15	0.01542	0.00823	0.00439
16	0.00907	0.00559	0.00207
17	0.00703	0.00312	0.00108
18	0.00546	0.00152	0.00066
19	0.00264	0.00060	0.00034
20	0.00159	0.00041	0.00013
21	0.00057	0.00020	0.00009
22	0.00034	0.00011	0.00002
23	0.00028	0.00003	0
24	0.00011	0.00005	0
25	0.00007	0.00002	0

The choice of selecting the function is quite arbitrary but for natural wind data gathered from many spots in the world, Weibull (Ernst Hjalmar Waloddi Weibull (18 June 1887 – 12 October 1979) Swedish engineer, scientist, and mathematician) suggested the following exponential function known as *Weibull probability density function*:

$$f(u) = \frac{k}{c}\left(\frac{u}{c}\right)^{k-1} \exp\left[-\left(\frac{u}{c}\right)^{k}\right] \qquad (k > 0, u > 0, c > 1). \qquad (2.22)$$

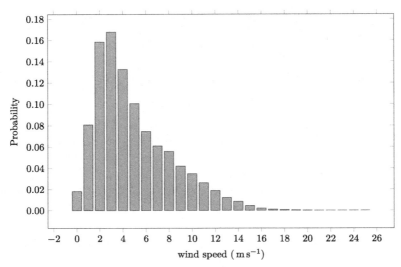

FIGURE 2.2 Histogram of wind data at Torbate Jaam for a whole year at height of 10 m above the ground.

The function has only two parameters that should be found for each wind table. They are known as the shape and scale parameters and are respectively denoted by k and c. These parameters can be found by using conventional curve fitting methods. But the problem is that Weibull PDF is a nonlinear function and the conventional curve-fitting algorithms result in a nonlinear system of equations. Hence to obtain the two parameters, you need to use nonlinear algorithms.

As explained, the parameters can be found using conventional methods, though, by applying some mathematics, we can find them very quickly by having statistical information about wind data in a region. Note that the average velocity, or $\bar{u}$, in terms of any PDF is defined as

$$\bar{u} = \int_0^\infty u f(u) du. \tag{2.23}$$

Using Weibull PDF, it becomes

$$\bar{u} = \int_0^\infty \frac{uk}{c} \left(\frac{u}{c}\right)^{k-1} \exp\left[-\left(\frac{u}{c}\right)^k\right] du. \tag{2.24}$$

To compute the integral, it is better to change the integration variables in the form of

$$x = \left(\frac{u}{c}\right)^k. \tag{2.25}$$

Thus the mean velocity is

$$\bar{u} = c \int_0^\infty x^{1/k} e^{-x} dx. \tag{2.26}$$

This integral can be evaluated using the gamma function defined as

$$\Gamma(y) = \int_0^\infty e^{-x} x^{y-1} dx. \tag{2.27}$$

It is clear that by replacing $y = 1 + 1/k$, the expression of Eq. (2.26) is exactly the same as Eq. (2.27). In other words,

$$\bar{u} = c\Gamma\left(1 + \frac{1}{k}\right). \tag{2.28}$$

To have a better understanding, it is better to plot $c/\bar{u}$ using Eq. (2.28). This value depends only on k:

$$\frac{c}{\bar{u}} = \frac{1}{\Gamma\left(1 + \dfrac{1}{k}\right)}. \tag{2.29}$$

A plot of Eq. (2.29) is shown in Fig. 2.3, showing that for $k > 1.6$, which is usually the case in wind data all over the world, the ratio $c/\bar{u} \approx 1.12$ is almost constant. Hence, for a quick estimation of the scale parameter, we can choose

$$c = 1.12\bar{u}. \tag{2.30}$$

Justus (1978) showed that the following equation gives a good estimate for k:

$$k = \left(\frac{s}{\bar{u}}\right)^{-1.086}, \tag{2.31}$$

in which s is the standard deviation.

Summing up, for obtaining k and c, we have three choices:

1. Applying conventional curve fitting methods and solving a nonlinear system of equations (not recommended);
2. Obtaining k using Eq. (2.31) and setting $c = 1.12\bar{u}$;
3. Obtaining k using Eq. (2.31), but obtaining c by means of Eq. (2.29).

It is worth mentioning that other methods are available in the open literature for calculation of k and c for a specific region but the methods discussed here are quite fine.

Example 2.13. Referring to Fig. 2.2, calculate k and c by two different methods and plot the data and Weibull PDFs in the same figure. Discuss the results.

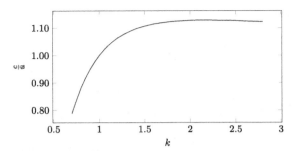

FIGURE 2.3 Variation of Weibull scale parameter with respect to shape parameter.

Answer. The mean wind speed value for the provided data was obtained as

$$\bar{u} = 4.9058599695586 \,\mathrm{m\,s^{-1}}$$

(note that to obtain good estimates for k and c, we have to keep many digits) and the standard deviation was $s = 3.273995$. Consequently, k can be easily obtained using Eq. (2.31):

$$k = \left(\frac{s}{\bar{u}}\right)^{-1.086} = \left(\frac{3.273995}{4.9058599695586}\right)^{-1.086} = 1.55146.$$

Using Eq. (2.30), we have

$$c = 1.12\bar{u} = 1.12 \times 4.9058599695586 = 5.494563.$$

We can also make use of Eq. (2.29) for finding c. This equation yields

$$c = \frac{\bar{u}}{\Gamma\left(1 + \frac{1}{k}\right)} = \frac{4.9058599695586}{\Gamma\left(1 + \frac{1}{1.55146}\right)} = 5.45517.$$

The measured probability and two fitted curves are plotted in Fig. 2.4. As can be seen, the difference between the two methods is quite negligible. □

The above example shows that obtaining k using Eq. (2.31) and setting $c = 1.12\bar{u}$ is quite reasonable and there is no need to calculate c by Eq. (2.29).

Example 2.14. Prove that the obtained Weibull function can be used as a PDF.

Answer. A function can be used as a PDF if it satisfies Eq. (2.21). For one of the obtained functions, we have:

$$\int_0^\infty \frac{1.55146}{5.45517} \left(\frac{u}{5.45517}\right)^{1.55146-1} \exp\left(-\left(\frac{u}{5.45517}\right)^{1.55146}\right) du$$

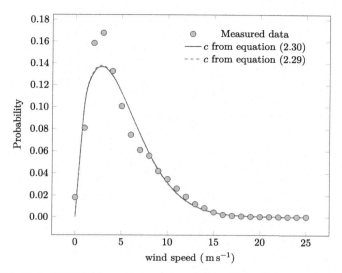

FIGURE 2.4 Weibull PDF for data at Torbate Jaam at height of 10 m above the ground.

$$= \left[-1.0 \exp(-0.0719 u^{1.55146}) \right]_0^{\infty} = 1.0,$$

hence the function can be used as a PDF.   □

2.2.3 Wind rose chart

For installation of a single wind turbine, the wind direction is not important because, if the turbine is a VAWT, it is not sensitive to wind direction, and if it is an HAWT, the yawing system rotates the turbine towards the wind. The wind direction is important when we are dealing with a wind farm.

Wind direction is measured clockwise in degrees, and with zero degrees pointing toward the north. It can also be expressed in cardinal directions denoted by initials such as N, E, S, and W. A full map of direction, together with cardinal directions and equivalent degrees, is plotted in Fig. 2.5.

When talking about wind direction, we name the wind according to its origin. For example, when we have the north wind, it means that the wind flows from the north to the south. Similarly, the west wind blows from the west to the east. Such a naming system is quite conventional in practice. The naming convention can be extended to cardinal directions, too. For example, we may have NNE wind which blows from NNE to SSW. These winds and their corresponding directions are shown in Fig. 2.5 by arrows.

For analyzing the wind direction, we can use wind rose charts or diagrams. The wind rose is a polar diagram in which the probability distribution of the wind direction is shown in polar coordinates as shown in Fig. 2.6. The radial distance indicates the probability and the angular coordinate represents the wind

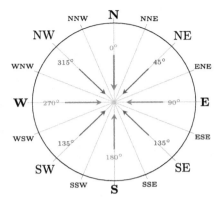

FIGURE 2.5 Wind direction terms and definitions.

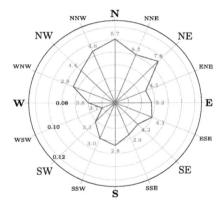

FIGURE 2.6 A sample wind rose diagram.

direction. As mentioned above, zero angle points toward the north, and the positive angular direction is clockwise, such that E direction is equal to 90°. The data are usually expressed for 16 cardinal directions.

It is conventional to superimpose the averaged wind speed on the wind rose. This adds more value to the chart and makes it more applicable for decision making. For example, Fig. 2.6 indicates that the north wind in the region has the probability of 10% and the average northern wind speed is $5.7\,\mathrm{m\,s^{-1}}$. Similarly, the western wind (blowing from the west to the east) has a probability of about 4%, while the average velocity is only $3.8\,\mathrm{m\,s^{-1}}$. Therefore, if we are going to install a wind farm in this region, it would be much better to arrange the turbines in a row extending from the west to the east because most of the time the wind blows either from the north to the south or from the south to the north. Or it would be even better if the turbines are aligned along the WNW–ESE axis since the mean wind velocity perpendicular to this axis is higher than for the W–E axis.

Example 2.15. For wind data of city Torbate Jaam, draw the wind rose.

Answer. The data of Torbate Jaam is available on SATBA (2020) in which the wind direction is tabulated at the height of 30 m. Hence we use data at that level.

The following table is a summary of the processed data. For each wind angle or direction, the frequency is given. Using the available data in the datasheet, the probability of the wind direction can be easily found by dividing each frequency by the total number of rows, n. The average wind speed is also calculated using the data available at the height of 30 m.

Angle	Frequency	Probability	Ave. Speed	Angle	Frequency	Probability	Ave. Speed
0	6434	0.126963	4.66388	22.5	15015	0.296294	8.48404
45	4683	0.0924106	6.81666	67.5	1722	0.0339806	3.51283
90	2210	0.0436104	3.81109	112.5	2251	0.0444194	3.94634
135	2456	0.0484648	3.69483	157.5	2752	0.0543058	3.62337
180	2379	0.0469453	3.58747	202.5	1508	0.0297577	3.4888
225	1148	0.0226537	2.80444	247.5	1048	0.0206804	2.67414
270	1177	0.023226	2.83756	292.5	1299	0.0256334	3.09523
315	1858	0.0366643	3.39849	337.5	2736	0.0539901	3.59415

According to the above table, a wind rose diagram is drawn in Fig. 2.7a and, for more clarity, a zoomed view is also drawn in Fig. 2.7b. The actual data indicates that in this region the wind usually blows from NNE to SSW, and the probability of blowing in other directions is very low. So, if we are going to install a wind farm, the turbines should definitely be arranged along the WNW–ESE axis. The mean average speed also indicates that the NNE wind (blowing from NNE to SSW) is also very powerful!

As a final remark, it is not necessary to calculate the average wind data at the same level where the wind direction is available because the direction of the wind does not vary too much with height. Therefore, we can show the average wind velocity on the wind rose chart, at any available height. □

Wind rose is a good diagram for studying the wind profile of a region. For example, studying Fig. 2.7 shows that in the selected region, for 30% of a year, the wind blows from NNE toward SSW with an average speed of 8.5 m s^{-1}. Also if we add 12% N-wind and 10% NE-wind, it shows that for about 52% of a year, the wind blows in a more or less the same direction with an average speed of 7.3 m s^{-1}. The weighted average speed in these three directions is calculated using probability as the weight function.

2.3 Wind power

The power of wind is a function of its velocity and density (Burton et al., 2001). Up to now, we have discussed wind speed and how to deal with it in statistical

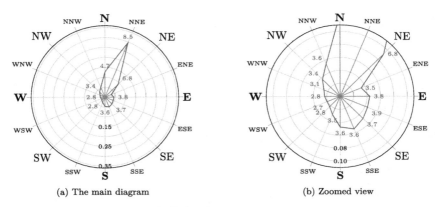

(a) The main diagram   (b) Zoomed view

FIGURE 2.7 Wind rose diagram for Torbate Jaam.

terms. The relation of wind power to its velocity and density is also important when a turbine is selected for a specific region. Since the density of air depends on the altitude (vertical distance from the sea level) and also the humidity of the place, we have to pay special attention. For example, assume that we select a 1 MW turbine from a manufacturer. We have to carefully study its technical notes to make sure under which circumstances the turbine reaches its rated capacity. The standard of calibration states that the technical data be given for $\rho = 1.225 \, \text{kg m}^{-3}$, but the datasheet of wind turbines contains technical data that relates their power to their density and wind velocity. If we install the turbine in a high altitude location, where the air density is less than the recommended value, we may not reach the full capacity of the turbine.

Before analyzing any wind turbine, we have to understand the relation of the wind speed and density to its power. In the following subsections, we discuss the issue in more detail. It should also be noted that we are not able to capture all the available wind power, in other words, the captured wind power is less than 100%. Just like thermal machines whose efficiencies are limited to the Carnot cycle, the share of power that can be captured has a limit that is called Betz's limit. This topic is also discussed in the following subsections.

2.3.1 Power of a wind element

When a fluid element flows, the power of that element is a function of its velocity (Hansen, 2008). Assume that an element with a constant cross-section area A moves with a constant velocity u in the x direction as shown in Fig. 2.8. The kinetic energy of the element is

$$U = \frac{1}{2}mu^2. \tag{2.32}$$

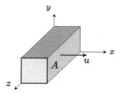

FIGURE 2.8 A fluid element moving with constant velocity, u.

As shown in Fig. 2.8, the volume of the fluid element is equal to

$$\mathcal{V} = A \times x, \tag{2.33}$$

where $\mathcal{V}$ is the volume of the element, A is the constant cross-section area of the element shown as the shaded area, and x is the thickness of the element. Using Eq. (2.33), the mass of the element is

$$m = \rho \mathcal{V} = \rho A \times x. \tag{2.34}$$

Substituting Eq. (2.34) into Eq. (2.32) yields

$$U = \frac{1}{2}\rho A u^2 \times x. \tag{2.35}$$

Having the energy of the element, its power can be found simply by differentiating Eq. (2.35) with respect to time, assuming that the density of air is constant, which is the case in subsonic regimes,

$$P = \frac{dU}{dt} = \frac{1}{2}\rho A u^2 \times \frac{dx}{dt} = \frac{1}{2}\rho A u^3. \tag{2.36}$$

Eq. (2.36) indicates that the power of wind is directly proportional to density and also proportional to the cube of u. This means that doubling the wind velocity results in an eight-fold increase in its power. This property causes lots of problems in designing wind turbines because we are dealing with a construction that should work from, let's say, 5 to about 25 m s^{-1}. So, the turbine should be designed and manufactured such that its friction is small enough for making it possible to operate at $u = 5$ m s^{-1}, and the whole structure should withstand the power of wind when $u = 5$ m s^{-1}, that is, 125 times higher than the lowest power.

2.3.2 Power of an ideal turbine, actuator disc model

Up to now, we found that the power of wind is proportional to the third power of its velocity. Now the question is how much power a wind turbine can extract from the wind. In other words, how large the efficiency of a wind turbine

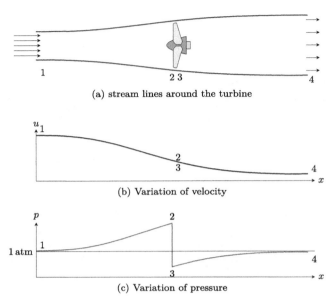

(a) stream lines around the turbine

(b) Variation of velocity

(c) Variation of pressure

FIGURE 2.9 An ideal wind turbine in an ideal flow or ADM.

is (Wagner and Mathur, 2013). It is a good idea to define a maximum theoretical value for efficiency to answer this question. The theoretical value indicates that no wind turbine can exceed the limit. From this viewpoint, the maximum theoretical efficiency can be compared with the Carnot cycle's efficiency that defines the maximum theoretical efficiency that a thermal power engine can achieve. Although the Carnot engine delivers the highest achievable efficiency, from a practical point of view, no one can make such a theoretical engine.

A theoretical model exists for the wind turbines named as *actuator disc,* or AD (Muyeen et al., 2008). An AD is a disc whose diameter is the same as the turbine's diameter and it has zero thickness. The turbine has no blade and does not rotate. Therefore, when the wind passes through the disc, it does not generate any rotation or turbulent pattern to the wind. The wind flows smoothly through the disc and moves away without getting any disturbance from the disc. Hence, the disc makes no dissipation of energy and consequently will have the highest efficiency. The only effect of AD is that it extracts a part of wind's power and delivers it as its own electrical power. Thus, the wind's power after the AD is less than its power before the disc.

Fig. 2.9 shows an ideal wind turbine in which the wind blows toward the turbine from infinity at point 1. The AD model is used instead of an actual turbine because we want to obtain the maximum efficiency. Since the flow regime is subsonic, the existence of the AD is sensed by the wind, and its velocity reduces as it moves closer to the turbine at location 2. For an ideal inviscid flow, we can use the Bernoulli equation between points 1 and 2 because there is no external force in the flow and the flow is inviscid and ideal. However, when it passes the AD, an amount of wind energy is captured by the turbine. Hence we cannot use the Bernoulli equation between points 2 and 3. These points are very close to each other due to the fact that the width of the turbine is negligible compared to the flow length. This is consistent with the assumption of a zero-width AD. From point 3 onward, the flow is again dominated by the Bernoulli equation because there is no external force in it, and it is again an ideal flow.

In a real case, the viscosity plays an important role in making a wake behind the wind turbine, and the velocity just behind the turbine drops suddenly. But in an ideal flow, there is no vortex behind the turbine. Also, as mentioned, points 2 and 3 are located in the same place. Hence the areas at points 2 and 3 are the same. Since there is no variation in area, the continuity states that the velocities at points 2 and 3 are equal.

The variations of pressure and velocity from point 1 to 4 are plotted in Fig. 2.9. Since point 1 is located far from the turbine, its pressure is 1 atm, and its velocity is u_1. By approaching the turbine, the wind velocity decreases, and, according to the Bernoulli equation, its pressure increases to the values above the atmospheric pressure. From point 2 to 3, we don't have any velocity change since the cross-section area does not change; hence as discussed above, due to continuity, the velocity remains the same. But at this point, the energy of wind drops because a part of its energy is delivered to the turbine. Therefore, according to the Bernoulli equation, we have to have a sudden decrease in pressure, as is illustrated in Fig. 2.9c. After that, the pressure of the wind reaches the atmospheric pressure because point 4 is far from the turbine. Once again, according to the Bernoulli equation, when the pressure increases to the atmospheric value, its velocity continues to decrease to a moderate value as shown in Fig. 2.9b.

In an ideal flow, there is no friction or any other energy dissipating mechanism. Hence, the total energy of the wind at point 1 must be conserved. Consequently, the balance of energy states that the power of wind at point 1, P_{w1}, is equal to the sum of turbine power, P_T, and wind power at point 4, P_{w4}, or

$$P_T = P_{w1} - P_{w4}. \tag{2.37}$$

Using Eq. (2.36), we can write

$$P_T = \frac{1}{2}\rho \left(A_1 u_1^3 - A_4 u_4^3 \right). \tag{2.38}$$

Note that the atmospheric properties such as ρ are constant. Or better said, it is assumed to be constant since we are dealing with a wind farm that is located in a specific region. Normally, the height or humidity variations in a farm are not very high.

2.3.3 Betz's limit

Eq. (2.38) is used to obtain the turbine power if we have enough information about the wind. In other words, having A_1, A_4, u_1, and u_4, we can calculate the turbine power. In general, it is not an easy task to obtain all these values. But from theoretical calculations, we can find the maximum theoretical value for P_T.

The exerted trust on the actuator disc or the turbine, T, can be found either from momentum theory or static balance on the disc. The momentum theory states that

$$T = u_1(\rho A u)_1 - u_4(\rho A u)_4, \qquad (2.39)$$

where the indices refer to the locations shown in Fig. 2.9. For a steady state flow, $(\rho A u)_1 = (\rho A u)_4 = \dot{m}$, where $\dot{m}$ denotes the mass flow rate. Hence Eq. (2.39) can be rewritten as

$$T = \dot{m}(u_1 - u_4). \qquad (2.40)$$

From location 1 to 2, we can use the Bernoulli equation:

$$p_1 + \frac{1}{2}\rho u_1^2 = p_2 + \frac{1}{2}\rho u_2^2, \qquad (2.41)$$

and also from location 3 to 4 we get

$$p_3 + \frac{1}{2}\rho u_3^2 = p_4 + \frac{1}{2}\rho u_4^2. \qquad (2.42)$$

On the other hand, we can calculate T by making a static balance around the AD that yields

$$T = A_2(p_2 - p_3). \qquad (2.43)$$

Now if we solve for $(p_2 - p_3)$ from Eqs. (2.41) and (2.42) and substitute it into Eq. (2.43), we find

$$T = \frac{1}{2}\rho A_2 \left(u_1^2 - u_4^2 \right). \qquad (2.44)$$

Equating Eqs. (2.40) and (2.44) yields

$$u_2 = \frac{u_1 + u_4}{2}. \qquad (2.45)$$

When the wind reaches the actuator disc, its velocity decreases. The fractional decrease is called the *induction factor* and is calculated as

$$a = \frac{u_1 - u_2}{u_1}. \qquad (2.46)$$

Eq. (2.46) can be rearranged in the following form:

$$u_2 = u_1(1 - a),\quad (2.47)$$

and from Eq. (2.45) we find

$$u_4 = u_1(1 - 2a).\quad (2.48)$$

It is more convenient to express the turbine thrust in its dimensionless form, which is also known as the thrust coefficient, or c_T. The thrust coefficient can be obtained simply by dividing Eq. (2.44) by $\frac{1}{2}\rho u_1^2 A_2$ (note that velocity is used at point 1, and the area at the rotor plane, or point 2):

$$c_T = \frac{\frac{1}{2}\rho A_2 \left(u_1^2 - u_4^2\right)}{\frac{1}{2}\rho u_1^2 A_2} = \frac{\frac{1}{2}\rho A_2 (u_1 + u_4)(u_1 - u_4)}{\frac{1}{2}\rho u_1^2 A_2}.\quad (2.49)$$

Using Eqs. (2.45) and (2.46), c_T is found to be a function of the induction factor expressed by

$$c_T = 4a(1 - a).\quad (2.50)$$

Up to now, the thrust on the turbine was obtained. We know that the power is equal to the thrust multiplied by velocity. Therefore, at the location of the turbine, the power is calculated multiplying Eq. (2.44) by u_2, which yields

$$P = \frac{1}{2}\rho A_2 \left(u_1^2 - u_4^2\right) u_2.\quad (2.51)$$

Substituting u_2 and u_4 from Eqs. (2.47) and (2.48), we get

$$P = \frac{1}{2}\rho A_2 u_1^3 \times 4a\,(1 - a)^2.\quad (2.52)$$

Again, it is better to use the dimensionless form of power, which is also an expression for the efficiency of the turbine. This parameter is also known as the *coefficient of performance*) and defined as the fraction of the power extracted by the turbine over the power of wind at infinity with an area equal to the rotor plane. In mathematical terms, it is

$$c_p = \frac{P}{\frac{1}{2}\rho A_2 u_1^3}.\quad (2.53)$$

Since the density of air differs from place to place, the standard states that c_p is usually calculated with $\rho = 1.225$, which corresponds to the density of air at standard pressure and $T = 15°C$.

Applying Eq. (2.52) yields

$$c_p = 4a(1 - a)^2.\quad (2.54)$$

Eq. (2.54) means that the coefficient of performance of a wind turbine is a function of its induction factor. This equation is suitable for finding the maximum turbine's power, since it is a one-variable function. To find the maximum value for c_p, we can take a derivative of Eq. (2.54) with respect to a and equate the result to zero. The operations yields $a = 1/3$. In other words, the maximum power can be extracted from the wind when

$$u_2 = u_3 = \frac{2}{3}u_1, \qquad u_4 = \frac{1}{3}u_1, \tag{2.55}$$

and consequently,

$$A_2 = A_3 = \frac{3}{2}A_1, \qquad A_4 = 3A_1. \tag{2.56}$$

Substituting $a = 1/3$ in Eq. (2.54), we can find the maximum achievable c_p as

$$c_{p,\max} = \frac{16}{27} = 0.5926. \tag{2.57}$$

It means that at best we can harvest only 16/27 of the wind power:

$$P_{\max} = c_{p,\max}\frac{1}{2}\rho A_2 u_1^3 = \frac{16}{27}\frac{1}{2}\rho A_2 u_1^3. \tag{2.58}$$

When using Eq. (2.58), just note that wind power is a fictitious power since the wind area is calculated at the rotor plane, A_2, while its velocity is at infinity, u_1. By keeping this fact in mind, the value $\frac{16}{27} \approx 60\%$ is called the Betz's limit after German physicist Albert Betz, who first published the law in 1919.

Betz's limit identifies the maximum power that can be extracted from the wind. Betz's limit states that we cannot build a wind turbine that is capable of extracting more than 60% of wind power if we calculate the wind power by its velocity at infinity but the area at the rotor plane.

2.4 Efficiency of wind turbine components

There are three main sources of energy loss in a wind turbine. These three sources are the turbine blades, gearbox, and generator, as shown in Fig. 2.10. Each of these components has its own efficiency, which strongly affects the final output, or electrical power.

The power of wind, P_w, first should be captured by the blades and then be converted to mechanical power on the low-speed shaft, or P_m. As stated before, the maximum theoretical value for power conversion is defined by Betz's limit. The mechanical power is transmitted by the gearbox to the high-speed shaft. Since the gearbox has its own efficiency denoted by η_m, the power on the high-speed shaft, P_t, differs from P_m. Finally, the transmitted power on the high-speed shaft is converted to electricity, P_e, via a generator. The conversion efficiency is shown by η_g.

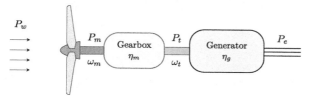

FIGURE 2.10 Wind turbine power transmission.

The overall efficiency of a wind turbine, η_O, is the product of all the above mentioned efficiencies, or

$$\eta_O = c_p \eta_m \eta_g. \tag{2.59}$$

It is quite evident that the input power of the wind turbine, or wind power, varies with the wind speed. Also, it is obvious that the efficiency of each component varies with its input power. Consequently, the efficiency of a wind turbine strongly depends on the wind speed. In the following subsections, these efficiencies are discussed in detail for each component.

2.4.1 Efficiency of blades or coefficient of performance

A typical turbine power generation curve is shown in Fig. 2.11a. In this figure, the wind power is plotted versus u, as well as the wind turbine gained power. The power of wind is directly proportional to the cube of the wind velocity and increases as the wind speed increases. However, for a wind turbine to be able to operate, the wind speed should meet a certain threshold. In other words, a minimum amount of power is required to overcome the turbine friction and internal resistance. Hence, up to a certain value, the gained power of the wind turbine is zero. At a specific velocity, the wind turbine starts rotating and generates power. This velocity is called the cut-in velocity which is designated by u_c in the figure. The output power of the wind turbine increases as the wind velocity increases until it reaches the designed value, or the rated power of the wind turbine. The wind velocity at which the wind turbine begins to produce its rated power is called the rated velocity and is denoted by u_R. From then on, the power of wind increases further, but we have to control the wind turbine so that it does not exceed the rated power. Otherwise, the generator becomes too hot, and it may melt or explode. Therefore, for the wind speeds greater than u_R, we see that the turbine power curve is flat, and we do not extract more power from the wind than the turbine's rated power. We can control the wind turbine up to a certain wind speed, at which the wind power becomes so larger that the controlling systems are not able to maintain the turbine at the desired operational conditions. At such wind speeds, we have to temporarily shut down the turbine. The velocity at which the turbine should be shut down is called the furling velocity and is designated as u_F.

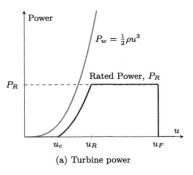

(a) Turbine power

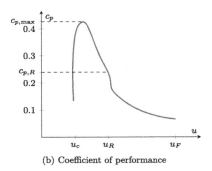

(b) Coefficient of performance

FIGURE 2.11 Turbine power generation curve.

In Fig. 2.11b, the coefficient of performance of such a turbine is plotted. As expected, at velocities less than u_c and more than u_f, we have $c_p = 0$ because the turbine is shut down and produces no power. From u_c to u_R turbine starts producing power and c_p increases to its maximum value, then it declines to c_{pR} at u_R. Since the controlling systems do not let the turbine produce more power at the velocities above u_R, a sharp decrease is seen for c_p. The curve shows that not only it is not able to reach Betz's limit, but also most of the time, a wind turbine works far from its own maximum coefficient of performance.

The c_p curve shown in Fig. 2.11b is valid only for a single rotational speed. If the rotational speed of the turbine changes, the gained power changes, and it produces a different power curve; consequently, the c_p curve is not the same as before. Thus for studying a wind turbine, c_p curve should be plotted for different turbine's rotational speeds. This means a lot of computations or tests, which is not practical. To overcome the problem, a dimensionless parameter known as the *tip speed ratio* is defined as

$$\lambda = \frac{r\omega}{u}, \tag{2.60}$$

where r is the radius of the turbine blade, ω is the rotational speed, and u is the wind speed. By using λ, both effects, i.e., the wind speed and turbine rotational speed, are cast into a single variable, and we get a single c_p curve. Such a curve is shown in Fig. 2.12, in which, if the wind speed increases, λ moves toward zero. This is why in contrast to Fig. 2.11b, $c_{p,\max}$ is located to the right of $c_{p,R}$.

Example 2.16. For a 1 MW wind turbine with $u_R = 15$ and $u_F = 20\,\mathrm{m\,s^{-1}}$, plot c_p in its rated range.

Answer. Referring to Fig. 2.11a, from u_R to u_F, the gained power of the turbine remains constant while the wind power increases. This fact is true for pitch-regulated turbines in which the pitch-controlling mechanism tries to keep a constant output power. For stall-regulated turbines, the power varies within a

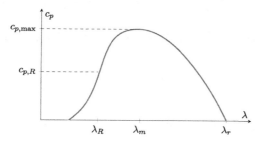

FIGURE 2.12 c_p curve versus λ.

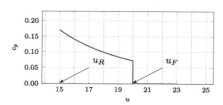

FIGURE 2.13 c_p plot of Example 2.16.

precalculated range such that the maximum power does not exceed the generator safety limits. See Micon M 530, Nordtank NTK 400, Nordtank NTK 500/41, and Wincon W755/48 in Appendix C for examples of stall-regulated turbines. A web search for 1 MW turbines (some samples are given in Appendix C) reveals that such turbines usually have a diameter of about 60 m. As mentioned, to obtain the coefficient of performance, according to standards, the air density is taken as $\rho = 1.225\,\text{kg}\,\text{m}^{-3}$. Thus c_p in the defined range is obtained from the following equation:

$$c_p = \frac{1,000,000}{\frac{1}{2}\rho A u^3} = \frac{577.4328}{u^3},$$

where $A = \pi \times 60^2/4$. This function is plotted in Fig. 2.13. Since we do not have enough information for this turbine, we cannot plot c_p for wind speeds less than u_R. Finally, after the wind speed exceeds u_F, the output power of the turbine becomes zero, since the turbine is stopped by controlling systems. Hence, c_p becomes zero afterwards as shown in the figure. □

2.4.2 Efficiency of gearbox

The angular velocity of the low-speed shaft is not usually high enough for power generation and should be increased by means of a gearbox. For reducing wearing and increasing the gears' service life, it is customary to design the gearbox such that each stage ratio does not exceed 6:1. Therefore, to increase the ratio to

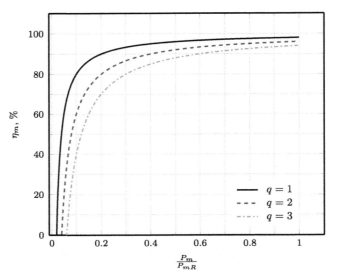

FIGURE 2.14 Gearbox efficiency.

more than 6:1, we can use two or three stages. Typically, a three-stage gearbox is enough for practical purposes since it can result in up to $6^3 = 216$-fold increase in the rotational speed. Depending on the design, 4-stage gearboxes also exist in the market.

Just like any other systems, gearboxes have their own characteristics and efficiencies. Thus the power on the high-speed shaft, P_t, is related to the mechanical power on the low-speed shaft, P_m, according to

$$P_t = \eta_m P_m. \tag{2.61}$$

The efficiency of gearboxes depends on many factors, including gear size, input power, quality of manufacturing, and many others. But, to have a rule of thumb for calculating the efficiency, it is convenient to assume that each gear stage dissipates about 2% of the rated power (Johnson, 1985). In mathematical formulas, we can write

$$\eta_m = \frac{P_t}{P_m} = \frac{P_m - (0.02)q\,P_{mR}}{P_m} = 1 - (0.02)q\,\frac{P_{mR}}{P_m}, \tag{2.62}$$

where q is the number of stages and subscript R denotes the rated value.

Fig. 2.14 illustrates the relation between the gearbox efficiency and its input power. From the figure, it can be concluded that if the mechanical input power drops to 70% of the rated power, the efficiency does not vary a lot, but for lower input power, the gearbox efficiency drops dramatically.

Example 2.17. For Nordtank NTK 400, calculate the number of required gear stages.

Answer. The properties of Nordtank NTK 400 are given in Section C.9. The rotor of this turbine rotates at 35 rpm while its generator requires 1800 rpm input speed. Therefore, the gearbox increases the rotational speed up to

$$g_r = \frac{1800}{35} = 51.428571429 \sim 52.$$

Using 2 stages, the maximum increase will be 36:1, but we need 52:1; hence we have to use a 3-stage gearbox. □

2.4.3 Efficiency of generator

Several items are contributing to generator efficiency, including hysteresis, eddy currents, windage losses, bearing friction which is of mechanical origin, and wire resistance. Like gearbox efficiency, we can use the following simple equation for generators (Johnson, 1985):

$$\eta_g = \frac{X - 0.5Y(1 - Y)(X^2 + 1)}{X}, \tag{2.63}$$

in which

$$X = \frac{P_t}{P_{tR}} \tag{2.64}$$

and

$$Y = 0.05 \left(\frac{10^6}{P_{eR}} \right)^{0.215}. \tag{2.65}$$

The above equations show that the generator efficiency is also a function of its input and rated power. A very close similarity exists between Eqs. (2.63) and (2.62). Fig. 2.15 shows the generator efficiency versus P_t/P_{tR} where the rated electrical power is a parameter. It is evident that, as the size of the generator increases, it becomes more efficient. Moreover, just like gearbox efficiency, the efficiency of the generators is almost constant in partial loads up to values of about 40% of its rated power. If the input power further decreases, the efficiency drops dramatically.

2.4.4 Overall efficiency

The wind turbine overall efficiency is expressed by Eq. (2.59). In analyzing the efficiency of the gearbox and generator, it was observed that the efficiency of these components drops dramatically in partial loads. This means that if a wind turbine is located in a place where it is mostly operational in partial loads, the

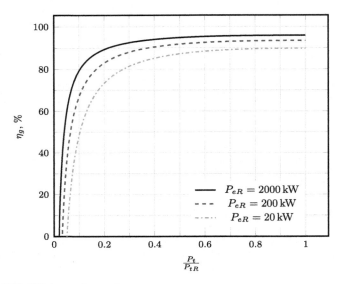

FIGURE 2.15 Efficiency of generator.

overall efficiency becomes way too small such that all the input power may be lost!

The following examples discuss the efficiency of different parts and their contribution in the overall efficiency of the wind turbine. The examples are prepared for both pitch- and stall-regulated models. As it will be see, the procedure for both types is the same, just the key parameters must be properly chosen.

Example 2.18. As an example for pitch-regulated turbines, consider Vestas V90 whose power curve and c_p are given in Appendix C. Calculate η_m and η_g versus the wind velocity. Plot c_p, η_m, η_g, and η_O on a single diagram and discuss the results.

Answer. The specifications of Vestas V90 Gridstreamer is given in Section C.18. This turbine operates from $u_c = 4.0\,\mathrm{m\,s^{-1}}$ up to $u_F = 25\,\mathrm{m\,s^{-1}}$. The rated wind speed for this turbine is $u_c = 13.5\,\mathrm{m\,s^{-1}}$, at which the rated electrical power is $P_{eR} = 2030\,\mathrm{kW}$, as can be see in its power curve. Since the turbine is pitch-controlled, the power curve remains flat from u_R to u_F. Finally, he gearbox contains one planetary stage and two helical stages. Thus it contains three stages, or $q = 3$.

The first step to solve the problem is to obtain necessary parameters at rated conditions. From the power curve, we see that at u_R we have

$$P_w = \frac{1}{2}\rho A u^3 = \frac{1.225}{2} \times 6362 \times 13.5^3 = 9{,}587{,}404\,\mathrm{W}.$$

In calculation of the wind power, the air density is chosen as $\rho = 1.225$, as the standard states, and the turbine's swept area is $A = 6362$, according to the available datasheet. At the rated condition, we can find

$$c_p = 0.227$$

from the power curve. Hence, P_{mR} is

$$P_{mR} = c_p P_w = 0.227 \times 9,587,404 = 2176340\,\text{W}.$$

The gearbox efficiency, η_m, is simply

$$\eta_m = 1 - 0.02q = 1 - 0.02 \times 3 = 0.94,$$

since at the rated condition $P_m = P_{mR}$. Consequently, the rated transmission power is

$$P_{tR} = \eta_m P_{mR} = 0.94 \times 2176340 = 2045760\,\text{W}.$$

At rated conditions, $P_t = P_{tR}$, hence $X = 1$. Also from rated electrical power, we have

$$Y = 0.05 \left(\frac{10^6}{P_{eR}} \right)^{0.215} = 0.05 \left(\frac{10^6}{2030000} \right)^{0.215} = 0.0429.$$

Substituting X and Y into Eq. (2.63), the efficiency of the generator at rated conditions becomes

$$\eta_g = \frac{1 - 0.5 \times 0.0429(1 - 0.0429)(1 + 1)}{1} = 0.959,$$

and finally, the overall efficiency becomes

$$\eta_O = c_p \eta_m \eta_g = 0.227 \times 0.94 \times 0.959 = 0.2046 = 20.46\%.$$

The overall efficiency can be used to calculate the electrical rated power as

$$P_{eR,cal} = \eta_O P_w = 0.2046 \times 9587404 = 1961688\,\text{W},$$

which is very close to the rated power obtained from the power curve. The error of calculation is

$$Error = \frac{P_{eR} - P_{eR,cal}}{P_{eR}} = \frac{2030000 - 1961688}{2030000} = 0.033 = 3.3\%.$$

Now that the rated values are all known, calculating these parameters at different velocities is quite simple and straightforward. The results are tabulated in Table 2.5 and also plotted in Figs. 2.16 and 2.17. $\qquad\square$

TABLE 2.5 Results of Example 2.18.

u m s^{-1}	P_w W	c_p	P_m W	η_m	P_t W	X	η_g	η_O	$P_{e,calc}$ W	$P_{e,curve}$ W	Error %
5	487091	0.490	238674	0.453	108094	0.053	0.610	0.135	65940	190000	65.3
6	841693	0.505	425055	0.693	294474	0.144	0.854	0.299	251567	354000	28.9
7	1336577	0.506	676281	0.807	545701	0.267	0.917	0.375	500673	582000	14.0
8	1995123	0.488	973979	0.866	843399	0.412	0.942	0.398	794218	883000	10.1
9	2840713	0.473	1342464	0.903	1211883	0.592	0.953	0.407	1155096	1240000	6.8
10	3896725	0.437	1703765	0.923	1573185	0.769	0.957	0.387	1506290	1604000	6.1
11	5186541	0.388	2014245	0.935	1883665	0.921	0.959	0.348	1805990	1893000	4.6
12	6733541	0.315	2123085	0.938	1992505	0.974	0.959	0.284	1910593	2005000	4.7
13	8561105	0.251	2151834	0.939	2021253	0.988	0.959	0.226	1938182	2027000	4.4
13.5	9587405	0.227	2176341	0.940	2045760	1.000	0.959	0.205	1961688	2030000	3.4
14	10692613	0.205	2196983	0.941	2066402	1.010	0.959	0.185	1981477	2030000	2.4
15	13151447	0.167	2198696	0.941	2068115	1.011	0.959	0.151	1983119	2030000	2.3
16	15960986	0.137	2193055	0.940	2062475	1.008	0.959	0.124	1977713	2030000	2.6
17	19144610	0.115	2196535	0.941	2065954	1.010	0.959	0.103	1981048	2030000	2.4
18	22725700	0.097	2200422	0.941	2069841	1.012	0.959	0.087	1984773	2030000	2.2
19	26727637	0.082	2202425	0.941	2071845	1.013	0.959	0.074	1986694	2030000	2.1
20	31173800	0.071	2211940	0.941	2081360	1.017	0.959	0.064	1995812	2030000	1.7
21	36087570	0.061	2206526	0.941	2075945	1.015	0.959	0.055	1990624	2030000	1.9
22	41492328	0.053	2191276	0.940	2060696	1.007	0.959	0.048	1976008	2030000	2.7
23	47411453	0.046	2163214	0.940	2032634	0.994	0.959	0.041	1949099	2030000	4.0
24	53868326	0.040	2145791	0.939	2015210	0.985	0.959	0.036	1932384	2030000	4.8
25	60886328	0.035	2145759	0.939	2015179	0.985	0.959	0.032	1932354	2030000	4.8

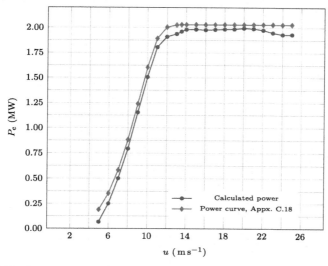

FIGURE 2.16 Calculated power of Example 2.18.

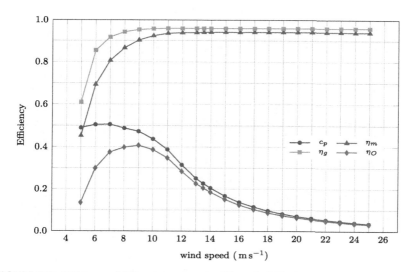

FIGURE 2.17 Efficiency of different components of Example 2.18.

The above calculations are very sensitive to the precision of different efficiencies. A slight change in c_p results in a significant change in P_e. Thus, having precise data is crucial in the calculation of power curves. Unfortunately, in many datasheets, the values of c_p are given using two or three decimal digits, which are not accurate enough. The same argument is valid for the efficiency of the gearbox and generator. Obviously, the more accurate the data, the more pre-

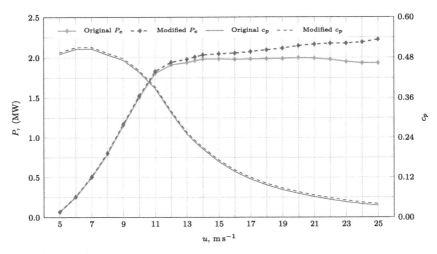

FIGURE 2.18 Comparison of power and c_p of Example 2.19.

cise the calculations. The following example shows the importance of accurate values.

Example 2.19. Investigate the effect of variation of c_p on the power curve.

Answer. To investigate the effect, we increase c_p by 0.005 and recalculate the power curve. The procedure is the same as described in the previous example. The results are shown in Fig. 2.18. As it can be seen, a slight change in the c_p curve has made lots of changes in the final calculated P_e.   □

There is no difference between simulation of stall- and pitch-controlled turbines. The only difference is that for a stall-regulated wind turbine, the efficiency of the gearbox or the generator may exceed the rated values. The following example investigates such cases.

Example 2.20. As an example of a stall-regulated turbine, compute the efficiency of different parts of Wincon W755/48 from Appendix C.14.

Answer. The procedure is the same as before. The first step is to find the rated values. From Appendix C.14, we see that Wincon W755/48 is a turbine with $P_{eR} = 755\,\text{kW}$, $A = 1810\,\text{m}^2$, $q = 3$, and $u_R = 14\,\text{m s}^{-1}$. The wind power at the rated state is

$$P_w = \frac{1}{2}\rho A u^3 = \frac{1.225}{2} \times 1810 \times 14^3 = 3{,}042{,}067\,\text{W}.$$

Also from the power curve and at the rated condition,

$$c_p = 0.25;$$

hence, P_{mR} becomes

$$P_{mR} = c_p P_w = 0.25 \times 3,042,067 = 760516\,\text{W}.$$

The gearbox efficiency, η_m, again is obtained as

$$\eta_m = 1 - 0.02q = 1 - 0.02 \times 3 = 0.94;$$

consequently, the rated transmission power is

$$P_{tR} = \eta_m P_{mR} = 0.94 \times 760516 = 714885\,\text{W}.$$

At rated conditions, $P_t = P_{tR}$, hence $X = 1$. Also from rated electrical power, we have

$$Y = 0.05 \left(\frac{10^6}{P_{eR}} \right)^{0.215} = 0.05 \left(\frac{10^6}{755000} \right)^{0.215} = 0.0531.$$

Substituting X and Y into Eq. (2.63), the efficiency of the generator at rated conditions becomes

$$\eta_g = \frac{1 - 0.5 \times 0.0531(1 - 0.0531)(1 + 1)}{1} = 0.949,$$

and finally, the overall efficiency becomes

$$\eta_O = c_p \eta_m \eta_g = 0.25 \times 0.94 \times 0.949 = 0.2232 = 22.32\%.$$

The overall efficiency can be used to calculate the electrical rated power as

$$P_{eR,cal} = \eta_O P_w = 0.2232 \times 3,042,067 = 678989\,\text{W}.$$

It seems that the present calculation predicted a lower value for P_{eR}. The error of calculation is

$$Error = \frac{P_{eR} - P_{eR,cal}}{P_{eR}} = \frac{755000 - 678989}{755000} = 0.100 = 10\%.$$

As stated before, accurate data is important for a precise result. It seems that the values for c_p given in Appendix C.14 do not have enough accuracy. If we add small values to the data, then the calculation may have more accuracy. For example, if we add only 10% to c_p at the rated power (i.e., $c_p = 0.275$) and recalculate the rated power, we get

$$\eta_O = 0.23945$$

and

$$P_{eR,cal} = \eta_O P_w = 0.23945 \times 3,042,067 = 728408\,\text{W},$$

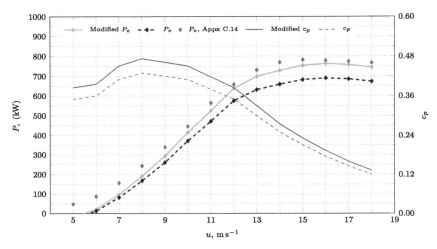

FIGURE 2.19 Power and c_p of Example 2.20.

which is more accurate than before. The error of calculation is

$$Error = \frac{P_{eR} - P_{eR,cal}}{P_{eR}} = \frac{755000 - 728408}{755000} = 0.0521 = 5.21\%.$$

Fig. 2.19 shows the whole calculations over the all range of wind speed. As it can be seen, increasing c_p by 10% has made a great improvement to the simulations. The efficiency of different parts of the turbine is plotted in Fig. 2.20 as before. □

In some cases, the power curve contains P_e, but there is no data available for c_p. For example, in Appendix C, there are many different examples of such turbines. In these situations, the same procedure can be followed to estimate the efficiency of different parts and obtain the overall efficiency. But since c_p is not available, a trial-and-error procedure must be carried on. The procedure is like this:

Step 1 Guess a typical value for c_p which, according to Betz's limit, must definitely be less than 0.6.

Step 2 Having c_p, follow the same procedure discussed above and find P_e.

Step 3 If the calculated P_e matches the provided data, the guessed c_p is correct. Otherwise, change the value of c_p and return to Step 1.

The above procedure gives a good estimate of the efficiency of different parts. But it may not be accurate since the formulas used for obtaining the efficiency of different parts give typical values. The above procedure is best understood by the following example.

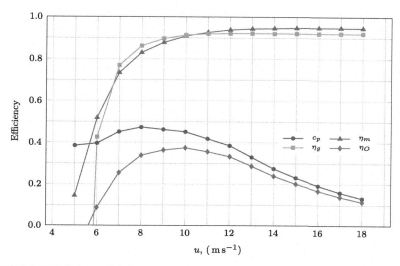

FIGURE 2.20 Efficiency of different components of Example 2.20.

Example 2.21. The power curve for Aerodyn-8.0MW is given in Appendix C.25. Calculate c_p, η_m, η_g, and η_O. Plot the results and discuss.

Answer. The swept area of Aerodyn-8.0MW is $22,167.0\,\mathrm{m^2}$. This turbine has a two-bladed rotor with downwind configuration. In contrast to the previous examples, Aerodyn-8.0MW uses a two-stage gearbox; thus $q = 2$. Finally, the power curve shows that the rated wind speed is $u_R = 12.5\,\mathrm{m\,s^{-1}}$, at which the rated power is $P_{eR} = 8\,\mathrm{MW}$.

The first step is to obtain the rated values. As we do not have any value for c_p, we follow the above-mentioned trial-and-error procedure. First, we try $c_p = 0.25$ and perform the calculations. The following table summarizes the results. As it can be seen, the calculated $P_e = 6167362$ is too small compared to the real value. Therefore, we choose $c_p = 0.4$ and repeat the calculations. The result, $P_e = 9867779$, indicates that the calculated output power is greater than $P_{eR} = 8\,\mathrm{MW}$. From these values and by performing linear regression, we find that $c_p = 0.324287$ must give proper values. Recalculation of the parameters shows that the output power reaches a correct value. Therefore, we conclude that at the rated velocity, $c_p = 0.324287$.

No.	c_p	P_{mR} (W)	P_{tR} (W)	P_{eR} (W)
1	0.25	6629534	6364353	6167362
2	0.40	10607255	10182965	9867779
3	0.324287	8599487	8255508	7999981

Up to now, the rated values are known. The next step is to calculate these values at other wind velocities. The results are shown in Figs. 2.21 and 2.22. As

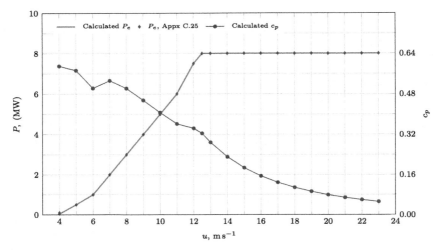

FIGURE 2.21 Power and c_p of Example 2.21.

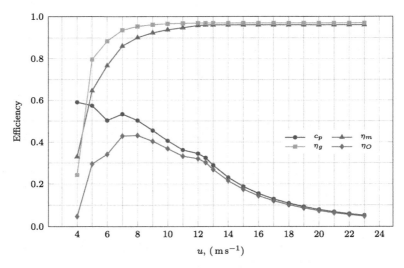

FIGURE 2.22 Efficiency of different components of Example 2.21.

it is clear from the figures, c_p values are chosen so that the power curve gives the same values as the tabulated ones. Hence, the calculated power curve is the same as data taken from Appendix C.25. The efficiency of gearbox shown in Fig. 2.22 is higher than those of previous examples, since Aerodyn-8.0MW has a two-stage gearbox. Finally, the generator's efficiency is also higher since the rated power of the turbine is higher than that of all the previous examples. □

2.5 Yearly gained energy of a wind turbine

The power curve of a wind turbine, as shown in Fig. 2.17, indicates that a wind turbine does not operate at its rated power at different wind speeds. For a pitch-regulated turbine, the power output is always equal to the rated power when the wind speed exceeds the rated velocity. This situation is clearly seen for these types of turbines (see Vestas V29, Mapna MWT2.5-103-I, and Aerodyn SCD 8.0/168 in Appendix C just to have some examples) where their power is constant after the wind speed is higher than their respective rated velocity. However, for lower wind speeds, their power is less than their rated values. This means that at specific intervals, the wind turbine does generate less power as expected.

The situation is more complicated for stall-regulated turbines since they do not have a constant power curve even when the wind exceeds the rated value. For some examples, see the power curves of Micon M 530, Nordtank NTK 500/41, and Wincon W755/48 in the same appendix. These turbines generate more power than their rated value when the wind speed is higher than the rated velocity. Therefore, these turbines generate either less power relative to their rated power when the wind speed is low, or higher power when it is high.

From the above statements, we conclude that the total gained energy of a wind turbine is not obtained simply by multiplying the rated power to the operational hours. Since the turbine power is a function of wind velocity, its produced energy also becomes a function of the wind speed. Therefore, the energy output of a wind turbine at a specific velocity is obtained using the wind turbine power at that velocity multiplied by the fraction of time in which the turbine operates at the same specific velocity. The total amount of time can be easily calculated by the following equation:

$$t(u) = 8760 \times f(u), \tag{2.66}$$

where $t(u)$ is the total time during which the wind blows with velocity u, and $f(u)$ is the PDF obtained from weather data. Multiplying Eq. (2.66) to the turbine power gives the appropriate equation for obtaining the produced energy as a function of velocity:

$$E(u) = P_e(u)t(u) = 8760 P_e(u) f(u). \tag{2.67}$$

Consequently, the net gained energy of a wind turbine is obtained using the following equation:

$$E = 8760 \int_0^\infty P_e(u) f(u) du. \tag{2.68}$$

Since the turbine power is zero for the velocities less than the cut-in velocity and over the cut-out velocity, the above integral would be equal to

$$E = 8760 \int_{u_c}^{u_F} P_e(u) f(u) du. \tag{2.69}$$

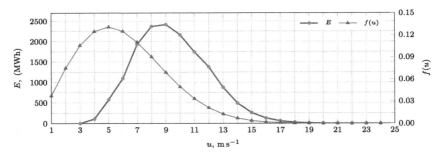

FIGURE 2.23 Hourly produced energy of the turbine of Example 2.19.

Example 2.22. Assume that an Aerodyn-8.0MW turbine is located in a place whose Weibull PDF is characterized by $k = 2.1$ and $\bar{u} = 6 \text{ m s}^{-1}$. Calculate the net energy produced by this wind turbine.

Answer. The power curve of Aerodyn 8.0MW is shown in Fig. 2.21. From Eq. (2.29), the shape parameter of Weibull PDF is found to be

$$c = \frac{\bar{u}}{\Gamma(1 + 1/k)} = \frac{6}{\Gamma(1 + 1/2.1)} = 6.77 \text{ m s}^{-1}.$$

Having k and c, the PDF is fully determined. Therefore, the following table can be calculated:

Velocity (m s^{-1})	3	4	5	6	...	22	23	24
Power (kW)	0	100	500	1000	...	8000	8000	8000
Probability	0.10561	0.12474	0.13084	0.12497	...	7×10^{-06}	2×10^{-06}	8×10^{-07}
Energy (kJ)	0	109276	573099	1094698	...	558	184	57

The result is better understood by plotting it on a chart as shown in Fig. 2.23, in which the velocity-based energy production, i.e., Eq. (2.67), is plotted superimposed with the Weibull PDF. As the figure indicates, the maximum probability of the wind speed occurs at $u = 5 \text{ m s}^{-1}$; however, the maximum energy production occurs at $u = 9 \text{ m s}^{-1}$. Also, it can be observed from the figure that the high power at higher wind velocities does not mean that the turbine produces more energy. This is because the energy output is a function of both power and PDF of wind velocity.

Finally, the net gained energy of this turbine can be found by integrating in Eq. (2.68). Since the data are given in a tabulated format, the integration can be easily completed using the trapezoidal rule:

$$E = \frac{\delta u}{2} \left(E(3) + 2 \sum_{u=4}^{23} E(u) + E(24) \right) = 15572.9 \text{ MWh}.$$

Note that in the trapezoidal rule, δu is the velocity step. In the present example, the velocity step is $\delta u = 1$ since the energy is obtained at $1\,\mathrm{m\,s^{-1}}$ intervals. ☐

2.6 The capacity factor of a wind turbine

Any energy-producing sector is designed for a particular power. However, it cannot produce its rated power 24 hours a day, 365 days a year. For many reasons, the power plants generate less power during a year. For example, they must be stopped for regular maintenance purposes. Or, they may stop producing energy due to the lack of their feed. Consequently, the net yearly produced energy of power plants is not equal to the product of its rated power and the total number of hours in a year (i.e., 8760 hours).

According to the above statement, the power plant efficiency, better known as the *capacity factor*, is defined as the ratio of the actually produced energy to the theoretical produced energy, as if the power plant was able to produce its rated power the whole year. Mathematically, the capacity factor is

$$C_F = \frac{E_{act}}{E_{theo}}, \tag{2.70}$$

where C_F is the capacity factor, and E_{act} and E_{theo}, respectively, are the actual and theoretical energies.

Wind turbines also follow the same condition, and during a year, they cannot produce their full-rated power. In its normal operation, the turbine power may be less than its rated value due to the low wind velocity. Also, the turbine stops producing energy at wind velocities less than its cut-in and over its cut-out velocities. In addition to its normal operation, a wind turbine requires regular maintenance like every other power plant. Therefore, the capacity factor of a wind turbine must be determined using Eq. (2.70).

Example 2.23. Compute the capacity factor of Example 2.22.

Answer. The rated power turbine of Example 2.22 is $P_{eR} = 8\,\mathrm{MW}$, which results in $u = 12.5\,\mathrm{m\,s^{-1}}$. Therefore, the theoretical energy of this turbine is

$$E_{theo} = 8760 \times P_{eR} = 8760 \times 8 = 70080\,\mathrm{MWh}.$$

But this turbine produces only $E_{act} = 15572.9\,\mathrm{MWh}$ as was calculated in Example 2.22. Therefore, its capacity factor is

$$C_F = \frac{E_{act}}{E_{theo}} = \frac{15572.9}{70080} = 0.222.$$

This means that the turbine produces 22.2% of its theoretical energy during a year. ☐

2.7 Summary

In this chapter, a brief introduction to the wind, its power, and gained power by a single turbine is introduced. It is shown that, in general, we need to have good wind data for any place where wind turbines or farms are to be installed. Then the efficiency of a wind turbine is discussed, and it is emphasized that not all the available wind power can be captured by a turbine. Therefore, the efficiency of the wind turbine and its major components becomes important. These efficiencies are discussed in more detail, and the overall efficiency of a practical wind turbine is obtained. Finally, the capacity factor a of wind turbine is introduced, and the way it can be calculated is discussed and shown by some examples.

2.8 Problems

1. Using Eq. (2.4), tabulate the variation of pressure and density with respect to altitude. Plot the values on a single graph.
2. For data given in Table 2.1, calculate α for all the rows between the heights of 10 and 30 m and obtain an average value for exponent α.
3. Repeat the previous problem, but this time use the data between the heights of 10 and 40 m and obtain an average value for exponent α. Compare the results of these problems.
4. Many wind datasheets are available, including wind data for Iran from (SATBA, 2020). Use the data and obtain a proper exponent for different cities. Compare the results and discuss.
5. Acquire the wind data from (SATBA, 2020) for cities of Kahak, Khaf, Zabol, and Manjil. Convert the ungrouped measured data to grouped dataset for each city and different available heights.
6. For the grouped data obtained in the previous problem, calculate the mean wind speed and corresponding standard deviation at different heights.
7. Plot the probability distribution of wind data for cities of Problem 5.
8. For each height, obtain the Weibull shape and scale parameters for cities of Problem 5 and compare the probability distribution obtained in Problem 7 and the Weibull PDF.
9. From data of Siemens SWT-4.0-130 in Section C.22 calculate the input rotational speed of its generator.
10. The datasheet of Siemens SWT-6.0-154 (Section C.23) shows that the rotor speed and generator input speed are both 11 rpm. Discuss the gearbox type.
11. For eno energy eno 100, whose data is given in Section C.19, calculate the number of gearbox stages. This turbine is a 2.2 MW turbine whose power curve is given in Fig. C.17. For the whole range of wind speed, calculate and plot the efficiency of its gearbox.
12. For Vestas V90 calculate and plot the power on the high-speed shaft versus wind speed.
13. For the previous problem, calculate and plot the electrical output power.

14. For Enercon E-18, calculate efficiency of gearbox and generator versus the wind speed. Plot the power curve and compare it with the practical data, i.e., Fig. C.2. Note that you need the c_p curve, so use Fig. C.2.

15. Repeat the previous problem for Enercon E-30, Nordtank NTK 500/41, Wincon W755/48, Vestas V90, and eno energy eno 100.

16. Calculate the net energy output of an Enercon E-16 that is installed in a field where its Weibull PDF parameters are $k = 1.8$ and $c = 7\,\mathrm{m\,s^{-1}}$. Plot the results and compute its capacity factor.

17. Study the effect of variation of Weibull shape parameter, k, on the net yearly gained energy of the turbine of Problem 16. Plot the net energy versus k.

18. Repeat the above study this time for the variation of c.

19. Do the same study as requested in Problems 16 to 18 for different wind turbines given in Appendix C.

References

Burton, Tony, Sharpe, David, Jenkins, Nick, Bossanyi, Ervin, 2001. Wind Energy Handbook, vol. 2. Wiley Online Library.

Hansen, M.O., 2008. Aerodynamic of Wind Turbines. Earthscan, London & Sterling, VA.

Heumann, Christian, Schomaker, Michael, et al., 2016. Introduction to Statistics and Data Analysis. Springer.

Johnson, Gary L., 1985. Wind Energy Systems. Citeseer.

Justus, C.G., 1978. Winds and Wind System Performance.

Mann, Prem S., 2007. Introductory Statistics. John Wiley & Sons.

Manwell, James F., McGowan, Jon G., Rogers, Anthony L., 2010. Wind Energy Explained: Theory, Design and Application. John Wiley & Sons.

Mathew, Sathyajith, 2006. Wind Energy: Fundamentals, Resource Analysis and Economics. Springer.

Muyeen, S.M., Tamura, Junji, Murata, Toshiaki, 2008. Stability Augmentation of a Grid-Connected Wind Farm. Springer Science & Business Media.

Ng, Chong, Ran, Li, 2016. Offshore Wind Farms: Technologies, Design and Operation. Woodhead Publishing.

SATBA, 2020. Map of wind and solar energy of Iran. http://www.satba.gov.ir/en/regions. (Accessed 20 April 2020).

Wagner, H.J., Mathur, J., 2013. Introduction to Wind Energy Systems, Green Energy and Technology.

Wikipedia, 2021. Roughness length. https://en.wikipedia.org/wiki/Roughness_length. (Accessed 12 May 2021).

Chapter 3

Basics of aerodynamics

In previous chapters, we studied the wind itself and the relation of wind power to its velocity, density, and area. We also talked about the maximum achievable power, or Betz's limit. A review of some conventional wind turbines also revealed that the wind turbines operate far from Betz's limit in practice. In many situations, we also try to reduce the energy gain since the power of wind increases with the increase of wind speed, but we have to maintain the generator power to a rated limit.

Now that we have enough information about the wind, it is time to explain how a wind turbine captures wind power. We know from Betz's limit that the maximum efficiency of a wind turbine is about 60%. Moreover, we found that the turbines operate quite far from that limit in the previous chapter. Therefore, in the present chapter, we seek methods that enable us to find the captured power.

In general, there are two approaches to harvesting wind power. The turbines use either drag or lift to convert the kinetic energy of wind into a rotational mechanical work. Whether it is a VAWT or HAWT, we can benefit from both drag and lift, depending on the design.

But what are the drag and lift forces? By definition, the drag force is a resistive force acting on an object moving in a fluid. The direction of the drag force is opposite to the direction of the oncoming fluid, or in other words, it is in the same direction as the relative velocity between the object and the fluid. The lift is the force exerted on the object by the fluid whose direction is normal to the relative velocity between the object and fluid. By these definitions, the drag and lift forces are always perpendicular to each other. Note that the relative wind direction determines the direction of these forces. Also, note that the relative wind direction differs from the direction of the wind at the far-field.

An important thing to remember is that the direction of drag and lift is not related to the position of the object. Moreover, their direction does not vary by rotation of the object since the object's rotation does not change the fluid's relative direction. Thus, the direction of the drag force is always in the same direction as the relative wind velocity, and the direction of the lift is always normal to the direction of the relative velocity. The term *relative velocity* is very tricky here since it plays an important role in wind turbine calculations. This parameter will be discussed in detail in the upcoming sections.

Having a good understanding of aerodynamic forces is crucial in determining and calculating the power gained by a wind turbine; thus, we give a brief

Fundamentals of Wind Farm Aerodynamic Layout Design. https://doi.org/10.1016/B978-0-12-823016-9.00009-7

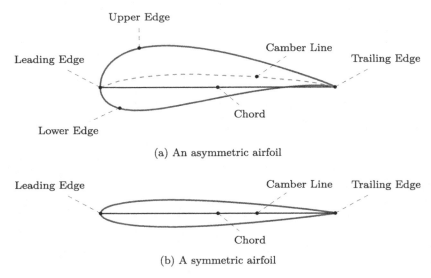

(a) An asymmetric airfoil

(b) A symmetric airfoil

FIGURE 3.1 Definitions of an airfoil.

introduction and discuss different important aspects of fluid motion and aerodynamics in this chapter. It is quite evident that the detailed design of wind turbines requires more information about aerodynamics and related topics. In this chapter, we first discuss the concept of airfoils and introduce different famous airfoils used in wind turbine industries. Second, we talk about the blade and the forces that are exerted on them. Third, we discuss different aspects of aerodynamic phenomena that are important when dealing with wind turbines. Finally, we give a brief introduction about the famous *blade element momentum* method which is suitable for calculating the gained power of a wind turbine.

3.1 Airfoils

Airfoil is a cross-section of a wing or blade; thus, it is a two-dimensional shape. A typical airfoil and its main terminology are shown in Fig. 3.1. When an airfoil is located in a fluid flow, as shown in the figure, its first point that hits the wind is called the *leading edge* (assuming that wind blows from left to right). Similarly, the last point of the airfoil on which the wind leaves the airfoil is called the *trailing edge*. The straight line that connects these two points is called the *chord*. The dashed line shown on the figure is called the *camber line* which is a line such that the thickness of the airfoil is equal on its both sides. Note that the thickness is measured perpendicular to the camber line.

In a symmetric airfoil, the camber line and chord are identical. Such a case is shown in Fig. 3.1b. Thus we can conclude that if an airfoil is not symmetric, its chord and camber line are not the same. Other than that, other things are the same in both cases.

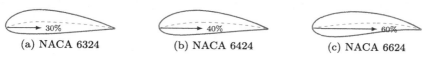

(a) NACA 6324   (b) NACA 6424   (c) NACA 6624

FIGURE 3.2 Airfoils of Example 3.1.

The chord of an airfoil is important because, as we will see in the future, it is used to nondimensionalize the airfoil's lengths. By nondimensionalizing, all similar airfoils show similar behavior. This fact is used in determining the aerodynamic properties of airfoils and airfoil families.

Many different airfoil families are commonly used in wind turbine industry. These families are optimized for wind turbines and are well documented. Some of these families are covered in Appendix B, but here some of them are briefly discussed so that a better understanding of airfoils and their behavior is gained.

3.1.1 NACA series

Since the beginning of the invention of airplanes, the related companies decided to make standards for making airfoils. One of the pioneers was NACA which was renamed NASA after a while. For making different airplanes, NACA introduced many different airfoils in some different families, namely NACA 4-digit, 5-digit, 6-digit, and some others. Among them, the NACA 4-digit family was used in the early wind industry, too.

Each NACA 4-digit airfoil, as its name indicates, has four different digits. For example, we have NACA0012 and NACA4628, and any other four-digit configuration. If we name the airfoils as NACA$mptt$, then each letter has the following meaning:

- The first digit, m, describes the maximum camber as a percentage of the chord.
- The second digit, p, indicates the location of the maximum camber from the leading edge in tenths of the chord.
- The last two digits, tt, indicate the maximum thickness of the airfoil as a percentage of the chord.

Example 3.1. Discuss NACA6324, NACA6424, and NACA6624 airfoils.

Answer. All the sections are from the NACA 4-digit family. The maximum camber thickness of all the sections is 0.06 of the chord (according to their first digit), but the maximum camber's location differs from section to section. As shown in Fig. 3.2, the maximum camber is located at 30%, 40%, and 60% of the chord for NACA6324, NACA6424, and NACA6624, respectively; their second digit expresses the camber maximum locations. Finally, for all the sections, the maximum thickness is 24% of their chord according to their last two digits. □

Example 3.2. Discuss NACA0012, NACA0024, and NACA0036 airfoils.

(a) NACA 0012   (b) NACA 0024   (c) NACA 0036

FIGURE 3.3 Airfoils of Example 3.2.

Answer. These sections are all symmetric because, as their first digit indicates, the maximum camber is 0. In other words, a camber is a straight line that obviously is located on the chord. The last two digits determine the thickness of the airfoil. Thus the higher the number, the thicker the airfoil. The maximum thickness of these airfoils is 12%, 24%, and 36% of their chord length, respectively. These sections are drawn in Fig. 3.3 for comparison. □

Although NACA airfoils have been used in the aircraft industry for many years, many wind turbine makers used them for making turbine blades. However, the turbine manufacturers found that these airfoils are not well optimized for wind turbines. For this reason, many industrial sectors and universities, such as NREL, DU, and many others, started to develop optimized airfoil sections that show better performance in wind turbines. NACA airfoils are still in use by some manufacturers, specially for small-scale turbines.

3.1.2 NREL

In the 1980s, the National Renewable Energy Lab (NREL) in the USA started to introduce new classes of airfoils specially designed for wind turbines. As mentioned before, NACA airfoils were not well optimized for wind turbines in which blade rotation is also a matter of concern. For example, the performance of NACA23XXX airfoils deteriorates rapidly with increasing thickness. Moreover, they generate noise in large turbines.

After analyzing the data, NREL has suggested seven airfoil families suitable for different HAWT blade sizes. These families are tabulated in Table 3.1.

3.1.3 Other types

Other than NACA and NREL airfoils, many manufacturers have developed their own airfoils. The most famous are:

Risø The family was developed by Risø National Laboratory for use on wind turbines. These researches and the data were transferred to Denmark Technical University (DTU).

DU The family was developed and tested by the Faculty of Aerospace Engineering of Delft University.

FFA This family was developed by The Aeronautical Research Institute of Sweden with cooperation with KTH university.

TABLE 3.1 NREL airfoil families (NREL, 2021).

Blade length (meters)	Thickness category	Root	Primary	Tip
1–3	Thick	S835	S833	S834
3–10	Thick	S823	–	S822
10–20	Thin	S807	S805A	S806A
	Thin	S808	S805A	S806A
	Thick	S821	S819	S820
20–30	Thick	S811	S809	S810
	Thick	S814	S812	S813
	Thick	S815	S812	S813
20–40	–	S814	S825	S826
	–	S815	S825	S826
	–	–	–	S829
30–50	Thick	S818	S816	S817
40–50	Thick	S818	S830	S831
	Thick	S818	S830	S832
	Thick	S818	S827	S828

FX The family was designed by Althaus and Wortmann, and the experiment was carried out in the Laminar Wind Tunnel at the Institute for Aerodynamics and Gasdynamics in Stuttgart.

These are not the only available families. Other sections and section families are available. Also, the airfoil development is still under research to find even more suitable geometries.

3.2 Aerodynamic forces on an airfoil

When an airfoil is located in a wind stream, the wind flows around it from both the upper and lower edges (Ng and Ran, 2016; Mathew, 2006). Such a situation is illustrated in Fig. 3.4 for a symmetric airfoil. The angle between the airfoil chord and the free-stream wind velocity is called the *angle of attack* and usually is denoted by α. When $\alpha = 0$, like the situation shown in Fig. 3.4a, the flow is completely symmetric around the airfoil. Therefore, the pressure field around the section is symmetric; hence it produces an equal pressure on both the upper and lower edges. The result of such a symmetric pressure field is a zero-lift condition. The only effect of wind on the airfoil is a drag force that is in the same direction as the upstream wind velocity.

If we rotate the airfoil, such as shown in Fig. 3.4b, the flow field on the upper and lower edges becomes different. As shown in the figure, the air moves faster on the upper edge than the lower edge because it travels a much longer path at

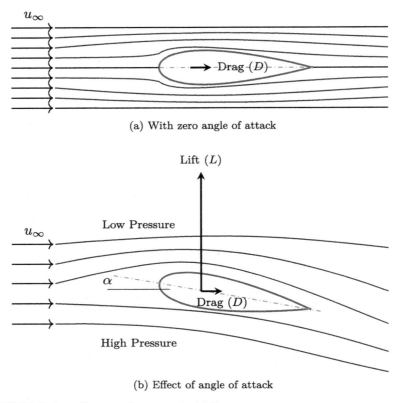

(a) With zero angle of attack

(b) Effect of angle of attack

FIGURE 3.4 Streamlines around a symmetric airfoil.

the same time. Therefore, according to Bernoulli's law, the air pressure is lower on the upper edge and higher on the lower edge. The result is a vertical force perpendicular to the wind direction called *lift*. Since airfoils are usually very thin, the lift is much larger than its drag.

For asymmetric airfoils, the same argument is also correct, but the only difference is that for asymmetric sections, the airfoil produces lift even at $\alpha = 0$. This is because, as shown in Fig. 3.5, the streamlines around the airfoil are not symmetric, resulting in an asymmetric pressure field around the airfoil.

The lift of an airfoil increases as its angle of attack increases. For thin airfoils, the slope of lift versus α is near 2π as long as the streamlines are attached to the airfoil. As α increases, at a certain value (depending on the airfoil shape and thickness), the streamlines cannot maintain their path attached to the upper edge, and a phenomenon known as *separation* occurs. At this stage, streamlines start making vortexes over the upper edge, resulting in a high-pressure zone on the top side. Consequently, a large negative lift pushes the airfoil downward; thus, the lift suddenly drops. This phenomenon is called the *airfoil stall*. The

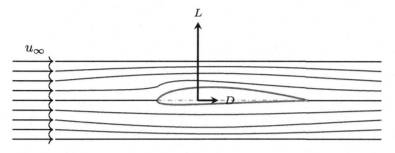

FIGURE 3.5 Streamlines around an asymmetric airfoil.

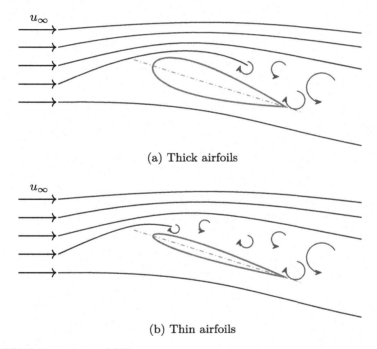

(a) Thick airfoils

(b) Thin airfoils

FIGURE 3.6 Separation on airfoils.

stall is a phenomenon in which the airfoil cannot produce more lift by increasing the angle of attack, and its lift decreases dramatically.

Stall in airfoils strongly depends on their thickness and shape. The thickness of the section affects the way the separation moves with respect to the angle of attack. By increasing the angle of attack, the separation point moves from tail to tip on the upper edge. For a better understanding, the separation is illustrated in Fig. 3.6 for both thick and thin airfoils. By increasing the angle of attack, for thicker airfoils, the separation takes place near the tail and smoothly moves to

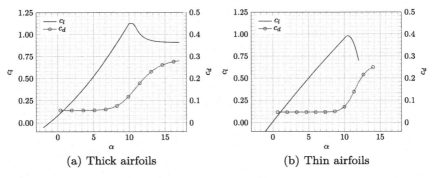

FIGURE 3.7 Lift and drag curves versus angle of attack.

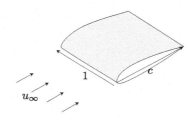

FIGURE 3.8 Illustration of projected area in calculation of lift and drag coefficients.

tip, but for thin airfoils, the separation moves very fast. Thus for thick airfoils in the stall zone, the lift decreases smoothly, while for thin airfoils, the lift drop is very sharp. Fig. 3.7 illustrates the difference for both cases. In this figure, instead of lift and drag, the *lift* and *drag coefficients* are plotted. These are the dimensionless lift and drag coefficients calculated from

$$c_l = \frac{L}{\frac{1}{2}\rho u_\infty^2 A}, \tag{3.1}$$

$$c_d = \frac{D}{\frac{1}{2}\rho u_\infty^2 A}, \tag{3.2}$$

in which ρ is the density of air, and A is an arbitrary area. Usually, A is set as

$$A = c \times 1, \tag{3.3}$$

where c is the chord length, and 1 means a unit length perpendicular to the airfoil section. In other words, Eq. (3.3) means that the lift (or drag) coefficient is a measure of lift (or drag) exerted on a unit length of a wing (Burton et al., 2001; Johnson, 1985). This arbitrary area is illustrated in Fig. 3.8.

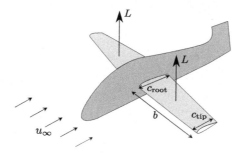

FIGURE 3.9 Lift of a moving plane.

Applying Eq. (3.3) to Eqs. (3.1) and (3.2) gives the following relations for c_l and c_d, respectively:

$$c_l = \frac{L}{\frac{1}{2}\rho u_\infty^2 c}, \tag{3.4}$$

$$c_d = \frac{D}{\frac{1}{2}\rho u_\infty^2 c}. \tag{3.5}$$

Returning to Fig. 3.7, two different airfoils are compared. Fig. 3.7a illustrates c_l and c_d versus α for a typical thick airfoil, while Fig. 3.7b demonstrates the same data for a thin section. The figure, once again, shows the different behavior of thick and thin airfoils in the stall, i.e., smooth reduction in the lift for the thick and sudden drop in the thin airfoils. For both sections, at the stall, the drag coefficient increases dramatically.

Example 3.3. Fig. 3.9 shows a plane that moves with $u = 100\,\mathrm{m\,s^{-1}}$. For this plane, $c_{\mathrm{tip}} = c_{\mathrm{root}} = 0.75\,\mathrm{m}$, and the wing length is $b = 8\,\mathrm{m}$. If the angle of attack of the wings is set to $\alpha = 9°$ by the pilot, calculate the total lift.

Answer. For the situation of the present problem, Eq. (3.1) gives the lift as

$$L = \frac{1}{2}\rho u_\infty^2 c \times b c_l.$$

We assume that the plane's wing is fabricated using airfoil of Fig. 3.7a. Hence, at $\alpha = 9$ we have $c_l = 1$. Substituting the known values, we get

$$L = \frac{1}{2}1.225 \times 100^2 \times 0.75 \times 8 = 36750\,\mathrm{N},$$

which is equal to 3750 kg. For both wings, the total lift is $L_{\mathrm{tot}} = 73500\,\mathrm{N}$ equal to 7500 kg (assuming $g = 9.8$). It means that at this condition, the plane can lift a total mass of 7500 kg. □

Example 3.4. Repeat the above example, if $c_{\mathrm{tip}} = 0.75\,\mathrm{m}$ and $c_{\mathrm{root}} = 1.5\,\mathrm{m}$.

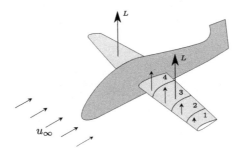

FIGURE 3.10 Lift distribution on variable cross-section wing.

Answer. Since the cross-section of the wings is not uniform, we have a distribution of lift on the wings as is shown in Fig. 3.10. According to Eq. (3.1), the lift distribution is:

$$dL(x) = \frac{1}{2}\rho u_\infty^2 c(x)c_l \times dx,$$

where x is the local coordinate starting from the tip pointing to the root. Since the lift distribution is a function of the chord and as the chord variation is linear, there is a linear lift distribution on each wing as illustrated for different sections in Fig. 3.10. The above equation must be integrated from tip to root to obtain the total lift of a single wing, which yields

$$L = \int_0^8 dL(x) = \int_0^8 \frac{1}{2}\rho u_\infty^2 c(x)c_l \times dx.$$

The chord function is

$$c(x) = 0.09375x + 0.75.$$

Therefore, by inserting all the known parameters into the above equation and performing integration, the lift of one wing becomes

$$L = \frac{1}{2}1.225 \times 10^4 \left[0.09375\frac{x^2}{2} + 0.75x\right]_0^8 = 55125\,\text{N},$$

which is equal to 5625 kg. For both wings, the total lift is $L_{\text{tot}} = 110250\,\text{N}$ equal to 11250 kg. $\square$

The above examples show the situation which is valid for an airplane. However, as we will see in the next sections, the case is not the same in a wind turbine, and we cannot assume that all the cross-sections have a constant angle of attack. Therefore, to better understand the situation, we have to know the differences between an airplane and a wind turbine. In the next section, the differences are covered in more detail.

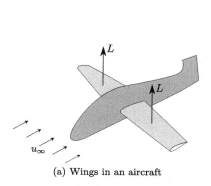

(a) Wings in an aircraft

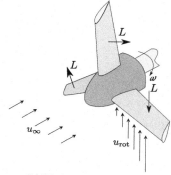

(b) Blades in a wind turbine

FIGURE 3.11 Using lift for different purposes.

3.3 Aerodynamic forces on a blade

As stated before, an airfoil is a two-dimensional cross-section of a wing or blade. For an aircraft, as shown in Fig. 3.11a, the lift for both wings is in the same direction. The result is a force that acts oppositely to the weight of the aircraft. This will cause the plane to overcome the weight and gradually fly. In a wind turbine, several blades are attached to a hub in order to rotate a shaft. The lift is almost perpendicular to each blade as illustrated in Fig. 3.11b.

There is a big difference between the aerodynamic phenomena around the wings of an aircraft and blades of a wind turbine (Manwell et al., 2010). When the speed of an aircraft is constant, the wings experience a uniform wind speed from root to tip. This is because the wind speed for all the wing sections is equal to the aircraft's velocity relative to the atmospheric wind. But in the case of wind turbines, even if the wind velocity is constant and even if the blade rotation is kept constant, the net wind speed that reaches each blade section is not uniform. The reason is that the blade rotation produces a wind flow in the same rotational plane of the blades but in the opposite direction. As it is well known, velocity is a vector quantity; hence the net sum of two different velocity vectors is obtained using vector summation. In the case of a wind turbine's blade, there are two different vector components at the blades' location, i.e., the far-field wind velocity and the rotational velocity induced by blade rotation as shown in Fig. 3.11b. Consequently, the net wind velocity that reaches the blade is the vector sum of the produced rotational wind and the far-field wind. The situation becomes more complex if we notice that the linear velocity of the blade increases from root to tip; thus, the net wind speed is not uniform along the blade. This situation is illustrated in Fig. 3.11b where the rotational velocity shown by arrows is in the opposite direction of the blade rotation. Note that in order to make it less complex, the rotational velocity is shown only for one blade.

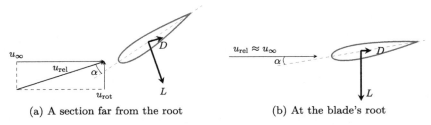

(a) A section far from the root   (b) At the blade's root

FIGURE 3.12 Effect of rotational velocity on the direction of the relative wind.

The rotational velocity plays an important role in turbine aerodynamic. As mentioned above, this velocity affects the far-field velocity and creates a relative velocity passing over the blade. Fig. 3.12 illustrates this effect for two different sections. In this figure, the blade is moving downward; hence the applied lift is toward the ground. At sections far from the root, as shown in Fig. 3.12a, the relative velocity, u_{rel}, is obtained by vector summation of the far-field velocity, u_∞, and the rotational velocity, u_{rot}. According to the definition of lift, we should first pay attention to the fact that it is always perpendicular to the far-field velocity. This is true for a plane, but the far-field velocity must be replaced by the relative velocity for a rotating blade because the relative velocity is the actual velocity that passes over the blade. In this case, as shown in the figure, the lift is perpendicular to u_{rel}. The second thing to note is that the drag is also in the same direction as u_{rel}, as illustrated. Finally, it is quite evident that both the lift and drag contribute to rotation because both of them have components on the rotational surface. While the lift component is trying to rotate the blade in its proper direction, the drag component tries to resist and reduces the lift effect. This is why designers try to minimize the drag and maximize the lift in the construction of airfoils.

Since the blade is moving downward, the section should be rotated such that the angle of attack, α, becomes negative to produce a downward lift. The larger the rotational velocity, the higher the section rotation should be. Otherwise, the lift becomes positive and resists the rotation. As we move toward the root, the rotational velocity becomes less and less. At the root section, the rotational speed is very small so that in some cases one can neglect it. Such a situation is shown in Fig. 3.12b where the relative velocity almost coincides with the far-field velocity. It is quite clear from both figures that the relative velocity angle differs from section to section, and for maintaining each section in good condition, or a reasonable range for α, we need to design a proper rotation angle for that section. In other words, in contrast to an aircraft wing, a wind turbine's blade is *twisted* from root to tip. The twisted blade is also shown in Fig. 3.11b in which the blade chord angle differs from tip to root.

The concept of relative wind velocity makes wind turbines become more complicated than aircraft. Some differences can be summarized as follows:

- In an aircraft, at cruise condition, the angle of attack is the same along the wings, but for a wind turbine blade, the angle of attack differs from section to section.
- If the far-field velocity of an aircraft changes, the angle of attack remains unchanged. But any change in the far-field velocity results in a different relative velocity, which results in changing the angle of attack for all sections.
- For a steady wind velocity, the wind velocity that passes over an aircraft's wing is uniform from root to tip. But in a wind turbine's blade, the relative wind varies from root to tip even in a steady far-field wind. Therefore, for each section, the relative wind velocity must be obtained.
- In an aircraft, there is no rotational speed; however, in a wind turbine, the rotational speed of the blades makes rotational velocity very important in analyzing the blades.

3.4 Generated vortex behind a wind turbine

The final phenomenon that must be considered for the wind turbine blades is the concept of the *induced velocity*. It is the velocity that is induced by the turbine's existence and its rotation. A wind turbine can be considered an obstacle to the wind. Hence, it opposes the find flow and reduces the far-field velocity as the wind reaches the turbine. Moreover, the blade's rotation creates rotational patterns in the passing wind. Therefore, when the wind passes a wind turbine, a rotational velocity is added to its velocity vector. These two effects are known as the induced velocity components.

The induced velocity has two components, one opposite to the far-field wind velocity, u_∞, and the other is in the same direction as the rotational velocity. The induced velocity modifies both u_∞ and u_{rot}; therefore, it changes the relative wind velocity and angle of attack. Consequently, since the generated lift and drag of the blade depend on these two factors, understanding and calculating induced velocity is very important.

In a subsonic regime, any obstacle in the wind's path generates pressure waves that oppose the wind. The result is the velocity reduction when the wind moves toward that obstacle. This phenomenon happens in wind turbines in the same manner, and the existence of the turbine induces a horizontal wind component against the upstream wind. This component is named $u_{ind,a}$, and schematically is shown in Fig. 3.13a. The axial induced velocity obviously is less than the far-field velocity and is commonly defined as a portion of u_∞. Mathematically, we can write

$$u_{ind,a} = au_\infty, \tag{3.6}$$

where a is called the *axial induction factor*. This factor is not uniform along the blade, and, as shown in Fig. 3.13a, it has a nonuniform distribution. Therefore, for a specific blade, a is not constant and should be obtained as a function of the blade radius.

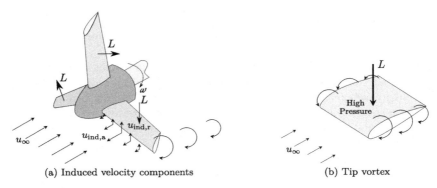

(a) Induced velocity components (b) Tip vortex

FIGURE 3.13 Illustration of induced velocity and tip vortex.

In addition to the axial induction velocity, just like an aircraft wing, a blade generates a tip vortex due to the pressure difference between the upper and lower blade surfaces. This phenomenon is illustrated in Fig. 3.13b, in which a downward moving blade is shown. In this case, the lift is toward the ground (downward), meaning that the pressure at the upper surface of the blade is higher than at the lower surface. At the tip and root of the blade, this pressure gradient pushes the air from the upper surface to the lower surface, as indicated by circular arrows. The result is the generation of tip vortex that extends behind the wind turbine, resulting in a very complex turbulent wake. The extension of tip vortexes is shown in Fig. 3.13a, where, for the sake of clarity, it is drawn only on the tip of one blade. Similar vortexes are generated on the tips of other blades and on each blade's root where the blade comes to an end.

From both illustrations of Fig. 3.13, it is easily concluded that the effect of tip vortex is a vertical induced velocity known as the *rotational induced velocity*. In contrast to the axial induced velocity, the rotational induced velocity is exactly in the same direction as the rotational velocity; hence it will be added to the rotational velocity. In the same manner that dealt with the axial velocity, the rotational induced velocity is calculated in terms of rotational velocity,

$$u_{\text{ind},r} = a' u_{\text{rot}}, \tag{3.7}$$

where a' is called the rotational induction factor, and as it is expected and illustrated in Fig. 3.13, it is not constant even for a specific blade.

Wind turbine aerodynamics is a very complicated phenomenon, and the calculation of turbine power is hard and tedious work. For example, obtaining induction factors, a and a', is not as easy as expected. As stated above, these factors are not constant even for a single blade under steady-state condition. If the operating conditions change, these factors will change. In general, the induction factors greatly depend on the geometrical parameters such as the length of the blade, used airfoils, local chord length, and local twist. In addition, they depend on operational conditions such as the wind velocity, turbine rotational speed, and, less importantly, on the ambient properties such as temperature, humidity level, and atmospheric pressure.

The normal way of study is through experimental tests and also numerical calculations. Hopefully, different applications exist for the simulation of Navier–Stokes equations. These codes give accurate results comparable to the actual data. However, they require lots of numerical calculations, hence are costly.

In addition to the numerical methods, some simple and available algorithms exist that enable us to calculate the characteristics of a full turbine. One of the most famous is called the blade element momentum, or BEM, method. This method is quite simple, and its implementation is straightforward. In the following section, BEM method is discussed. The method is very fast and can simulate a wind turbine in a fraction of a second.

Example 3.5. An Enercon E-16 operates at its rated condition. If for this turbine, the axial induction factor is $a = 0.25$ and the rotational induction factor is $a' = 0.15$, calculate the actual wind components in the axial and radial directions.

Answer. Enercon E-16 is an onshore turbine whose characteristics are given in Section C.1. The diameter of this turbine is $D = 16.2\,m$, and its rotor speed is 50 rpm. The turbine generates 55 kW power at a rated wind speed of $12\,m\,s^{-1}$.

The axial induced velocity reduces the far-field wind speed. Hence, we can write

$$u_{act,a} = u_\infty - u_{ind,a},$$

in which the induced velocity is calculated from Eq. (3.6). Substituting the available data in the above equation yields

$$u_{act,a} = 12 - 0.25 \times 12 = 9\,m\,s^{-1}.$$

The actual velocity in the axial direction is equal for the blades from their root to tip. But the radial velocity differs from root to tip due to its rotational nature. The rotational speed of the blade is

$$u_{act,r} = r\omega - u_{ind,r},$$

in which the induced velocity is calculated using Eq. (3.7). Substituting the available data in the above equation yields

$$u_{act,r} = 50\frac{\pi}{30}r - 50\frac{\pi}{30} \times 0.15r = 4.45r.$$

For this turbine, the radial induced velocity is a linear function of radius. It is zero at the axis of rotation and $36\,m\,s^{-1}$ at the tip radius ($r_{tip} = 8.1\,m$). $\square$

The axial and radial induction factors are assumed constant in the above example. However, these values are not uniform across a blade and have a different distribution profile. Therefore, even for a steady-state condition, the induced velocity in both directions will not be uniform (even for the axial direction). Unfortunately, these induction factors are unknown and must be calculated a priori.

3.5 Blade element method

Although the turbine aerodynamics is very complex, with some simplifications, one can make a simple and almost accurate estimation of the turbine's generated power. In the calculation of aerodynamic forces, the most challenging part is the calculation of induction factors. These factors are not easy to obtain. But assume that in some magical manner we have a reasonable estimate of induction factors in each section of a blade. In that case, we can calculate the local angle of attack, and consequently the local lift and drag on that section. In general, for obtaining the induction factors, we have to use either complex and time-consuming CFD models or simple procedures such as BEM method. In this section, the concept of BEM is discussed in more detail (Hansen, 2008; Hamedi et al., 2015).

BEM procedure is illustrated in Fig. 3.14. In this method, a blade is divided into several elements as shown in Fig. 3.14a. The more elements, the more accurate the method. However, just like many other numerical algorithms, the results converge to a unique solution for a specific number of elements, hence increasing the number of elements while consuming more computational resources and increasing the calculation time will not give more accurate results. Therefore, we have to find the correct number of elements for each individual problem. For each element, we assume that:

- The airfoil is known, thus the lift and drag curves versus alpha (the curves similar to those in Fig. 3.7) are given.
- In addition to the aerodynamic curves, the geometrical parameters of the airfoil are also known. Therefore, the local chord, twist angle, thickness of the element, and radial distance of the element from the axis of rotation, r, are known.
- Each element has its own lift and drag, and the elements do not have any effect on each other.
- The aerodynamic forces are applied to the center of each element.
- The exerted force differs from element to element, and we assume that these forces vary linearly from one element to another.

Since the elements do not affect each other, the aerodynamic force of each element is calculated separately. Fig. 3.14b shows the cross-section of a representative element shown in Fig. 3.14a. In BEM method, we first assume that the induction factors, a and a', are known. Thus, we can calculate the relative wind velocity, u_{rel}, as shown. The axial component of u_{rel} is obtained by subtracting axial induced velocity from the far-field velocity, and its rotational component

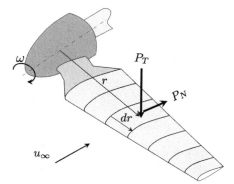

(a) Dividing a blade into elements

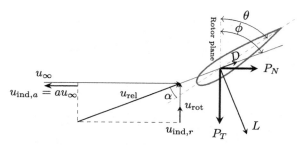

(b) Effect of induced velocity on the relative wind

FIGURE 3.14   Schematic concept of blade element momentum.

is obtained by the addition of rotational induction velocity to the rotational velocity. Mathematically we have

$$u_a = u_\infty - au_\infty = u_\infty(1 - a), \tag{3.8}$$
$$u_r = u_{\text{rot}} + a'u_{\text{rot}} = u_{\text{rot}}(1 + a'). \tag{3.9}$$

The angle of attack, or α, is defined as the angle between the relative velocity and the local chord. In studying Fig. 3.14b, in addition to α, two other angles are shown. The first angle is θ, which is called the *local pitch angle* and defined as the angle of the local chord and the rotation plane. The other angle is ϕ, which is defined as

$$\phi = \alpha + \theta. \tag{3.10}$$

Note that θ has two components, namely $\theta = \theta_p + \beta$, where θ_p is called the *pitch angle,* that is, the angle between the tip chord and the rotor plane; and β is known as the *twist,* which is measured relative to the tip chord. It means that it is zero at the tip and has its maximum value at the root. By this definition, the angle of attack is

$$\alpha = \phi - \beta - \theta_p. \tag{3.11}$$

Also, from Fig. 3.14b we see that

$$\tan(\phi) = \frac{u_\infty(1-a)}{r\omega(1+a')}. \tag{3.12}$$

According to the definition of ϕ and Fig. 3.14b, we see that

$$u_{\text{rel}} \sin \phi = u_\infty(1-a), \tag{3.13}$$

$$u_{\text{rel}} \cos \phi = u_{\text{rot}}(1+a') = \omega r(1+a'). \tag{3.14}$$

By having α and the properties of the airfoil such as c_l and c_d curves (curves such as those shown in Fig. 3.7), we can find L and D. Just note that, when using Eqs. (3.4) and (3.5), u_∞ should be replaced by u_{rel}. Therefore we have

$$L = \frac{1}{2}\rho u_{\text{rel}}^2 c_l, \tag{3.15}$$

$$D = \frac{1}{2}\rho u_{\text{rel}}^2 c_d. \tag{3.16}$$

From the figure, it is clear that both the lift and drag contribute to the tangential and normal forces. The tangential force, or P_T, is in the rotor plane and causes the blade to rotate, but the normal force, P_N, is in the direction of the far-field wind velocity, or u_∞, and is an unfavorable force causing the wind turbine to bend. The geometrical relations yield

$$P_T = L \sin(\phi) - D \cos(\phi), \tag{3.17}$$

$$P_N = L \cos(\phi) + D \sin(\phi). \tag{3.18}$$

These forces can be expressed in a dimensionless form using the far-field momentum, $\frac{1}{2}\rho u_{\text{rel}}^2$, that yields

$$c_t = \frac{P_T}{\frac{1}{2}\rho u_{\text{rel}}^2} = c_l \sin(\phi) - c_d \cos(\phi), \tag{3.19}$$

$$c_n = \frac{P_N}{\frac{1}{2}\rho u_{\text{rel}}^2} = c_l \cos(\phi) + c_d \sin(\phi). \tag{3.20}$$

Having P_T and the radial distance from the axis of rotation, r, the torque of each element shown in Fig. 3.14a is obtained using

$$dM = r N_B P_T dr, \tag{3.21}$$

in which N_B is the number of blades and dr is the width of that element. In addition to the torque, the thrust of normal force can also be calculated using

$$dT = N_B P_N dr. \tag{3.22}$$

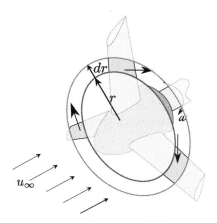

FIGURE 3.15 Annular control volume for calculating dM and dT.

The torque and thrust obtained by Eqs. (3.21) and (3.22) can be obtained using the famous Euler equation in pumps and turbines and momentum theory, respectively. These parameters for the annular control volume shown in Fig. 3.15 are

$$dM = 4\pi r^3 \rho V_{\text{rel}} \omega (1 - a) a' dr \qquad (3.23)$$

and

$$dT = 4\pi r \rho V_{\text{rel}}^2 a (1 - a) dr. \qquad (3.24)$$

In deriving Eqs. (3.23) and (3.24), it is assumed that the whole control volume shown in Fig. 3.15 gains power or the whole control volume is made of blades. I reality, the surface of the control volume is open to the wind in the intervals between the blades. Therefore, these intervals do not capture any power, and the wind blows through their open surface. Consequently, the assumption of Eqs. (3.23) and (3.24) can be translated to the infinite number of blades which is not correct. Prandtl noticed the problem and made a correction to this fact, which is called *Prandtl's tip loss correction*. According to this correction, the above equations are modified as follows:

$$dM = 4\pi r^3 \rho V_{\text{rel}} \omega (1 - a) a' F dr \qquad (3.25)$$

and

$$dT = 4\pi r \rho V_{\text{rel}}^2 a (1 - a) F dr. \qquad (3.26)$$

The correction factor, F, is defined as:

$$F = \frac{2}{\pi} \cos^{-1} \exp(-f), \qquad (3.27)$$

where

$$f = \frac{N_B}{2} \frac{R - r}{r \sin \phi}. \tag{3.28}$$

As shown in Fig. 3.15, for the annular control volume, the air cannot flow through the entire control volume since a portion of the control volume is filled with the blades. The fraction of the control volume that is filled with the solid blade to the whole annular volume is called solidity and is given by the following relation:

$$\sigma(r) = \frac{c(r)N_B}{2\pi r}, \tag{3.29}$$

where $c(r)$ is the local chord that is a function of r and N_B is the number of blades as before.

Equating Eqs. (3.21) and (3.25) and also Eqs. (3.22) and (3.26), the induction factors can be obtained. Applying Eqs. (3.17) to (3.20), using the definitions (3.13) and (3.14), and the concept of solidity (Eq. (3.29)), yields the following equations for a and a':

$$a = \frac{1}{\dfrac{4F \sin^2 \phi}{\sigma(r)c_n} + 1}, \tag{3.30}$$

$$a' = \frac{1}{\dfrac{4F \sin \phi \cos \phi}{\sigma(r)c_t} - 1}. \tag{3.31}$$

The main objective of the BEM is to find proper values for a and a' for each element shown in Fig. 3.14a. In this algorithm, the shown elements are considered one by one, and as stated before, the elements do not affect each other. For each element, a and a' are calculated separately using an 8-step algorithm shown in Fig. 3.16. The algorithm starts with guessing initial values for a and a'. Since these values are between zero and 0.4, $a = a' = 0$ would be considered a good guess. The algorithm continues until a and a' are calculated again in Step 6. If a and a' differ from the guessed values, it returns back to Step 2 and repeats the whole procedure. This time, a and a' are not zero, and their values are taken from Step 6. The process iterates until a and a' converge.

At the end of the algorithm, we should check whether $a < 0.4$ or not. If the axial induction factor becomes approximately larger than 0.4, Glauert showed that the classical momentum theory that was used for obtaining Eq. (3.25) and (3.26) fails to give a good answer. The largest possible value for a is named as a_c. The maximum value of $a_c = 0.4$ is quite an approximation, and in many research articles different values are reported. Some researchers suggested $a_c = \frac{1}{3}$ and some others suggested different values. To correct the axial induction value, if $a > a_c$, empirical expressions for c_t must be used. The details of the method can be found in many references (Hosseini et al., 2019; Moradtabrizi

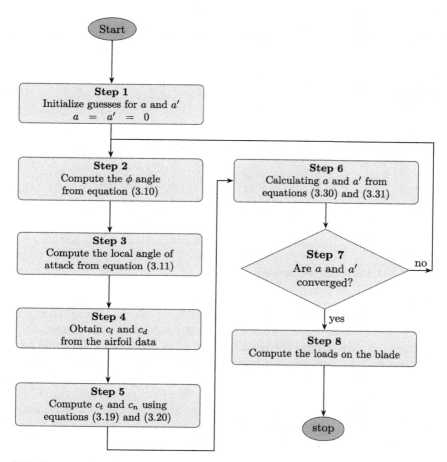

FIGURE 3.16 BEM flowchart.

et al., 2016; Nili et al., 2016; Tahani et al., 2014; Goupee et al., 2015; Grant, 2005; Ivanell et al., 2007; Madsen et al., 2012; Merrill, 2011; Sathyajith, 2006; Piana, 2019; Troldborg et al., 2009), but the final result is that, at the end of the flowchart of Fig. 3.16, the following modification must be done:

- If $a < a_c$, the value of a remains the same as obtained from Eq. (3.30).
- If $a > a_c$, we have to replace a by

$$a = \frac{1}{2}\left(2 + K(1 - 2a_c) - \sqrt{(K(1 - 2a_c) + 2)^2 + 4(Ka_c^2 - 1)}\right), \quad (3.32)$$

where

$$K = \frac{4F \sin^2 \phi}{\sigma(r)c_n}. \quad (3.33)$$

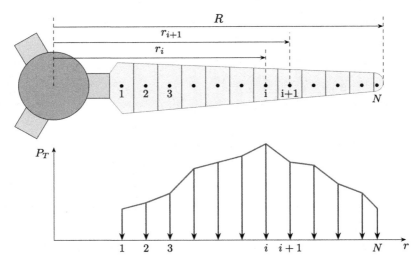

FIGURE 3.17 Calculation of axial torque.

The BEM algorithm discussed here enables us to calculate the aerodynamic forces on the blades. These forces are required for estimating the bending momentum and axial torque. The bending momentum is useful for structural calculations both for the rotor and tower. Therefore, it is understandable that P_N is the only responsible component for bending momentum, and P_T is the force required for the calculation of the axial torque. Note that, according to Fig. 3.14a, both P_N and P_T are applied on the center of the blade element. This fact is illustrated in Fig. 3.17 from another view. We can assume that the tangential forces on the elements vary linearly from one element to the next to calculate the axial torque. Hence from the ith to the $(i + 1)$th element, we can write

$$P_T = A_i r + B_i, \qquad (3.34)$$

where

$$A_i = \frac{P_{T,i+1} - P_{T,i}}{r_{i+1} - r_i}, \qquad (3.35)$$

$$B_i = \frac{P_{T,i} r_{i+1} - P_{T,i+1} r_i}{r_{i+1} - r_i}. \qquad (3.36)$$

Eq. (3.34) evaluates the linear variation of tangential force from element i to $i + 1$. Since it is a variable load, the axial torque of a infinitesimal element of length dr is obtained using

$$dM = r P_T dr = (A_i r^2 + B_i r) dr. \qquad (3.37)$$

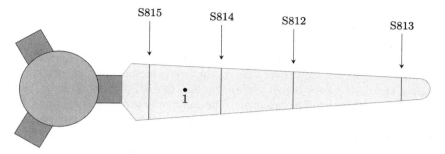

FIGURE 3.18 Construction of a turbine blade with different airfoils.

Consequently, the contribution $M_{i,i+1}$ to the total shaft torque can be easily calculated by integrating Eq. (3.37). The result is

$$M_{i,i+1} = \left[\frac{1}{3}A_i r^3 + \frac{1}{2}B_i r^2\right]_{r_i}^{r_{i+1}} = \frac{1}{3}A_i\left(r_{i+1}^3 - r_i^3\right) + \frac{1}{2}B_i\left(r_{i+1}^2 - r_i^2\right). \quad (3.38)$$

Eq. (3.38) gives the axial torque resulted from P_T from element i to $i+1$ for one blade. The total turbine axial torque is the sum of the torques of all the elements for all the blades. This means that M_{tot} is

$$M_{\text{tot}} = N_B \sum_{i=1}^{N-1} M_{i,i+1}, \quad (3.39)$$

in which N_B is the number of blades and the summation takes place from $i=1$ to $i = N - 1$.

3.6 Blades with different airfoils

A blade is designed by choosing the appropriate airfoil as its cross-section. But it does not mean that the whole blade must have a unique airfoil type. For small-scale blades, a single airfoil can be used to construct the entire blade. However, for large turbines, different airfoils are used in different sections. For example, thick airfoils are used at the root section because the root bears the mechanical loads. Hence, it must be mechanically more resistant.

Fig. 3.18 shows a multiairfoil blade in which different airfoils are used at different blade's radii. NREL proposed the shown configuration for wind turbines with generator sizes between 150 to 400 kW. For these turbines, the blade length is usually between 10 and 15 meters.

As indicated in the figure, the airfoils are only used in certain sections. Therefore, using aerodynamic coefficients is valid only at the shown specific radii. For other radii, the airfoils are unknown; hence we cannot use the tabulated data for c_l and c_d of the airfoils. For example, at point 1, the aerodynamic coefficients are neither for S815 nor for S814, but the values are somewhere in-between the two.

From one airfoil to another, the blade is lofted with different scales. For example, it can be linearly lofted or lofted by a nonlinear profile. In any case, the interval cross-section between the two airfoils is neither the starting nor the ending airfoil. It means that if we want to use the BEM method, the c_l and c_d data of neither airfoil are adequate. The solution is to use interpolations. In other words, at these intervals, aerodynamic characteristics must be interpolated between the data of the two ending airfoils.

3.7 Simulation of wind turbines

To show the capability of BEM method on simulation of wind turbines, we use this algorithm (see Fig. 3.16) to simulate stall- and pitch-regulated wind turbines. The turbines' specifications are summarized in Appendix C and the aerodynamic coefficients are taken from Appendix B.

3.7.1 Stall-regulated wind turbines

Nordtank NTK 500/41 (Nordtank, 2021) is a stall-regulated wind turbine whose specifications are given in Section C.11. From Fig. C.11 it is clear that the output power of this turbine is automatically controlled by aerodynamic design since it is not fixed after the wind speed exceeds the rated velocity, i.e., 14.0 m s^{-1}.

The blades' characteristics of NTK 500/41 are tabulated in Table C.1. In this table the chord and twist of the blades are given with respect to the radius. Unfortunately, there is no exact information about the used airfoils for this turbine. Hence, we need to choose an arbitrary blade section for simulation. The maximum rotational velocity of its rotor is 27.1 rpm.

Example 3.6. To show the applicability of BEM method on simulation of stall-regulated wind turbines, simulate NTK-500/41 and compare the power curve with real data.

Answer. The specification of NTK-500/41 is given in Section C.11. As it can be seen, the blade specifications are tabulated in Table C.1. In this table, the chord length and twist angles are given at different radii. But unfortunately, there is no information for the used airfoil; hence, we need to choose a proper airfoil for simulation. Here we choose FX66-S196-V1, whose characteristics are given in Appendix B. In other examples, the effect of airfoils will be investigated.

The simulation process is shown in Fig. 3.16. But for simulation of a real case at any desirable wind velocity, we need to have adequate information about the rated conditions since the efficiency of different components depends on the rated values such as P_{mR}, P_{tR}, and P_{eR}. Therefore, the first step in the wind turbine simulation is to simulate the turbine at its rated velocity.

The blade is divided into 17 sections. Therefore, the BEM procedure must be conducted over all 17 sections. We assume that all the sections have the same

TABLE 3.2 Results of BEM algorithm at rated velocity $u_R = 14\,\mathrm{m\,s^{-1}}$.

Section No.	α (degree)	c_l	c_d	a	a'
1	25.3427	0.65762	0.01944	0.0391027	0.04087
2	23.5543	0.65762	0.01944	0.0418759	0.02948
3	22.3548	0.65762	0.01944	0.0443096	0.02234
4	21.5888	0.65762	0.01944	0.0467599	0.01763
5	21.1001	0.65762	0.01944	0.0490986	0.01432
6	18.8037	1.18388	0.01944	0.0968319	0.02186
7	17.5324	1.24345	0.01944	0.106489	0.01940
8	16.5915	1.24087	0.01944	0.109756	0.01647
9	15.9169	1.23180	0.01944	0.111958	0.01400
10	15.1422	1.22468	0.01944	0.113958	0.01197
11	14.5419	1.22632	0.01944	0.116864	0.01033
12	14.1455	1.23209	0.01944	0.120537	0.00894
13	13.7736	1.24126	0.01944	0.125583	0.00773
14	13.2908	1.25870	0.01944	0.134668	0.00671
15	12.5856	1.29463	0.01944	0.148326	0.00566
16	11.7785	1.34723	0.01944	0.16862	0.00429
17	11.1018	1.39501	0.01944	0.189385	0.00206

airfoil; therefore, Eqs. (B.17) and (B.18) are used to obtain the lift and drag coefficients. However, as Fig. B.9 illustrates, these equations are valid for specific ranges. Thus a constant value is used, where the angle of attack is out of the range. This fact can be better understood by analyzing the results of simulation shown in Table 3.2. As it can be seen, the angle of attack for different sections varies from 11 to 25 degrees, while the curve data for FX66-S196-V1 is valid up to 11 degrees for drag and 18 degrees for lift. Hence, as it can be seen, the lift and drag coefficients remain constant when α exceeds the limit. Although this assumption imposes some errors in calculations, the results show that BEM could give reasonable results. Table 3.2 also shows the variation of a and a' in different sections. These data are plotted in Fig. 3.19 to better visualize the induction factor distribution. The jump in the induction factors is attributed to the curve-fitting profiles used in c_l and c_d. This examples shows that proper curve-fitting is crucial in aerodynamic calculations.

The datasheet of the turbine shows that $P_{eR} = 500\,\mathrm{kW}$, hence this value is used to obtain the generator's efficiency. The result of the algorithm is shown in Fig. 3.16 gives the following data for the rated values:

$$P_{mR} = 548.68\,\mathrm{kW},$$
$$P_{tR} = 515.76\,\mathrm{kW}.$$

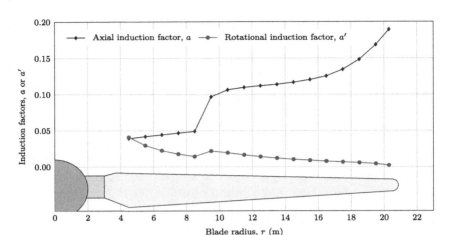

FIGURE 3.19 Distribution of the axial and rotational induction factors.

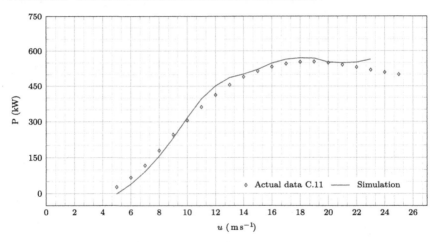

FIGURE 3.20 Simulation of Example 3.6, using FX66-S196-V1 airfoil.

Now that the rated values are calclated, we can use the BEM algorithm to calculate the mechanical power at any desirable velocity. For obtaining the electrical power, we have to calculate the efficiency of the gearbox and the generator. These efficiencies are given by Eqs. (2.62) and (2.63), and can be easily calculated using the rated parameters (i.e., P_{mR}, P_{tR}, and P_{eR}). The result of calculations is shown in Fig. 3.20. In this figure, the electrical power is compared to the power curve taken from the turbine's catalog. The result indicates that BEM was able to capture the power curve in a wide range accurately. □

The above example shows that BEM is able to simulate the stall-regulated wind turbines. Note that the airfoil data used for simulation was not accurate

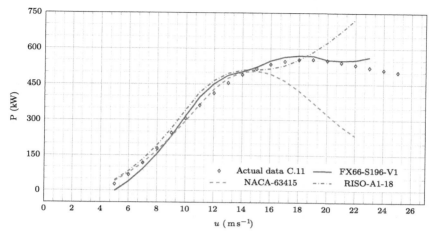

FIGURE 3.21 Simulation of Example 3.6, using different airfoils.

enough since we have no data for a wide range. The more accurate the input parameters, the more accurate the results.

In addition to the accurate data, it must be noted that exact information about the used airfoil is crucial for obtaining accurate results. In the following example, the effect of proper airfoil on the simulation results is investigated.

Example 3.7. Repeat the previous example and simulate the same turbine, using different airfoils, and compare the results.

Answer. Appendix B contains the information for different airfoils. To have a good comparison, we select NACA-63415 (see Fig. B.1), and RISO-A1-18 (see Fig. B.6) to simulate NTK-500/41. The procedure is the same as described in Example 3.6, hence is not repeated again.

The results are plotted in Fig. 3.21. The results show that NACA-63415 predicts the best results for lower wind speeds, but it fails to capture a proper power for wind speeds over the rated velocity. Moreover, RISO-A1-18 overpredicts almost in all the ranges. For speeds higher than $18\,\mathrm{m\,s^{-1}}$, RISO-A1-18 results in a sharp increase, which does not agree with the catalog data.

The investigations reveal that the BEM method must be used with care, and proper input data must be provided. Otherwise, we may fail in obtaining a reasonable simulation. □

The above example shows that how aerodynamic coefficients affect the simulation results. It can be concluded that for a proper simulation, one must gather as much information as possible; otherwise, the simulation may be inaccurate. For the present simulations, a single airfoil is used for the whole blade; however, as explained, different airfoils are practically used to construct a blade.

Hence, if the exact geometry of the blade were available, a better result could be generated.

3.7.2 Pitch-controlled wind turbines

A pitch-controlled wind turbine can be simulated using BEM method in the same manner as we did for stall-regulated ones. The only difference is that for a stall-regulated wind turbine, the pitch angle, θ_p, in Eq. (3.11) is always zero. However, it is zero for a pitch-regulated turbine until the wind velocity is lower than the rated velocity. But when the wind speed exceeds the rated value, θ_p must be obtained so that the output power becomes equal to the rated power. One of the easiest ways for obtaining the proper pitch angle is the method of trial and error. The procedure starts with a guessed value for θ_p and performs the simulation. If the obtained power is not equal to the rated power, we modify the pitch angle and start the simulations again. The procedure repeats until the turbine's power becomes equal to its rated power.

Example 3.8. As a pitch-controlled wind turbine, select MWT2.5-103-I from Appendix C. Simulate the turbine and discuss the results.

Answer. The turbine operates with a horizontal axis 103 m rotor diameter whose maximum rotational speed is 14.5 rpm and its rated power is 2,500 kW. The power curve of MWT2.5-103-I is shown in Fig. C.18 and its main characteristics are listed in Table C.2.

The specification of blades is tabulated in Table C.2. These characteristics are enough for simulation of the wind turbine by BEM method except that there is no information about the blade profile or its used airfoils. Therefore, as we did for the previous simulation, we use some arbitrary blade sections and compare the results with the experimental data.

The power curve of MWT2.5-103-I (see Fig. C.18) clearly indicates that this turbine is equipped with a pitch controller since, after reaching the rated power, its power remains constant. For this reason, the BEM algorithm should be modified such that this condition is satisfied. In other words, if the obtained power of the turbine exceeds the rated power, the installation angle, or local pitch angle, θ_p in Eq. (3.11), should be modified, and the calculations repeated such that the power becomes the rated power.

The results of the simulation are shown in Figs. 3.22 and 3.23. The power curve obtained by simulation agrees well with the catalog data, although, as explained in the previous examples, the exact type of airfoils used for the construction of the blades is unknown. But the simulation results for two different airfoils, namely FX66-S196-V1 and NACA-63415, indicate that both airfoils are able to produce power at the rated level. For lower wind speeds, the power curve has a little deviation from the tabulated catalog data. This fact is more obvious by comparing the c_p curves as shown in Fig. 3.23. □

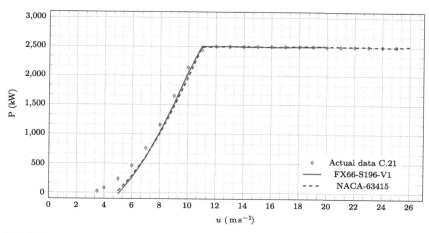

FIGURE 3.22 Simulation of pitch-regulated wind turbine using different airfoils.

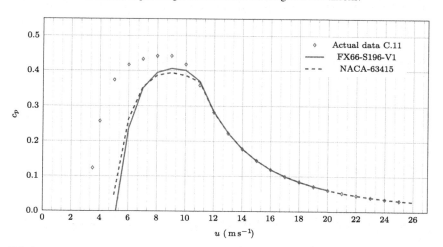

FIGURE 3.23 Comparison between different airfoils.

Example 3.9. Calculate the efficiency of different components of the Mapna turbine.

Answer. The efficiency of the gearbox and generator have been calculated using the relations explained in the previous chapter. The results are plotted in Fig. 3.24 for NACA-63415 airfoil. For comparison, c_p is also plotted in the same figure, as well as the overall efficiency. It indicates that the gearbox efficiency is lower than the generator efficiency in all ranges.   □

Example 3.10. How does a pitch-regulated turbine controls the output power?

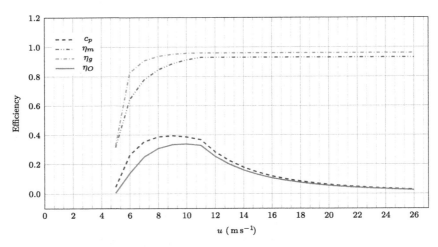

FIGURE 3.24 Efficiency of different parts.

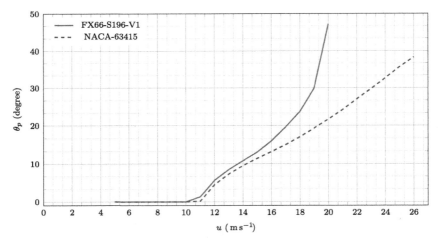

FIGURE 3.25 Variation of θ_p at different wind speed.

Answer. The answer to this question is better understood by studying Fig. 3.25, in which the pitch angle is plotted versus the wind speed. When the wind speed is less than the rated velocity, as can be seen in the figure, the pitch angle is zero. Hence, the turbine operates just like a normal turbine. But when the wind speed exceeds the rated power, the pitch angle increases. The increase in pitch angle causes a different angle of attack, α, for all the sections of the blade. Therefore, the turbine's captured power can be controlled by adjusting the pitch angle. The controller is designed so that the turbine's output power becomes constant from the rated velocity, u_R, to the furling, or cut-out, velocity, u_F.

Fig. 3.25 compares the pitch angles obtained for two different airfoils. Since the airfoils have different characteristics, for controlling the output power, different pitch angles are obtained. Unfortunately, for the present turbine, no experimental data is available to compare the results. □

3.8 Summary

The aerodynamic analysis of a wind turbine is a complicated task. It is a multi-disciplinary field in which many different physical aspects are involved. To have a deep understanding of the aerodynamics of wind turbines, one needs to have a vast theoretical background and experience. In the present chapter, the fundamentals of aerodynamics are discussed. The material of the chapter gives a good introduction to the basics of aerodynamics that are necessary for analyzing wind turbines. It starts with the concept of airfoils and continues to extend the physical phenomena to the aerodynamics of a three-dimensional blade. After that, the BEM method is introduced, which is a simple tool for the simulation of a wind turbine. Finally, BEM is used to illustrate its power for the simulation of stall- and pitch-controlled turbines.

The contents of the present chapter can be used for the simulation of a single wind turbine. As will be discussed in the future chapters, the same material are useful for the simulation of wind farms, which is the main concern of the present book. It is tried to solve some practical examples so that the material of the present chapter be better understood.

3.9 Problems

1. One of the tools that are used to analyze subsonic airfoils is called XFOIL (MIT, 2021). This open-source software is a cross-platform code and can be run under various operating systems. Using XFOIL, calculate lift and drag coefficients of airfoils given in Appendix B.
2. Another free tool for analyzing airfoils is called Airfoil Tools (Airfoil Tools, 2021), which is an online platform. Recalculate the airfoils of Appendix B, and compare the results with XFOIL.
3. According to the conventions of NACA 4-digit airfoils, discuss the following shapes: NACA0012, NACA0024, NACA0036, and NACA0048.
4. What is the difference between NACA2448 and NACA3648?
5. Investigate the usage of NACA 4-digit airfoils in the wind turbine industry.
6. Investigate the development of NACA airfoils that are suitable for wind turbines.
7. Investigate the development of FFA airfoils that are suitable for wind turbines.
8. Investigate the development of FX airfoils that are suitable for wind turbines.
9. Investigate the development of RISØ airfoils that are suitable for wind turbines.

10. Investigate the development of DU airfoils that are suitable for wind turbines.
11. Consider an airfoil is located in a free stream. How does the orientation of lift and drag change when the airfoil is oscillating?
12. Studying the airfoils of Appendix B, discuss the thickness of these airfoils with regard to their c_l.
13. Compare DU91-W2-250 B.15 and NACA63415 B.1. Which one seems to be thicker? Investigate your answer with real data.
14. Repeat Example 3.3, using different airfoils. Plot the obtained lifts versus the airfoil name and compare the results.
15. Do the same thing for the variable cross-section wing explained in Example 3.4.
16. In Problem 14, calculate the drag of the airplane wings. Compare the results for different airfoils.
17. Do the same thing for the variable cross-section wing explained in Example 3.4.
18. Write a computer code to implement the BEM method. Use the guidelines and codes of Appendix G.
19. Modify the code developed in Problem 18 to make the code suitable for pitch-regulated turbines.
20. The BEM method explained by Fig. 3.16 calculates c_p but not η_m and η_g. Apply the necessary equations for calculation the gearbox and generator efficiencies and apply them to the developed BEM (Problem 19) software.
21. Use the developed code of Problem 20 and calculate NTK-500/41 turbine with different airfoils of Appendix B. Which airfoils yield the correct value?
22. The simulation results depend on the functions that are fitted on the aerodynamic data. Choose arbitrary functions for estimating c_l and c_d data of different airfoils and resimulate NTK-500/41 turbine. Compare the results with the presented fitted curves.
23. Using the data obtained from XFOIL and Airfoil Tools (see Problems 1 and 2), repeat Problem 22.
24. Repeat Problems 21 to 24 for Mapna turbine.
25. Assume that the blade of NTK-500/41 turbine is made of FX66-S196-V1 and NACA63415. Obviously, the thicker airfoil is used at the root and the thinner at the tip section, and the blade is linearly lofted between the two sections. If the root section is located at $r_{root} = 4.5$ m and the tip airfoil at $r_{tip} = 18.5$ m, simulate the turbine and compare the results with previous simulations.
26. Repeat Problem 26 with different airfoils of Appendix B and compare the results.
27. Investigate the effect of multiple airfoils as explained in previous problems for Mapna turbine.

References

Airfoil Tools, 2021. Airfoil Tools website. http://airfoiltools.com/index. (Accessed 27 May 2021).

Burton, Tony, Sharpe, David, Jenkins, Nick, Bossanyi, Ervin, 2001. Wind Energy Handbook, vol. 2. Wiley Online Library.

Goupee, Andrew J., Kimball, Richard W., de Ridder, Erik-Jan, Helder, Joop, Robertson, Amy N., Jonkman, Jason M., et al., 2015. A calibrated blade-element/momentum theory aerodynamic model of the MARIN stock wind turbine. In: The Twenty-Fifth International Ocean and Polar Engineering Conference. International Society of Offshore and Polar Engineers.

Grant, Ingram, 2005. Wind turbine blade analysis using the blade element momentum method. Duncan University, School of Engineering, Durham University, Durham, NC. https://www.academia.edu/1749333/Wind_Turbine_Blade_Analysis_using_the_Blade_Element_Momentum_Method.

Hamedi, Razieh, Javaheri, Alireza, Dehghan, Omid, Torabi, Farschad, 2015. A semi-analytical model for velocity profile at wind turbine wake using blade element momentum. Energy Equipment and Systems 3 (1), 13–24.

Hansen, M.O., 2008. Aerodynamic of Wind Turbines. Earthscan, London & Sterling, VA.

Hosseini, Radmarz, Roohi, Reza, Ahmadi, Goodarz, 2019. Parametric study of a novel oscillatory wind turbine. Energy Equipment and Systems 7 (4), 377–387.

Ivanell, Stefan, Sørensen, Jens N., Henningson, Dan, 2007. Numerical computations of wind turbine wakes. In: Wind Energy. Springer, pp. 259–263.

Johnson, Gary L., 1985. Wind Energy Systems. Citeseer.

Madsen, H.Aa., Riziotis, V., Zahle, Frederik, Hansen, Martin Otto Laver, Snel, H., Grasso, F., Larsen, Torben J., Politis, E., Rasmussen, Flemming, 2012. Blade element momentum modeling of inflow with shear in comparison with advanced model results. Wind Energy 15 (1), 63–81.

Manwell, James F., McGowan, Jon G., Rogers, Anthony L., 2010. Wind Energy Explained: Theory, Design and Application. John Wiley & Sons.

Mathew, Sathyajith, 2006. Wind Energy: Fundamentals, Resource Analysis and Economics. Springer.

Merrill, Robert S., 2011. Nonlinear aerodynamic corrections to blade element momentum model with validation experiments.

MIT, 2021. Xfoil, subsonic airfoil development system. https://web.mit.edu/drela/Public/web/xfoil/. (Accessed 27 May 2021).

Moradtabrizi, Hamid, Bagheri, Edris, Nejat, Amir, Kaviani, Hamid, 2016. Aerodynamic optimization of a 5 megawatt wind turbine blade. Energy Equipment and Systems 4 (2), 133–145.

Ng, Chong, Ran, Li, 2016. Offshore Wind Farms: Technologies, Design and Operation. Woodhead Publishing.

Nili, Ahmadabadi Mahdi, Mokhtarinia, Farzad, Shirani, Mehdi, 2016. Performance improvement of a wind turbine blade using a developed inverse design method.

Nordtank, 2021. Nordtank NTK 500/41. https://en.wind-turbine-models.com/turbines/384-nordtank-ntk-500-41. (Accessed 12 May 2021).

NREL, 2021. Wind turbine airfoil families. https://wind.nrel.gov/airfoils/AirfoilFamilies.html. (Accessed 20 March 2021).

Piana, Claudio, 2019. Implementation of an actuator line method for aerodynamic analysis of horizontal axis wind turbines.

Sathyajith, Mathew, 2006. Wind Energy: Fundamentals, Resource Analysis and Economics. Springer Science & Business Media.

Tahani, Mojtaba, Sokhansefat, Tahmine, Rahmani, Kiana, Ahmadi, Pouria, 2014. Aerodynamic optimal design of wind turbine blades using genetic algorithm. Energy Equipment and Systems 2 (2), 185–193.

Troldborg, N., Sørensen, J., Mikkelsen, R., 2009. Actuator line modeling of wind turbine wakes. PhD thesis. Technical University of Denmark.

Chapter 4

Wind turbine wake and its role in farm design

The flow regime behind a wind turbine is turbulent and quite far from the ideal case that was studied in Section 2.3.2. In the ideal case, the velocity before and after the turbine remains constant, and the pressure suddenly drops. But in reality, as explained in Section 3.4, the axial and rotational induced velocities, viscosity, turbulent effects, and blade rotation generate a turbulent wake behind the turbine, which extends downstream. In this region, the wind speed is not the same as it was just in front of the turbine, and as we move along the wind direction, the velocity gradually recovers so that it reaches the far-field velocity at an infinite distance behind the turbine.

From a practical point of view, the wake is not an important matter for a single turbine. If a single turbine is installed in a region, then the investigation of wind is not needed. However, the wake is important when we are dealing with a wind farm where several wind turbines are going to be located in a finite field (Abkar et al., 2018; Blondel and Cathelain, 2020; Carbajo Fuertes et al., 2018; Sun and Yang, 2018, 2020; Vermeulen et al., 1979). As shown in Fig. 4.1, the flow field downstream a turbine is not as powerful as if it were upstream. Therefore, turbine-2, located downstream of turbine-1, experiences a lower wind velocity, hence producing less energy. Then the wake of turbine-2 is a complex combination of the wake of turbine-1 and turbine-2. Therefore, if another turbine is located downstream of turbine-2, it will produce even less energy. Consequently, the overall energy production of the field will not be very impressive.

In locating the turbines, whatever the layout, some turbines will be located in the wake of their upstream turbines, and we cannot avoid locating the turbines in the wake regions. However, according to the wind rose diagram of a specific area, we can choose the best layout to maximize the annual gained energy. As an example, in the city of Torbate Jaam, the dominant wind blows from NNE as shown in Fig. 2.7. According to the figure, 35% of the time, the wind blows from the NNE direction. Therefore, if two distinct wind turbines were to be installed in this city, installing them perpendicular to this direction would be much better. In other words, the connecting line between these turbines should extend from WNW to ESE. By this installation layout, 35% of time we are sure that the turbines are not located in the wake of each other. But if the connecting line is aligned in the NNE–SSW direction, most of the time, one of the turbines

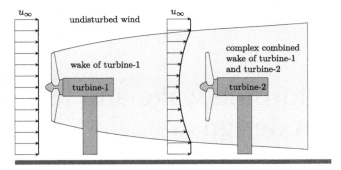

FIGURE 4.1 Effect of wake on downstream turbines.

lies in the wake of the other. The penalty of this layout design is that the turbine in the back produces less energy since it is operating at a lower wind velocity.

Although the solution seems to be simple, in practice, many different problems arise. In many cases, the wind rose does not show a dominant direction. Then the above solution does not work. In some cases, the field on which the farm is being installed, is limited by natural obstacles such as hills, mountains, lakes, etc. Therefore, the solution may not work for the region. In these situations, the wind farm layout design problem arises.

The above example shows the importance of wake and its evaluation in wind farm design. By increasing the number of wind turbines, the layout of the turbines becomes more and more important. To obtain an optimum layout design, advanced optimization algorithms should be incorporated for such a complex system. It is worth mentioning that layout design does not only depend on aerodynamic calculations but is affected by other important factors such as cabling, civil cost, and geographical features.

The main focus of the present chapter is on the physics of wake, its origin, and modeling methods. After that, the wake effect and its role in farm optimization are discussed.

4.1 Wake generation of a wind turbine

The induced velocity discussed in Chapter 3 and illustrated in Fig. 3.13 causes a sudden velocity drop just behind the turbine. Therefore, in a real case, the ideal behavior shown in Fig. 2.9 is not seen. The low-speed region just behind the wind turbine is called the *wake*. In this region, as shown in Fig. 4.2, the velocity profile just behind the turbine is affected by the turbine blades and nacelle. As shown in the figure, the downstream of the turbine is divided into the wake region, where the velocity profile is not uniform because of the effect wind turbine, and the undisturbed atmosphere. The further we go along the wind direction shown by the x axis, the more the velocity profile is recovered by the interaction with ambient.

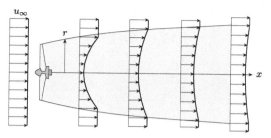

FIGURE 4.2 Behavior of wake.

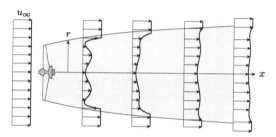

FIGURE 4.3 The complexity of the wake region.

The velocity in the wake region is very complicated because it is affected by axial and rotational induced velocity components. In addition, the tip vortex at the root and tip of the blades that are also shown in Fig. 3.13 makes the wake region much more complicated. Therefore, the velocity profile just behind the wind turbine is not as simple as shown in Fig. 4.2. Instead, the profile is best described by Fig. 4.3 in which, as we can see, the profile just behind the turbine is much more complex.

Regardless of the complexity of the wake, the wake can be considered as a *turbulent free-shear flow* because after the wake is released from the turbine, there is no source of momentum in the region, and the profile is moving in the atmosphere. In the topic of turbulent flows, the wake problem is a classic example of free-shear flow, and it is studied in more detail.

Wake profile is a function of two independent variables, namely the axial distance in the x direction and the radial distance r. As the wake flows in the x direction, the profile absorbs more mass from the ambient and expands in the r direction. The profile then becomes flatter, and eventually, far away from the turbine, it becomes the same as the upstream velocity.

As discussed in turbulent flow references, the wake shows *self similarity*. Although the wake profiles at different x locations are different, all the profiles at any distance x can be cast to a single profile by introducing proper scale parameters. For example, studying Fig. 4.2 reveals that all the profiles in the wake region have almost a parabolic shape. By introducing proper dimensionless parameters, it can be shown that all these parabolas can become the same.

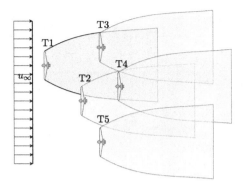

FIGURE 4.4 Effect of wake on downstream turbines.

The self-similarity solution and turbulent analysis will be discussed in the next chapter. It will be shown that the wake propagation can be accurately modeled by choosing proper scale parameters. Even if the wake is as complex as that shown in Fig. 4.3, the self-similarity analysis can be applied to obtain the wake profile at any location behind the wind turbine.

In a wind farm, the wake propagation becomes more and more complex because the wakes of different turbines may interact with each other and generate a complex region downstream. Such a situation is shown in Fig. 4.4 where a wind farm with five turbines is located in a specific field. The turbines downstream of the others may be located in different conditions. For example, T1 is the first turbine that faces the wind. Therefore, it experiences an undisturbed wind velocity. According to the figure, the same situation is true for T5, though the turbine is located downstream of T1. At the same time, as T1 and T5 are experiencing undisturbed wind, other turbines are located in different situations. T3 is located completely in the wake of T1, but T2 is partially in the wake of T1 and partially in the undisturbed wind. Finally, T4 is located in a combined wake of T1 and T2.

The above situation shows that in a wind farm, each turbine is operating in different operating conditions; for example, it operates under a different wind velocity and turbulence intensity. Therefore, not all the turbines in a single farm equally generate power because they are operating in different conditions. They may tolerate different mechanical stress, produce electricity with different frequencies, and many other factors.

The wind direction results in a completely different scenario, as shown in Fig. 4.5. Since each turbine rotates toward the wind, the whole wake interaction changes in pattern and even intensity. It means that the downstream turbines encounter different wind speeds, even if the speed of undisturbed velocity at far-field remains the same. These two examples show how complex the wind farm layout design can be and how important its evaluation is.

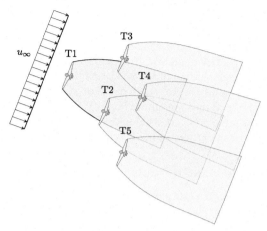

FIGURE 4.5 Effect of wind direction on downstream turbines.

To have a good estimate of the wake profile of each individual turbine, we need a proper wake model by which two main factors are to be obtained:

1. **Wake growth rate**

   The first parameter is the wake growth rate which is designated by $r(x)$. This factor can be used to track the wake and check if downstream turbines are located inside the wake or not.

2. **Velocity profile**

   From this property, the wind velocity at the location of a downstream turbine can be obtained. This velocity is then used to estimate the generated power of that turbine.

The advances in numerical methods, especially in the field of computational fluid dynamics (CFD), make it possible to accurately predict the wake region and the gained energy of the whole wind farm. However, the simulation of an entire wind farm with all the wind turbines and the modeling of geographical features and the terrain, though possible, require lots of numerical resources and are quite costly. Moreover, these types of simulations cannot be used in optimization algorithms due to their simulation time. For example, an ordinary optimization algorithm may require executing hundreds of simulations which is not possible if the solver is based on CFD models. In contrast to the CFD models, many different analytical models can estimate the wake and its propagation profile. These models are based on fundamental laws of fluid mechanics and also turbulent flow. The analytical models are not as accurate as those of the CFD, but their relative error is quite acceptable from an engineering perspective. These models are frequently used in many projects to optimize the wind farm layout. Some software has also incorporated the analytical models for layout design purposes. These applications use the model and integrate it with different optimization algorithms. The result is a layout in which the overall energy gained during a year is the maximum attainable value.

In the present chapter, some frequently-used models are introduced. These models have their own advantages and disadvantages. The study of these models gives an insightful understanding of the wake and the involved phenomena.

In the next chapter, a new wake model based on self-similarity solution is introduced and will be used throughout the book. The proposed model has higher accuracy and gives a more conformal velocity profile than the other conventional models. The details of the proposed model are given in the next chapter.

4.2 Conventional wake models

There are many different analytical models for wind turbine wake. Some of the models are pretty simple, and some are based on fluid dynamics fundamentals and even turbulence modeling. In order to have a better understanding of the wake and the involved parameters, here we have selected some famous models. Many researchers have used these models for different purposes. Some of the models are incorporated into famous wind simulators.

4.2.1 Jensen's model

Jensen (Jensen, 1983; Katic et al., 1986) carried on one of the first attempts towards wake modeling. He assumed that the wake behind a wind turbine has a simple hat shape as shown in Fig. 4.6. It is clear that in the real world, the blade rotation and viscosity effect creates a complicated wake just behind the wind turbine, and the present simple assumption is not valid. But in Jensen's model these effects are neglected for simplicity. The figure shows two different wake profiles, one of them is located just behind the turbine, shown by subscript w, and the other at an arbitrary distance, x. According to the Jensen's model, the velocity just behind the turbine reduces to u_w. Also, the velocity at any distance downstream the turbine is equal to u_1, which is greater than u_w. Jensen assumed that wake propagation has a linear shape with a slope equal to α. Therefore, according to the symbols shown in Fig. 4.6, the wake radius from r_w to $r(x)$ changes according to

$$r(x) = r_w + \alpha x. \tag{4.1}$$

Note that the initial wake radius differs from the turbine's radius or r_0 (see Fig. 4.6), but in Jensen's model, the difference is neglected. Hence, it is assumed that the initial wake radius is the same as the radius of the turbine, or

$$r_w = r_0. \tag{4.2}$$

Assuming continuity for the control volume shown by dashed lines in Fig. 4.6 yields

$$\pi r_w^2 u_w + \pi (r(x)^2 - r_w^2) u_\infty = \pi r(x)^2 u(x). \tag{4.3}$$

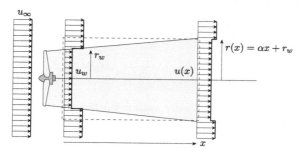

FIGURE 4.6 Simple linear model proposed by Jensen.

At this point, we have to give an estimate for the velocity just behind the turbine, u_w. As discussed in the previous chapter, the velocity just behind a turbine is determined by the value of the induction factor. This value is given by Eq. (3.8), which for clarity is repeated here,

$$u_w = u_\infty (1 - a). \tag{4.4}$$

Jensen assumed that $u_w = \frac{1}{3} u_\infty$, meaning that the induction factor is $a = \frac{2}{3}$. In practice, this is a large induction factor and may not always be true. Jensen pointed out this fact and mentioned that the value is uncertain and varies in different turbine loads.

Jensen's model was based on the same assumption, $u_w = \frac{1}{3} u_\infty$, resulting in the following equation for the velocity at any location x:

$$u(x) = u_\infty \left(1 - \frac{2}{3} \left(\frac{r_w}{r_w + \alpha x} \right)^2 \right), \tag{4.5}$$

where the wake expansion constant, or entrainment factor, α, was assumed to be around 0.05, but Frandsen (1992) gave the following relation for this parameter:

$$\alpha = \frac{0.5}{\ln(\frac{h}{z_0})}, \tag{4.6}$$

in which h is the hub height and z_0 is the surface roughness.

Example 4.1. For Vestas V80 turbines installed in Horns Rev 1 farm, obtain the entrainment factor. Then calculate the wake diameter of one turbine at the location of its first downstream turbine.

Answer. The data of Vestas V80 is given in Section C.17. The turbine is designed for offshore installation. Both its diameter and hub height are 80 m. According to the data of Appendix D.3, the distance between each turbine is $7D$, or 560 m.

From Table 2.2, we see that for a flat desert or rough sea, $z_0 = 0.001$ m. Thus, from Eq. (4.6) we have

$$\alpha = \frac{0.5}{\ln(\frac{h}{z_0})} = \frac{0.5}{\ln(\frac{80}{0.001})} = 0.0443.$$

Inserting this value into Eq. (4.1), the wake radius can be estimated as

$$r(x) = r_w + \alpha x = 40 + 0.0443 \times 560 = 64.8 \, \text{m}.$$

The same procedure may be conducted for other values of z_0. □

As we know, the assumption of $a = \frac{2}{3}$ is not always valid; therefore, if we use the general formulation, i.e., Eq. (4.5), the velocity at any location downstream of the turbine becomes

$$u(x) = u_\infty \left(1 - a\left(\frac{r_w}{r_w + \alpha x}\right)^2\right). \tag{4.7}$$

Then a is a factor that should be determined case by case. This factor is a function of turbine design, velocity field, and operational conditions.

In the original work of Jensen, the induction factor was chosen to be $a = \frac{1}{3}$ to make consistency with the Betz theory and to reach a maximum c_P based on one-dimensional momentum theory. In future studies, based on the real site data obtained by Vermeulen et al. (1979) and Højstrup (1983), Jensen amended the induction factor to $a = 0.324$. The field data were gathered from a Nibe-A turbine with 20 m diameter and assuming $8.1 \, \text{ms}^{-1}$ wind velocity at 100 m hub-height.

Although Jensen selected this value for a specific turbine under specific operating conditions, it created many misconceptions about the value. Many succeeding researchers used the same value for different purposes without noticing that the induction factor is not a unique value.

Although Jensen's model is quite simple and uses a linear expansion profile, it is still one of the frequently used models. This is because its simplicity makes it possible to be integrated with optimization algorithms. But in using the model, one should pay attention to the value of the induction factor because it is not a constant parameter for all the turbines under all operating conditions.

Example 4.2. Assume that two Siemens SWT-4.0-130, namely T1 and T2, are installed in the wind farm shown in Section D.1. Obtain the wake radius and wind velocity at the location of T2, according to Jensen's model. Plot the wind speed versus u_∞.

Answer. According to the data of Section C.22, the turbine diameter is $D = 130$ m and the hub height is $h = 89.5$ m. Since the turbine is manufactured for onshore installation, we can assume that it is installed in open flat terrain. For

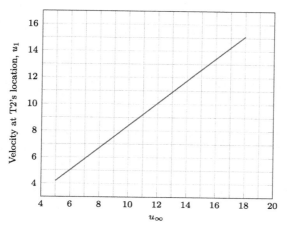

FIGURE 4.7 Result of Example 4.2.

such a field, the roughness $z_0 = 0.03$ m, according to Table 2.2. Hence, from Eq. (4.6) we have

$$\alpha = \frac{0.5}{\ln(\frac{h}{z_0})} = \frac{0.5}{\ln(\frac{89.5}{0.03})} = 0.0625.$$

To calculate the wake diameter, we need to have the distance between the two turbines. According to Fig. D.1, we assume that $x = 7$, which is conventional in many wind farms. Hence, the distance between the two turbines becomes $x = 7 \times 130 = 910$ m. Inserting this value into Eq. (4.1), the wake diameter can be estimated as

$$r(x) = r_w + \alpha x = 130 + 0.0625 \times 910 = 186.875 \text{ m}.$$

According to Jensen's model, the velocity at the location of T2 is obtained using Eq. (4.7). Also, in this model, it is assumed that $a = 0.324$ for all the conditions. Hence, we can calculate the wind speed to be

$$u_1 = u_\infty\left(1 - 0.324\left(\frac{130}{186.875}\right)^2\right) = 0.843 u_\infty.$$

The result shows that Jensen's model predicts a linear relationship between the upstream wind velocity and the velocity at the location of T2. This relation is plotted in Fig. 4.7. □

Example 4.3. Consider the same situation of Example 4.2. Discuss the variation of wind velocity at the location of T2 with respect to the distance between the two turbines.

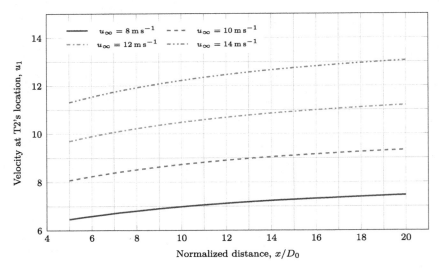

FIGURE 4.8 Result of Example 4.3.

Answer. If the horizontal distance between the two turbines changes, Jensen's model predicts the velocity at the location of T2 by Eq. (4.7). Applying the available data to the equation yields

$$u(x) = u_\infty \left(1 - 0.324 \left(\frac{130}{130 + 0.0597x}\right)^2\right).$$

The above equation is plotted for different u_∞ in Fig. 4.8. Jensen's model predicts that $u(x)$ reaches the wind velocity at infinity at very long spacing. □

4.2.2 Frandsen's model

Frandsen et al. (2006) believed that the wake model is related to the thrust coefficient, c_T. Therefore, they developed the following equation for the calculation of the thrust:

$$T = \rho A(x)u(x)(u_\infty - u(x)), \tag{4.8}$$

where $x = 0$ denotes the location of the turbine; hence, $u(0)$ means the velocity just at the turbine's rotating surface. The thrust can also be calculated according to the definition of thrust coefficient as shown by

$$T = \frac{1}{2}\rho A_0 u_\infty^2 c_T. \tag{4.9}$$

When writing Eq. (4.9), it is assumed that the velocity is at infinity, but the area is the swept area of the turbine. This fact was discussed in detail in the previous chapter.

Combining Eqs. (4.8) and (4.9), we get the following equation for the wake velocity expansion:

$$\frac{u(x)}{u_\infty} = \frac{1}{2} + \frac{1}{2}\sqrt{1 - 2c_T \left(\frac{r_0}{r(x)}\right)^2}. \tag{4.10}$$

According to Eq. (4.10), the velocity at distance x from the turbine is only a function of x and not r. In other words, Fandsen's model also assumes a hat-shape profile for the wake velocity as shown by Fig. 4.6. On the other hand, the model requires a relation for wake radius expansion, since in the above equation $r(x)$ is still unknown.

Frandsen further assumed that the initial wake area is related to the turbine's area according to the following relation:

$$\beta = \frac{A_w}{A_0} = \frac{1 - a/2}{1 - a}, \tag{4.11}$$

in which a is the induction factor. Also, the diameter of the initial wake is equal to $A_w = A(0)$. Equating Eqs. (4.8) and (4.9) and using Eq. (4.11), we get

$$c_T = a(2 - a) \longrightarrow a = 1 - \sqrt{1 - c_T}. \tag{4.12}$$

Note that solving Eq. (4.12) results in two different answers, one of which gives inadequate value for c_T. Substituting Eq. (4.12) in Eq. (4.11) yields

$$\beta = \left(\frac{r_w}{r_0}\right)^2 = \frac{1}{2}\left(\frac{1 + \sqrt{1 - c_T}}{\sqrt{1 - c_T}}\right). \tag{4.13}$$

Eq. (4.13) gives a relation for the wake radius just behind the wind turbine. In other words, the wake radius for $x = 0$. However, we still need an equation, or expression, for the wake radius for any distance x. Frandsen further discussed the matter and made use of the works done by Schlichting and Engelund on turbulence modeling with similarity solutions. Consequently, he assumed that the wake propagation profile is proportional to the cubic root of the distance. In other words,

$$r(x) \propto x^{1/3}. \tag{4.14}$$

Then using eddy viscosity concepts, he proposed the following equation for wake expansion:

$$D(x) = (\beta^{k/2} + \alpha s)^{\frac{1}{k}} D_0, \tag{4.15}$$

in which

$$s = \frac{x}{D_0}, \tag{4.16}$$

is the dimensionless wake distance. Also α is the entrainment factor discussed before. This factor can be obtained either from Eq. (4.6), or from the following equation proposed by Frandsen:

$$\alpha = \frac{\beta^{k/2}}{s} \left[(1 + 2\alpha_J s)^k - 1 \right], \tag{4.17}$$

where $\alpha_J = 0.05$ is the entrainment factor given by Jensen. This equation shows that the entrainment factor is not constant in Frandsen's model, even in a specified upstream wind velocity. It is a function of both c_T and the axial distance x. Therefore, we need to choose a specific distance to match the flow speeds and calculate α. This distance must be selected far away from the wind turbine. Typically, $s > 5$ would be a reasonable distance. Frandsen explained that the decay factor obtained from Eq. (4.17) is very large, on the order of $10\alpha_J$. The large values for α may result in a reasonable wake velocity at any distance x, but this predicts a very large wake radius that does not seem physical. This fact will be discussed in more detail through some examples.

As discussed before and also shown by Eq. (4.14), the wake expansion is proportional to the cubic root of x. This fact is a result of turbulent modeling via the similarity solution. Thus, according to this assumption and Schlichting, $k = 3$ is a proper value in Eq. (4.15). However, Frandsen mathematically explained that to have a nonvanishing and nonincreasing flow speed, the expansion of wake must have a square-root shape, or $D(x) \propto x^{1/2}$. Thus he used $k = 2$ instead of 3.

Frandsen's model is more advanced than Jensen's because it uses c_T, which contains the effect of wind speed and turbine operational conditions. Moreover, wake expansion is expressed according to turbulent flow fundamentals. But in many articles that used Frandsen's model, the authors chose $c_T = 0.88$ instead of the real c_T value. Surprisingly enough, this value comes from the same assumptions that Jensen made. We know that maximum c_T and the induction factor are related to each other according to Eq. (2.50). This relation shows that $c_T = 0.88$ is equal to $a = 0.324$. Again the induction factor is considered to be $a = 0.324$, which is consistent with the value used by Jensen. Unfortunately, this value does not apply to all the turbines under all possible operating conditions.

Example 4.4. For Eno Energy Eno 100 turbine (see Appendix C.19), obtain the initial wake radius, just behind the wind turbine, according to Frandsen's model.

Answer. According to Frandsen's model, the initial wake radius is obtained from Eq. (4.13). As can be seen, the radius is solely a function of c_T. Therefore, we need to obtain c_T for the turbine. On the other hand, c_T is a function of the wind velocity. Hence, it must be given by the manufacturer. However, a quick look at the data in Appendix C.19 reveals that we only have the c_p curve and not c_T. Thus, we need to first calculate c_T before proceeding any further.

From Eq. (2.54) we know that

$$c_p = 4a(1 - a)^2.$$

TABLE 4.1 Results of Example 4.4.

u_∞ (m s^{-1})	c_p	a	c_T	β	r_w
3	0.29	0.086969	0.3176	1.10528	52.82
4	0.41	0.137920	0.4755	1.19045	54.82
5	0.43	0.148139	0.5047	1.21050	55.28
6	0.44	0.153516	0.5197	1.22153	55.53
7	0.45	0.159095	0.5351	1.23334	55.80
8	0.44	0.153516	0.5197	1.22153	55.53
9	0.43	0.148139	0.5047	1.21050	55.28
10	0.39	0.128318	0.4474	1.17261	54.41
11	0.32	0.098419	0.3549	1.12254	53.23
12	0.26	0.076158	0.2814	1.08984	52.45
13	0.21	0.059331	0.2232	1.06732	51.91
14	0.17	0.046773	0.1783	1.05160	51.53
15	0.13	0.034892	0.1346	1.03750	51.18
16	0.11	0.029177	0.1133	1.03098	51.02
17	0.09	0.023600	0.0921	1.02477	50.86
18	0.08	0.020861	0.0817	1.02176	50.79
19	0.07	0.018153	0.0712	1.01883	50.72
20	0.06	0.015475	0.0609	1.01596	50.64
21	0.05	0.012826	0.0506	1.01316	50.57
22	0.04	0.010207	0.0404	1.01042	50.51
23	0.04	0.010207	0.0404	1.01042	50.51
24	0.03	0.007615	0.0302	1.00773	50.44
25	0.03	0.007615	0.0302	1.00773	50.44

The solution of this equation results in a at each wind velocity. Then c_T can be obtained using Eq. (2.50) as

$$c_T = 4a(1 - a).$$

Finally, Eq. (4.13) is used to calculate β and also the wake radius. In calculating r_w, we need to have the turbine's radius. This value is given in the catalog of the turbine, which is also available in Appendix C.19, namely $r_0 = 50.25$ m.

The results of calculations are summarized in Table 4.1 and also graphically shown in Figs. 4.9 and 4.10.

Note that the method used here for the calculation of c_T gives reasonable data. But, as we will discuss later, it is not quite accurate. ☐

Example 4.5. Calculate and plot the entrainment factor for the problem explained in Example 4.4.

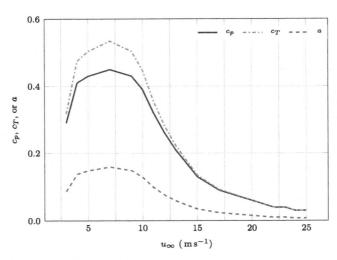

FIGURE 4.9 c_T and a of Example 4.4.

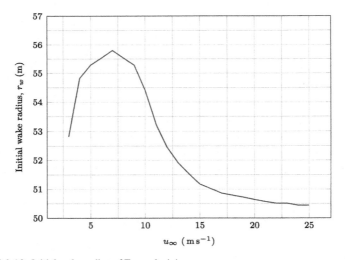

FIGURE 4.10 Initial wake radius of Example 4.4.

Answer. The entrainment factor is defined by Eq. (4.17). It is a function of β and x. As explained in the previous example, β itself is a function of c_T, or the upstream wind velocity. Therefore, for each upstream wind velocity, a different relation must be obtained. Moreover, as stated before, there are two different values for k, one suggested by Schlichting and the other by Frandsen. Here, we use both values to show the differences.

As shown by Fig. 4.11, using $k = 3$ predicts a much larger α factor. Comparing to Jensen's model, both values for k predict larger values even in wind

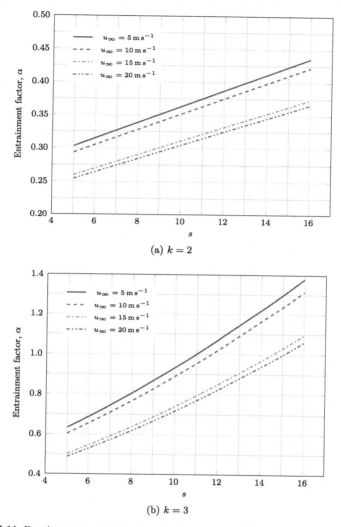

FIGURE 4.11 Entrainment or expansion factor for Example 4.5.

speeds close to Jensen's experiment. We can calculate α according to Eq. (4.6) using the data of Appendix C:

$$\alpha = \frac{0.5}{\ln(\frac{h}{z_0})} = \frac{0.5}{\ln(\frac{99}{0.03})} = 0.06171.$$

We assumed that the turbine is mounted in an open flat terrain covered with grass. As we can see, the predicted value for α by Eq. (4.17) is 10 times larger than that obtained from Eq. (4.6). $\qquad\square$

TABLE 4.2 Results of Example 4.6.

u_∞ (ms^{-1})	α, $k = 3$	α, $k = 2$
3	0.64956	0.28385
4	0.72607	0.29459
5	0.74449	0.29706
6	0.75469	0.29841
7	0.76566	0.29985
8	0.75469	0.29841
9	0.74449	0.29706
10	0.70981	0.29237
11	0.66483	0.28606
12	0.63600	0.28186
13	0.61638	0.27894
14	0.60282	0.27687
15	0.59074	0.27501
16	0.58518	0.27415
17	0.57989	0.27332
18	0.57735	0.27292
19	0.57486	0.27253
20	0.57244	0.27214
21	0.57007	0.27177
22	0.56775	0.27140
23	0.56775	0.27140
24	0.56549	0.27104
25	0.56549	0.27104

Example 4.6. Two Eno Energy Eno 100 turbines discussed in Example 4.4 are located in a farm as shown in Section D.1. Obtain the wake radius and wind velocity at the location of T2, according to Frandsen's model.

Answer. In Jensen's model, the wake radius is only a function of x. But in Frandsen's model, it is a function of c_T and the distance. Hence, since c_T varies with the upstream wind velocity, we will find different wake radii for different u_∞. Therefore, we can use the data obtained in Table 4.1 for the present calculations.

In addition to c_T, we must determine the location at which we will match the flow speeds to find α. As explained before, we can choose any distance in a far region. For the present calculations, we use $s = x/D_0 = 7$ and apply it to Eq. (4.17).

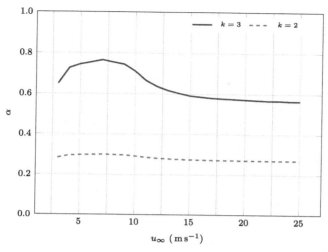

FIGURE 4.12 Entrainment or expansion factor for Example 4.6.

The results of calculations are tabulated in Table 4.2 and also plotted in Fig. 4.12. The results show that, for $k = 2$, α is almost a single value about $\alpha = 0.28$. But the case when $k = 3$ shows more variations. In both cases, α is much larger than the value obtained in the previous example.

The wake profile shown in Fig. 4.13a is compared with Jensen's model. As is evident in the figure, Frandsen's model has either a second- or third-order profile depending on the value of k. Both profiles show a higher value compared to Jensen's model at moderate distances, and eventually, Jensen's model predicts a higher value at very long distances. Frandsen's model shows a more consistent profile with the experimental tests. Also, it shows a separate profile for different upstream wind velocities, while Jensen's model gives a single profile for all the situations.

Finally, the wake profile with respect to the turbine's distance is plotted in Fig. 4.13b. This is a dimensionless plot showing $u(x)/u_\infty$. Again, as illustrated by the figure, the velocity build-up has different behavior with respect to u_∞. Comparing the results of Frandsen's model to Jensen's model (Fig. 4.8) reveals that Frandsen's model calculates a higher wind profile than Jensen's. Since Frandsen's model is more realistic, its results match the experimental and field tests better. □

4.2.3 Larsen model

Larsen (2009) assumed that wind turbine wake follows the thin shear layer approximation of the Navier–Stokes equations because the velocity gradient is more significant in the r direction compared to the x direction. Therefore, he used the axisymmetric form of the Navier–Stokes equations, and under the free-

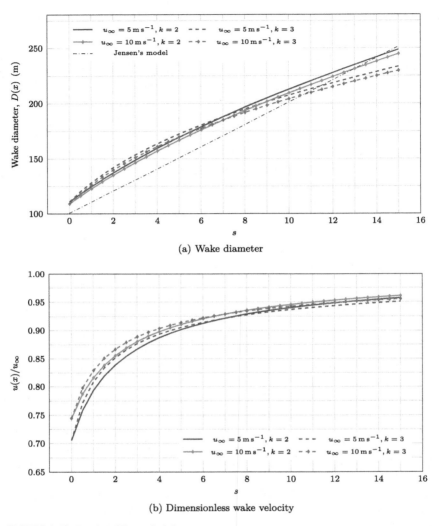

(a) Wake diameter

(b) Dimensionless wake velocity

FIGURE 4.13 Results of Example 4.6.

shear flow assumption, he calculated the wake profile as a function of x and r. Larsen assumed that the wake profile, $u(x, r)$, is composed of a mean flow velocity, $\bar{u}$, on which a perturbation profile, Δu, is superposed. In other words, the wake profile is

$$u(x, r) = \bar{u} + \Delta u(x, r). \tag{4.18}$$

In this model, the uniform wind velocity is called pseudouniform since the actual wind is not uniform and follows a logarithmic profile. However, a

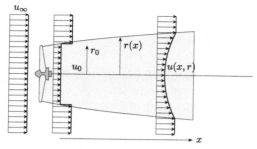

FIGURE 4.14 The wake model proposed by Larsen.

pseudouniform velocity is needed in order to obtain the velocity deficit, or perturbation.

Larsen obtained the perturbed velocity by solving the Navier–Stokes equations. To do this, he used the concept of self-similarity and incorporated Prandtl's mixing length theory for the calculation of turbulence terms. The solution of the governing equations requires two boundary conditions. In this model, the first boundary condition is defined right behind the wind turbine, and the other at a distance of $9.6D$ behind the wind turbine where D is the rotor diameter.

In addition to the wake velocity, Larsen used the wake modeling results obtained from turbulent wake models and gave a relation for the wake expansion radius. As explained in Frandsen's model, the wake expansion is proportional to $x^{\frac{1}{3}}$. The Larsen wake model and its expansion as shown in Fig. 4.14.

The relation is defined by

$$r(x) = \left(\frac{105c_1^2}{2\pi} \right)^{\frac{1}{5}} (c_T A(x + x_0))^{\frac{1}{3}}, \tag{4.19}$$

where x_0 is defined as

$$x_0 = \frac{9.6D}{\left(\frac{2R_{9.6}}{kD} \right)^3 - 1} \tag{4.20}$$

and c_1 is

$$c_1 = \left(\frac{kD}{2} \right)^{5/2} \left(\frac{105}{2\pi} \right)^{-1/2} (c_T A x_0)^{-5/6}. \tag{4.21}$$

In these equations, A is the rotor diameter, and k is a parameter that solely depends on the thrust coefficient, c_T, and is obtained using

$$k = \sqrt{\frac{m+1}{2}}, \tag{4.22}$$

in which m is given by

$$m = \frac{1}{\sqrt{1 - c_T}}. \tag{4.23}$$

Finally, $R_{9.6}$ is the wake radius at $9.6D$ downstream the turbine. The radius depends on the atmospheric turbulence intensity, I_a, and c_T. Larsen gave the following equation for the calculation of $R_{9.6}$:

$$R_{9.6} = a_1 \exp\left(a_2 c_T^2 + a_3 c_T + a_4\right)(b_1 I_a + 1) D, \tag{4.24}$$

where the coefficients are:

$$\begin{aligned} a_1 &= 0.435449861, & a_2 &= 0.797853685, \\ a_3 &= -0.124807893, & a_4 &= 0.136821858, \\ b_1 &= 15.6298. \end{aligned} \tag{4.25}$$

In addition to the wake diameter, its velocity is also important. The wake velocity profile is obtained using Eq. (4.18), where a pseudouniform flow is superposed on a velocity deficit profile. We know that the velocity is not uniform across the turbine swept area. Therefore, we have to obtain an averaged velocity out of the nonuniform profile. The first approximation is the volume-averaged velocity which is determined by

$$\bar{u} = \frac{1}{A} \int_A u \, dA. \tag{4.26}$$

However, this velocity is not consistent with the thrust coefficient of the wind turbine. In other words, we can calculate the turbine thrust with a nonuniform velocity and also calculate it with a pseudouniform velocity. The equation is

$$T = \frac{1}{2} \rho c_T \int_A u^2 \, dA = \frac{1}{2} c_T \bar{u}^2 A. \tag{4.27}$$

Equating these values yields the pseudouniform velocity as

$$\bar{u} = \sqrt{\frac{1}{A} \int_A u^2 \, dA}. \tag{4.28}$$

Consequently, Eq. (4.28) is used to obtain the pseudouniform velocity instead of Eq. (4.26).

Larsen defined the velocity deficit, Δu, as the sum of the first- and second-order profiles, i.e., Δu is

$$\Delta u(x, r) = \Delta u_1(x, r) + \Delta u_2(x, r). \tag{4.29}$$

The first-order profile dominates in practical cases, and the second-order profile can be neglected most of the time. Here, for the model to be complete, both terms are given.

The first-order profile is given by Larsen as

$$\Delta u_1(x,r) = -\frac{\bar{u}}{9} \left(c_T A(x+x_0)^{-2} \right)^{1/3}$$

$$\times \left[r(x)^{3/2} \left(3c_1^2 c_T A(x+x_0) \right)^{-1/2} - \left(\frac{35}{2\pi} \right)^{3/10} \left(3c_1^2 \right)^{-1/5} \right]^2.$$

(4.30)

All the parameters of the above equation are discussed before. Hence, the first-order profile can be easily calculated.

The second-order term is defined as

$$\Delta u_2(x,r) = \bar{u} \left(c_T A(x+x_0)^{-2} \right)^{2/3} \sum_{i=0}^{4} d_i z(x,r)^i,$$

(4.31)

with

$$z(x,r) = r^{3/2} \left(c_T A(x+x_0) \right)^{-1/2} \left(\frac{35}{2\pi} \right)^{-3/10} \left(3c_1^2 \right)^{-3/10}$$

(4.32)

and

$$d_0 = K \left(-1 - 3 \left(4 - 12 \left(6 + 27 \left(-4 + \frac{48}{40} \right) \frac{1}{19} \right) \frac{1}{4} \right) \frac{1}{5} \right) \frac{1}{8},$$

(4.33)

$$d_1 = K \left(4 - 12 \left(6 + 27 \left(-4 + \frac{48}{40} \right) \frac{1}{19} \right) \frac{1}{4} \right) \frac{1}{5},$$

(4.34)

$$d_2 = K \left(6 + 27 \left(-4 + \frac{48}{40} \right) \frac{1}{19} \right) \frac{1}{4},$$

(4.35)

$$d_3 = K \left(-4 + \frac{48}{40} \right) \frac{1}{19},$$

(4.36)

$$d_4 = K \frac{1}{40}.$$

(4.37)

In these equations, K is defined as

$$K = \frac{4}{81} \left[\left(\frac{35}{2\pi} \right)^{1/5} \left(3c_1^2 \right)^{-2/15} \right]^6.$$

(4.38)

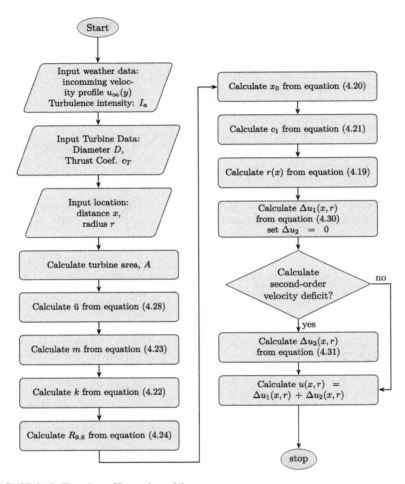

FIGURE 4.15 Flowchart of Larsen's model.

Fortunately, all the necessary parameters of the second-order term are also known and previously defined. Hence this term can be easily calculated.

Larsen's model seems more complicated than the other models, but the wake calculation is quite straightforward. Fig. 4.15 shows the flowchart of the model. As can be seen, all the wake parameters can be easily found if the input parameters are given. The input parameters are:

- Weather data, including the incoming wind velocity profile, $u_\infty(y)$, and turbulence intensity, I_a. Note that for the turbines that are located inside the wake of other turbines, the incoming velocity is not the same as the undisturbed far-field velocity. For these turbines, the incoming velocity is calculated using the wake velocity model.

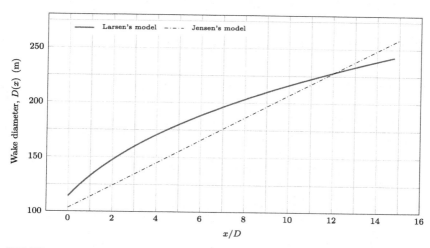

FIGURE 4.16 Wake diameter of Example 4.7.

• Turbine data, such as its diameter and thrust coefficient c_T. These values are used to determine the turbine's diameter and other parameters given in the flowchart.

• Wake location, including the axial distance downstream the turbine, x, and the radial distance, r.

Example 4.7. For Mapna MWT2.5-103-I turbine, calculate and plot the wake's radius.

Answer. The required data for Mapna MWT2.5-103-I is given in Section C.21. The turbine's radius is $R = 103.5$ m, and its power curve is given in Fig. C.18. Since the power curve contains c_p, we need to calculate c_T with the same procedure described in Example 4.4. The results are given in Table 4.3. Since Larsen's model depends on c_T, we solve the present problem only for one c_T. The calculation for other values would be the same. Hence, $c_T = 0.52283$, which is the maximum c_T corresponding to $u_\infty = 9$ m s^{-1}, is chosen as the reference value. Also, we need to specify the turbulence intensity. For the present calculations, we assume that the turbulence intensity is $I_a = 5\%$.

Now that all the input data are available, the calculations are quite straightforward. The following solution is of the same order as shown by the flowchart of Larsen's model (see Fig. 4.15).

Since we are modeling a uniform flow, the mean flow velocity equals the wind velocity at infinity. Hence, we have $\bar{u} = u_\infty$. Eq. (4.23) yields

$$m = \frac{1}{\sqrt{1 - c_T}} = \frac{1}{\sqrt{1 - 0.52283}} = 1.447,$$

TABLE 4.3 Results of Example 4.7.

u_∞ (m s^{-1})	c_p	a	c_T
3.5	0.123	0.03287	0.12718
4	0.257	0.07510	0.27787
5	0.373	0.12057	0.42413
6	0.417	0.14142	0.48568
7	0.433	0.14973	0.50925
8	0.442	0.15461	0.52283
9	0.442	0.15461	0.52283
10	0.418	0.14192	0.48713
11	0.358	0.11401	0.40407
12	0.282	0.08402	0.30786
13	0.222	0.06324	0.23698
14	0.178	0.04922	0.18721
15	0.145	0.03927	0.15092
16	0.119	0.03173	0.12289
17	0.099	0.02609	0.10165
18	0.084	0.02195	0.08588
19	0.071	0.01842	0.07233
20	0.061	0.01574	0.06197
21	0.053	0.01361	0.05373
22	0.046	0.01177	0.04654
23	0.040	0.01020	0.04041
24	0.035	0.00890	0.03531
25	0.031	0.00787	0.03124

and Eq. (4.22) yields

$$k = \sqrt{\frac{m+1}{2}} = \sqrt{\frac{1.447+1}{2}} = 1.106.$$

Then we can calculate $R_{9.6}$ from Eq. (4.24) as

$$R_{9.6} = a_1 \exp\left(a_2 c_T^2 + a_3 c_T + a_4\right)(b_1 I_a + 1) D = 153.98.$$

Eq. (4.20) gives the value of x_0, namely

$$x_0 = \frac{9.6D}{\left(\dfrac{2R_{9.6}}{kD}\right)^3 - 1} = 53.3.$$

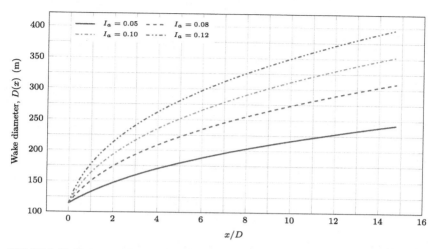

FIGURE 4.17 Wake diameter of Example 4.8.

The last parameter c_1 is obtained using Eq. (4.21):

$$c_1 = \left(\frac{kD}{2}\right)^{5/2}\left(\frac{105}{2\pi}\right)^{-1/2}(c_T A x_0)^{-5/6} = 0.2026.$$

Finally, the wake's radius can be computed by Eq. (4.19) as

$$r(x) = \left(\frac{105c_1^2}{2\pi}\right)^{\frac{1}{5}}(c_T A(x + x_0))^{\frac{1}{3}}.$$

As we can see, all the parameters of the equation have already been calculated. Since $r(x)$ is a function of x, it would be better to plot the wake's radius. The plot is shown in Fig. 4.16. Note that in this figure the diameter is plotted. □

Example 4.8. Investigate the effect of turbulence intensity on wake's diameter.

Answer. In Example 4.8, the turbulence intensity was assumed to be $I_a = 5\%$. It is quite clear that turbulence intensity has an increasing role in the wake's diameter. To investigate its role, the same example is calculated again with different intensities. The results are shown in Fig. 4.17. It is clear that by increasing the intensity, the wake's diameter increases. Also, it can be concluded that turbulence intensity has a strong effect on the wake's diameter. □

Example 4.9. Plot the velocity profile behind Mapna MWT2.5-103-I turbine at different distances.

Answer. For the results obtained in Example 4.8, we can calculate the wake profile according to Larsen's model. However, here we neglect the second-order

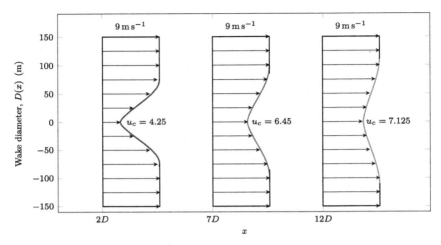

FIGURE 4.18 Wake expansion of Example 4.9.

deficit profile, i.e., Eq. (4.31), as suggested by Larsen and shown in the flowchart of Fig. 4.15. The results are shown in Fig. 4.18, in which the wake profile is plotted for three different distances, namely $2D$, $7D$, and $12D$. As can be seen, the velocity deficit at the center of the wake is becoming less and less as we move along the x direction. Meanwhile, the wake diameter is enlarging.

The simulation has been done for the turbulence intensity of $I_a = 0.05$. A higher turbulence intensity results in a larger wake diameter. However, it should be noted that according to the datasheet of the turbine, the hub height is 85 m. Consequently, if the wake diameter exceeds 170 m, the wake hits the ground and cannot maintain its classical shape. Since this phenomenon is not modeled, the wake radius grows boundlessly. In practice, for Mapna MWT2.5-103-I turbine, which is the case here, the wake hits the ground at around $7D$, as is shown in the figure. It should be noted that this behavior is not a flaw for Larsen's model since all the theoretical models behave the same. □

4.2.4 Ghadirian model

One of the main advantages of Frandsen's model is that the model is based on the induction factor. But this can also be considered as a disadvantage since the induction factor is unknown in most cases. The solutions proposed in solving the example problems of the present chapter are not very accurate since calculating c_T out of c_p according to the ideal theoretical relations is not accurate. Many researchers have been using Frandsen's model in their farm design tools or software. As mentioned above, they have used $a = 0.324$ for all their simulations. Ghadirian et al. (2014) pointed out the problem and tried to give a method for the calculation of the induction factor. They showed that the induction factor is a function of:

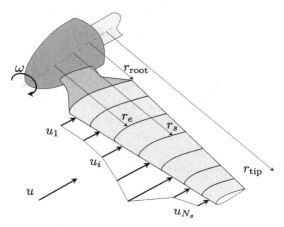

FIGURE 4.19 Ghadirian's model.

- Wind characteristics such as upstream wind velocity;
- Geometrical data such as the blade length, airfoils used to create the blade, and, of course, the local chord and twist;
- Operational parameters such as rotational speed.

As shown in Fig. 4.19, when the upstream wind reaches the blade, it converts into a nonuniform profile because each element produces a different induction value. Therefore, the induction factor has a nonuniform profile, and we cannot assign a single value for the whole blade. However, Frandsen's model requires a single value for the overall turbine. Ghadirian assumed that each element's induction factor is obtained using BEM (see Chapter 3). Then, he performed a mass balance on the turbine, meaning that the total mass of air that passes through the turbine swept area is equal to the sum of the air mass that passes through each element. Mathematically, the mass balance is

$$\dot{m}_{\text{total}} = \sum_{i=1}^{N_s} \dot{m}_{\text{ele}}, \tag{4.39}$$

in which $\dot{m}_{\text{total}}$ and $\dot{m}_{\text{ele}}$ are the total and element mass flow rates, and N_s is the number of blade elements.

The left-hand side of the above equation can be found from

$$\dot{m}_{\text{total}} = \rho A_{\text{tot}}(1 - a_{\text{ove}})u, \tag{4.40}$$

where a_{ove} is the overall induction factor that we are looking for; and A_{tot} is the turbine swept area of the blades,

$$A_{\text{tot}} = \pi(r_{\text{root}}^2 - r_{\text{tip}}^2). \tag{4.41}$$

The right-hand side of Eq. (4.40) is the sum of the mass that passes through each element, which is called the element mass flow rate. It can be calculated from the following equation:

$$\dot{m}_{\text{ele}} = \rho(1 - a_{\text{loc}})u\pi(r_e^2 - r_s^2). \tag{4.42}$$

In the above equation, a_{loc} is the local induction factor, and r_e and r_s are the radii of the end and start of the blade element, respectively. These parameters are shown in Fig. 4.19 for a single element. Moreover, each element area has an annular shape, as shown in the same figure.

Equating Eqs. (4.40) and (4.42) results in

$$(r_{\text{tip}}^2 - r_{\text{root}}^2)(1 - a_{\text{tot}})u = \sum_{1}^{N_s}(r_e^2 - r_s^2)(1 - a_{\text{loc}})u. \tag{4.43}$$

From this equation, the total induction factor can be obtained as

$$a_{\text{tot}} = 1 - \frac{\displaystyle\sum_{1}^{N_s}(r_e^2 - r_s^2)(1 - a_{\text{loc}})}{r_{\text{tip}}^2 - r_{\text{root}}^2}. \tag{4.44}$$

The obtained total induction factor can be used in Frandsen's model to obtain the wake data, such as its diameter and speed. Therefore, Ghadirian's model is just a modification to Frandsen's model. But this modification gives better predictions when the operating conditions change. As we know, the induction factor is a function of many parameters, including (a) the wind velocity, (b) the turbine's blade profile, and (c) the turbine's rotational velocity. These effects were studied in the previous chapter for different turbines and operational conditions. The present model is able to calculate a more realistic value for the axial induction factor used in Frandsen's model.

Example 4.10. Investigate the effect of upstream wind velocity on the total induction factor.

Answer. To investigate this effect, we simulate NTK-500/41 turbine once again. The turbine is simulated using the BEM method in Example 3.6. Hence, the details of the simulation are not repeated here. The axial induction factor for $u_\infty = 14$ is given in Table 3.2, but we need to recalculate the turbine characteristics for other velocities. For each simulation, as shown in Table 3.2, 17 induction factors are obtained, which correspond to each element. These values must be inserted into Eq. (4.44) to obtain the overall, or total, induction factor.

The final result is plotted in Fig. 4.20. As illustrated by the figure, the total axial induction factor decreases as the upstream velocity increases. This phenomenon is quite compatible with the concepts of low-speed aerodynamics,

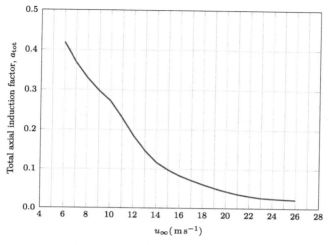

FIGURE 4.20 Effect of upstream wind velocity on a_{tot}.

while the higher the speed, the smaller the upstream influence. Consequently, the effect of the axial induction factor is greater in lower wind speeds. □

The above example shows that using a constant value (i.e., $a = 0.324$) is not justified, even for a specific wind turbine. This parameter is a strong function of the wind velocity and reaches very small values when u_∞ is higher than the rated velocity.

Example 4.11. Investigate the effect of airfoil profile on the total induction factor.

Answer. In Example 3.7, the turbine is simulated using three different airfoils. The same results are used here to calculate the total axial induction factor. Fig. 4.21 illustrates the obtained a_{tot} for different airfoils. As can be seen, a_{tot} is a function of the airfoil, too. But comparing Figs. 4.20 and 4.21, it seems that the upstream wind velocity has a stronger effect. □

Example 4.12. Investigate the effect of rotor angular velocity on the total axial induction factor.

Answer. To investigate the effect, NTK-500/41 is simulated under three different angular velocities, namely $\omega = 27.1$ rpm, which is its rated one, $\omega = 24.1$ rpm, and $\omega = 30.1$ rpm. The result is shown in Fig. 4.22. The figure shows that the rotor angular velocity also has a strong effect on the total induction factor. As the rotational velocity increases, the total induction factor increases. This relation is stronger at lower wind speeds. At very high upstream wind velocities, the effect becomes less and less. □

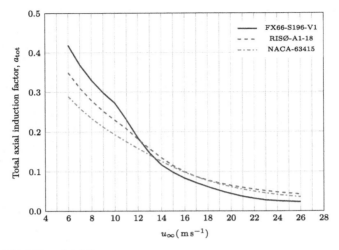

FIGURE 4.21 Effect of airfoil on a_{tot}.

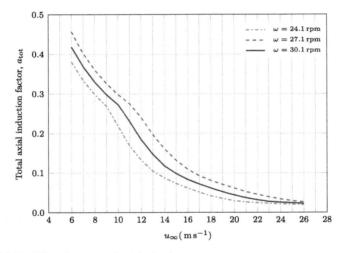

FIGURE 4.22 Effect of rotor angular velocity on a_{tot}.

The above examples show that the induction factor is a strong function of different parameters, hence a typical value cannot be used for different situations. The important feature of Ghadirian's modeling is that it can model the induction factor under different operational conditions and geometrical shapes. Moreover, since the total induction factor is obtained using a detailed calculation, the final results may be of greater accuracy.

Ghadirian's model requires a final remark. As stated, it uses the BEM method for the calculation of the total induction factor. We know that the BEM method calculates the thrust coefficient itself. Therefore, one can use the thrust coefficient obtained by the BEM method and use it as the input parameter to Frandsen's model. Also, the obtained values can be used for Larsen's model as well.

4.3 Gained energy of a farm

The gained energy of a farm in a year is the ultimate goal of a wind farm design. The best layout is that which produces the most energy throughout the year. Analyzing Figs. 4.4 and 4.5 reveals that wind turbines encounter different wind velocity at different times. The actual wind speed that applies to a wind turbine depends not only on the upstream wind velocity but also on the wind direction. As an example, consider turbine T5 in Fig. 4.4 where the wind blows from the west. In this situation, the turbine faces undisturbed upstream velocity, u_∞. However, when the wind direction changes, such as in Fig. 4.5, T5 is located partly at the wake of T2 and partly faces u_∞. Thus, even if the wind velocity is constant, the change in direction causes different actual speeds at T5's location. Consequently, T5 produces less energy when the wind blows from the northwest. The above example shows that each wind turbine produces a different amount of power at any moment; thus, the net farm power is the sum of all the turbines' power.

From the wake's physics, it is clear that the farther the wind turbines are located, the less critical the velocity deficit becomes. Hence, from the first estimation, the best practice is to install the turbines as far from each other as possible. However, the longer distance requires more land, more cabling, and higher operational costs. Therefore, when there is a limited area for installing the farm, we are not free to install the wind turbines as far apart as we like.

Obtaining the net energy of a wind farm is not as easy as it seems. There are many different parameters involved in this regard. These parameters can be summarized as follows.

- **Wind speed**
  The wind speed is the first effective parameter, and, as we know, it is a random parameter. Chapter 2 entirely covered this problem and used statistical relations to analyze the wind data. However, incorporating these statistical data and formulations in calculating the net farm's gained energy is not an easy task. Lots of simulations must be carried on in order to obtain the net gained energy.

- **Wind direction**
  As stated here, the wind direction is also an important parameter. Unfortunately, the wind direction also has statistical behavior; hence we have to take this fact into account. For this purpose, the wind rose diagrams that were discussed in Chapter 2 are the main tools. The simulations should be done according to the statistical concepts to ensure that the results are correct.

- **Wind turbines' location**
  As it is clear, the turbines' locations are significant in the calculation of the net yearly energy. This fact is discussed before and should be considered in simulations.
- **Homogeneity of the wind farm**
  The homogeneity of the wind farms means that all the installed turbines are of the same type and size. In actual wind farms, this may not be the case because the farm has been updated through time by installing new turbines that may have different characteristics than the original ones. Then, the simulation of these turbines requires more effort than a homogeneous farm.

Since the net output energy of a farm is the ultimate goal, it must be maximized. Usually, this fact must be handled with optimization algorithms, which are discussed in the next section. However, for any optimization, an accurate solver is necessary. According to the above-mentioned facts, any applicable solver must contain the operational factors discussed here.

4.4 Optimization

As the name states, optimization deals with maximizing a profit or minimizing a cost. Regarding the wind energy industry, many different problems involve optimization (Bazacliu et al., 2015; Chowdhury et al., 2012, 2013; González et al., 2011; Grady et al., 2005; Kiranoudis et al., 2001; Kusiak and Song, 2010; Lopez et al., 2019; Mosetti et al., 1994; Ozturk and Norman, 2004; Rajper and Amin, 2012; Son et al., 2014). For example, civil work requires optimization to reduce cost. The tower and structure of wind turbines encounter a similar problem. Finding the optimum size for gearbox, generator, shaft, and blade profile are other practical problems that require optimization. In wind farm design, the optimization methods try to find the best layout for the farm by which the farm's overall energy output is maximized. There are many optimization methods for different purposes. These methods can be categorized into the following categories:

- **Deterministic algorithms**
  These algorithms are used where the problem has a specific rule for marching toward the optimum value.
- **Stochastic algorithms**
  In complex problems, including the optimization of wind farms, the problem is so complex that one cannot find a straightforward rule for moving from an initial state to the optimum value. For these problems, the stochastic algorithms that are based on probabilistic rules give better answers.

For wind layout optimization, the researchers have successfully applied different methods such as genetic algorithm (GA), particle swarm optimization (PSO), ant colony (AC), and other stochastic algorithms. However, regardless of the method, an optimization algorithm includes the following main three parts:

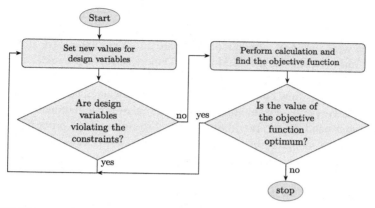

FIGURE 4.23 Optimization flowchart.

1. **Cost or objective function**

   The objective function (also known as the cost function) is the function that is going to be minimized (or maximized). Note that, in the optimization literature, there is no difference between maximizing or minimizing. In other words, if one maximizes a function $f(x)$, it means that the function $1/f(x)$ is minimized. For wind farm optimization, the farm's overall annual energy output is usually chosen as the objective function. One may also involve financial issues; hence the cost function would be maximizing the internal rate of return (IRR) or potential investments' profitability.

2. **Design variables**

   The cost function is usually a complicated and nonlinear function of many variables. Some of the variables cannot be adjusted by us, but some of them are adjustable. The adjustable variables are called the design variables. As an example of a wind farm layout design problem, the wind turbines' locations are design variables since, by moving the turbines over the field, the energy output of the farm varies. Also, the turbine locations can be adjusted as we wish. Note that although the wind velocity affects the overall energy output for this problem, it cannot be chosen as a design variable because we cannot alter the wind velocity.

3. **Constraints**

   The design variables have some constraints. For example, the turbines' locations are limited to the farm boundaries for wind farm layout design. In such a problem, the geographical issues such as rivers, hills, woods, and so on should also be considered the optimization constraints.

   The general flowchart of the optimization process is shown in Fig. 4.23. The flowchart does not contain the details of any specific algorithm and shows the overall procedure.

In addition to the simulation of the wind velocity field (using the methods mentioned in the present chapter), the optimization of wind farm layout also depends on wind turbines' simulation. There are different methods for simulation of wind turbines, such as the BEM and CFD models. It is quite understandable that the simulation methods based on CFD solvers give the best answer, and their results agree quite consistently with the experimental data. However, these models are very time-consuming and require lots of computational resources. On the other hand, BEM algorithms provide a straightforward procedure while giving accurate enough results to calculate turbines' power generation. It should be noted that other calculation methods can also be incorporated into optimization algorithms. For example, methods based on artificial intelligence (AI) and deep learning (DL) are quite suitable for this purpose.

Regardless of the method, since optimization algorithms require lots of iterations to find the optimum value, the process should be fast and computationally affordable. In this regard, fast algorithms such as BEM, AI, or DL are superior to CFD models. However, some modifications or assumptions may lead to fast CFD solvers that eventually make them quite reasonable for incorporating into optimization loops. For example, the actuator-disc-based CFD model that is covered in more detail in Section 6.3 reduces lots of computations, and the method is faster than full turbine modeling. At the same time, its results are more accurate than BEM, resulting in much better optimization. As another example, the numerical methods based on proper orthogonal decomposition (POD) methods can also be used to calculate a wind farm-generated power.

4.5 Summary

Understanding the physics of wake is very important in analyzing a wind farm. In the present chapter, the wake was studied, and different approaches commonly used for wake calculation were introduced. These models are widely used in various references and also in some industrial applications.

The methods discussed in this chapter are all based on the fundamental laws of fluid dynamics and turbulence. The results are analytical or semianalytical formulations that are easy to implement. However, for a more realistic simulation, one is encouraged to perform advanced numerical techniques used to solve the governing equations of fluid dynamics. These techniques are commonly known as CFD (computational fluid dynamics) and are implemented in many different programs. Some of these programs are commercial, including COMSOL Multiphysics or Ansys Fluent, and some of them are open-source such as OpenFoam and SU2. These programs can simulate the whole wind farm with great accuracy but with a substantial numerical cost. The CFD codes are quite suitable for the sectors that can afford the cost of supercomputers or large parallel clusters.

4.6 Problems

1. Enercon E-16 is a pitch-regulated wind turbine. For this turbine, calculate c_p and c_T using the power curve and the method discussed in Chapter 2. Then, use the obtained results to simulate the wake expansion and velocity according to Jensen's model.
2. The datasheet of Enercon E-18 (see Section C.2) contains c_p. Calculate c_T for this turbine and discuss about its wake according to Larsen's model.
3. Nordtank NTK 150 is an onshore turbine. Calculate and compare the entrainment factor for this turbine for different surface roughness. Then calculate its wake according to Jensen's and Frandsen's models.
4. Nordtank NTK 200 is a stall-regulated turbine given in Section C.4. Use the data to calculate and plot its c_T versus the upstream wind velocity. Then, choose some samples from the c_T curve and obtain its wake shape and velocity according to Larsen's model.
5. For the previous example, investigate the effect of turbulence intensity on the wake shape and velocity.
6. For Micon M 530 with the specifications given in Section C.7, calculate c_p and c_T. Then simulate the wake radius and velocity according to Larsen's model. Plot the wake profile for different distances downstream.
7. For the previous problem, compare the results with and without considering the second-order velocity deficit as explained by Larsen's model.
8. The power curve of Enercon E-30 is given in Section C.8, in which c_p is also available. Use Frandsen's model to predict the shape and velocity of its wake. Plot the results for $k = 2$ and $k = 3$.
9. Compare the wake radius and velocity obtained in the previous problem and Jensen's model.
10. Use Jensen's model for simulation of Nordtank NTK 400. Compare the model when $\alpha = 0.05$ with that obtained by Eq. (4.6). The data is available in Section C.9.
11. Use the data of Wincon W755/48 given in Section C.14 and calculate the wake of the turbine for some different upstream wind velocities.
12. Compare the results of the previous problem with Jensen's model.
13. Obtain c_p and c_T for Vergnet GEV HP 1000/62 and calculate its wake properties according to Frandsen's and Larsen's models.
14. Consider the previous problem. Obtain the distance at which the above two models fail to give a physical answer as the wake hits the ground.
15. Simulate Mapna MWT2.5-103-I with the BEM method. Obtain its c_p and a_{tot} from Ghadirian's model. Then obtain its wake properties according to Frandsen's model.
16. Do the previous problem and obtain the wake profile according to Larsen's model.
17. Investigate the effect of airfoil profile on the wake models, namely Frandsen's and Larsen's models.

18. For Mapna MWT2.5-103-I, investigate the effect of rotor rotational velocity on the wake profile and speed according to Larsen's model.
19. Again use Mapna MWT2.5-103-I turbine and investigate the effect of the second-order velocity deficit of Larsen's model.
20. For Mapna MWT2.5-103-I, investigate the effect of turbulence intensity on wake's profile and velocity.
21. Investigate the interaction of turbulence intensity and blade's profile. On which airfoil, the turbulence intensity has less effect?
22. Obtain c_T from the BEM model and redo Problems 15 and 16, then compare the results.

References

Abkar, Mahdi, Sørensen, Jens Nørkær, Porté-Agel, Fernando, 2018. An analytical model for the effect of vertical wind veer on wind turbine wakes. Energies 11 (7), 1838.

Bazacliu, Gabriel, Lazaroiu, George Cristian, Dumbrava, Virgil, 2015. Design of wind farm layout for maximum wind energy capture. UPB Scientific Bulletin, Series C: Electrical Engineering 77 (1), 269–276.

Blondel, Frédéric, Cathelain, Marie, 2020. An alternative form of the super-Gaussian wind turbine wake model. Wind Energy Science 5 (3), 1225–1236.

Carbajo Fuertes, Fernando, Markfort, Corey D., Porté-Agel, Fernando, 2018. Wind turbine wake characterization with nacelle-mounted wind lidars for analytical wake model validation. Remote Sensing 10 (5), 668.

Chowdhury, Souma, Zhang, Jie, Messac, Achille, Castillo, Luciano, 2012. Unrestricted wind farm layout optimization (UNFLO): investigating key factors influencing the maximum power generation. Renewable Energy 38 (1), 16–30.

Chowdhury, Souma, Zhang, Jie, Messac, Achille, Castillo, Luciano, 2013. Optimizing the arrangement and the selection of turbines for wind farms subject to varying wind conditions. Renewable Energy 52, 273–282.

Frandsen, Sten, 1992. On the wind speed reduction in the center of large clusters of wind turbines. Journal of Wind Engineering and Industrial Aerodynamics 39 (1–3), 251–265.

Frandsen, Sten, Barthelmie, Rebecca, Pryor, Sara, Rathmann, Ole, Larsen, Søren, Højstrup, Jørgen, Thøgersen, Morten, 2006. Analytical modelling of wind speed deficit in large offshore wind farms. Wind Energy: An International Journal for Progress and Applications in Wind Power Conversion Technology 9 (1–2), 39–53.

Ghadirian, A., Dehghan, M., Torabi, F., 2014. Considering induction factor using BEM method in wind farm layout optimization. Journal of Wind Engineering and Industrial Aerodynamics 129, 31–39.

González, J. Serrano, González Rodríguez, Á.G., Castro Mora, J., Burgos Payán, M., Riquelme Santos, J., 2011. Overall design optimization of wind farms. Renewable Energy 36 (7), 1973–1982.

Grady, S.A., Hussaini, M.Y., Abdullah, Makola M., 2005. Placement of wind turbines using genetic algorithms. Renewable Energy 30 (2), 259–270.

Højstrup, J., 1983. Nibe wake, part one. Technical report. Risø National Laboratory, Roskilde, Denmark.

Jensen, Niels Otto, 1983. A note on wind generator interaction.

Katic, I., Højstrup, J., Jensen, N.O., 1986. A Simple Model for Cluster Efficiency. European Wind Energy Association, Rome, Italy.

Kiranoudis, C.T., Voros, N.G., Maroulis, Z.B., 2001. Short-cut design of wind farms. Energy Policy 29 (7), 567–578.

Kusiak, Andrew, Song, Zhe, 2010. Design of wind farm layout for maximum wind energy capture. Renewable Energy 35 (3), 685–694.

Larsen, G.C., 2009. A simple stationary semi-analytical wake model. Technical University of Denmark.

Lopez, Daniel, Kuo, Jim, Li, Ni, 2019. A novel wake model for yawed wind turbines. Energy 178, 158–167.

Mosetti, G.P.C.D.B., Poloni, Carlo, Diviacco, B., 1994. Optimization of wind turbine positioning in large windfarms by means of a genetic algorithm. Journal of Wind Engineering and Industrial Aerodynamics 51 (1), 105–116.

Ozturk, U. Aytun, Norman, Bryan A., 2004. Heuristic methods for wind energy conversion system positioning. Electric Power Systems Research 70 (3), 179–185.

Rajper, Samina, Amin, Imran J., 2012. Optimization of wind turbine micrositing: a comparative study. Renewable and Sustainable Energy Reviews 16 (8), 5485–5492.

Son, Eunkuk, Lee, Seungmin, Hwang, Byeongho, Lee, Soogab, 2014. Characteristics of turbine spacing in a wind farm using an optimal design process. Renewable Energy 65, 245–249.

Sun, Haiying, Yang, Hongxing, 2018. Study on an innovative three-dimensional wind turbine wake model. Applied Energy 226, 483–493.

Sun, Haiying, Yang, Hongxing, 2020. Numerical investigation of the average wind speed of a single wind turbine and development of a novel three-dimensional multiple wind turbine wake model. Renewable Energy 147, 192–203.

Vermeulen, P.E.J., Builtjes, P., Dekker, J., Lammerts van Bueren, G., 1979. An experimental study of the wake behind a full scale vertical-axis wind turbine.

Chapter 5

Analytical model based on similarity solution

Turbulence is considered one of the most challenging phenomena for which a solid formulation does not exist. It is worth mentioning that most of the currents in nature are considered turbulent. For example, atmospheric flows, ocean currents, clouds' motion, river currents, combustion, and almost all the other fluid flows are considered turbulent. In fact, laminar flows rarely happen in nature. To become laminar, the flow must be either very slow or very viscous. In addition to the speed and viscosity, the dimension is also important. The flows in small tiny paths such as narrow tubes or channels become laminar. As a natural laminar flow, one can name the capillary flow inside the stems of plants.

It is well known that turbulence is dominant by two key parameters, namely (a) kinetics, and (b) viscosity. The flow kinetics is responsible for generating turbulence, but viscosity is the term that damps the vortexes generated by kinetic forces. These two forces are working against each other until one of them wins the game. If the kinetic forces are more substantial, turbulence occurs, and if viscosity dominates, then the flow becomes laminar. The ratio of inertial over viscous forces is called *Reynolds number* after the famous scientist Osborne Reynolds (1842–1912), who first defined the number as

$$Re = \frac{\rho U D}{\mu} = \frac{U D}{\nu},\tag{5.1}$$

where U is the flow velocity, D is the characteristic length, ρ is the fluid viscosity as usual, and μ is the dynamic viscosity. In this formulation, the fluid kinematic viscosity is defined as the ratio $\nu = \mu/\rho$ and is an indication of viscous forces, while $U D$ is an indication of inertial forces. Hence, Eq. (5.1) defines the ratio of inertial over viscous forces. By this definition, the higher the Reynolds number, the more turbulent the flow.

Turbulence is quite complex, and there is no unanimous theory that describes its exact involved physical phenomena (Davidson, 1997; Launder and Spalding, 1972; Paofsky and Dutton, 1984). However, turbulence is tackled by different scientists from different perspectives. Before thinking about the mathematical models of turbulent flows, it is better to understand turbulence physics. The things that we have to know about the turbulent flows can be summarized as follows:

Fundamentals of Wind Farm Aerodynamic Layout Design. https://doi.org/10.1016/B978-0-12-823016-9.00011-5

Turbulent flow is a flow There is no turbulent fluid since turbulence is not a property of the material. It is a property of fluid flow. Regardless of the fluid type, if its velocity increases, it becomes turbulent.

Continuum Although turbulence is very complex, it is a continuum phenomenon and is governed by fluid dynamics equations such as the Navier–Stokes equations.

Irregularity Turbulence is quite irregular, and one predefined pattern cannot describe its nature. For example, the river current is not the same for all the rivers or even for those in a specific place. More importantly, in an atmospheric flow, the flow patterns are entirely random.

Diffusivity A turbulent flow tends to mix fluid properties, whether it is momentum, heat, or any other property. The mixing mechanism is present because of the existence of eddies which act as a natural mixer inside the flow.

Dissipation As stated before, a turbulent flow experiences two different forces, namely kinetic and viscous forces. The viscous forces increase as the shear stress increases, and the shear stress increases by kinetic forces. These mechanisms constantly interact with each other, resulting in a dissipative nature. As a result, the bulk flow energy should always be compensated to keep the fluid moving.

Large Reynolds number Turbulence happens where the kinetic forces are far more significant than viscous forces. This happens for large Reynolds numbers as explained by Eq. (5.1).

For a wind turbine, Reynolds number is usually expressed by

$$Re = \frac{\rho u_{rel} c}{\mu} = \frac{u_{rel} c}{\nu}, \tag{5.2}$$

in which u_{rel} is the relative velocity shown in Fig. 3.10, and c is the local chord length. Since both parameters vary from root to tip, the Reynolds number will not be uniform, too. But it should be noted that since the kinematic viscosity of air is small (of order 10^{-5}), the flow will be turbulent. For external flows, the transition Reynolds number at which the flow is considered turbulent is about $Re = 500,000$.

Example 5.1. For a wind turbine with chord length $c = 1\,\mathrm{m}$ that operates at $T = 25°C$, calculate the Reynolds number for $u_\infty = 10\,\mathrm{m\,s^{-1}}$.

Answer. The kinematic viscosity of atmosphere at $T = 25°C$ is

$$\nu = 1.55 \times 10^{-5}\,\mathrm{m^2\,s^{-1}}.$$

If we neglect the blade rotation (since we have no information), then the Reynolds number is

$$Re = \frac{u_{rel} c}{\nu} == \frac{10 \times 1}{1.55 \times 10^{-5}} = 6.45 \times 10^5.$$

It shows that the flow is entirely turbulent even if we neglect the rotational induced velocity. It is quite understandable that if we add the rotational speed, the Reynolds number will become even larger. ☐

5.1 Turbulent free-shear wake

There are lots of different classified turbulent flows. One class of turbulent flows is called *free-shear* flows. In this class, the flow is far from solid boundaries; hence there is no viscous shear-layer inside the flow. In the absence of the shear layer, turbulence happens due to the velocity difference inside the bulk flow.

One of the examples of free-shear flow, is called *Wake*. This happens when the flow leaves an obstacle and continues to downstream. The wake flow is one of the classical forms of free-shear flow and hopefully, wind turbine wakes are also categorized in this type of flow.

5.2 Self-similarity method

To obtain a simple model for wind turbine wakes, one can use the similarity solution since the wind wake is actually a free-shear flow. Therefore, once the wake passes the turbine, it will be influenced by turbulence mixing and diffusion the same way as described before. Although the terrain and atmospheric wind direction change influence the wind turbine wakes, they can be neglected since we seek a simple model. It is quite understandable that the whole farm should be simulated numerically for a more realistic model, which is not the purpose of the current modeling.

As it will be discussed in the present section, a self-similarity solution requires a reference profile that will further expand in the flow direction. The reference profile is the wake profile just behind the wind turbine where the flow contains large vortexes induced by blade rotation, axial induced velocity, shear flow developed from nacelle and tower surface, upstream wind velocity, and other physical phenomena. These facts reveal that the initial wake is very complicated and cannot be obtained easily. However, we can have an estimate for it by some simplifications. For the present modeling, we use the BEM theory to obtain an initial axisymmetric wake profile. This initial wake is not uniform along the turbine blade because BEM is able to calculate the wake velocity for each radius.

The BEM can give a wake profile for the blade section; however, the similarity solution requires a full initial wake, including the nacelle section. The induced wake behind the nacelle depends on its shape and size, but we can estimate its generated wake through fluid dynamics' fundamental laws. In the present modeling, the nacelle's induced wake is modeled using a second- or third-order polynomial with proper boundary conditions to make it consistent with physics.

In summary, the present modeling includes the following steps for wake modeling:

1. Use the BEM algorithm with operational parameters and obtain a wake profile just behind the blades;
2. Estimate a second- or third-order profile for nacelle according to the fluid dynamics laws and the profile obtained in the previous step;
3. Use the similarity solution to advance along the wind direction and obtain the flow profile at any distance.

For a wind farm, the wakes should be combined, which will be discussed later in this chapter.

According to the free-shear wake models and using scale analysis, for axisymmetric free shear wakes the following equation describes the momentum equation in the polar coordinate system and along the wind direction (Pope, 2000; Torabi and Hamedi, 2015; Hamedi et al., 2015):

$$u\frac{\partial(U-u)}{\partial x}+\frac{1}{r}\frac{\partial r\overline{uv}}{\partial r}=0, \tag{5.3}$$

in which u, U, x, r, and $\overline{uv}$ are wind speed at infinity, wind speed at any distance inside the wake, the axial coordinate in the wind direction, the radial distance from the turbine rotational axis, and Reynolds stress term, respectively.

The velocity U can be expressed by

$$\frac{u-U}{U_s}=f(\frac{r}{l_{ch}}), \tag{5.4}$$

and the Reynolds stress term, $\overline{uv}$, is calculated using

$$-\overline{uv}=U_s^2 g(\frac{r}{l_{ch}}). \tag{5.5}$$

In these equations, U_s is the difference between the far-field wind velocity and the local wake velocity just on the axis or $r=0$, and l_{ch} is a characteristic length that must be obtained.

In Eqs. (5.4) and (5.5), f is an unknown function for having a good estimate of the velocity at the wake's centerline and similarly, the function g is used to give a good estimate for the Reynolds stress term. These two functions are to be obtained using the following statements. Using Eqs. (5.4) and (5.5) in Eq. (5.3) yields

$$-\frac{ul_{ch}}{U_s^2}\frac{dU_s}{dx}f+\frac{u}{U_s}\frac{dl_{ch}}{dx}\eta f'=g'+\frac{g}{\eta}, \tag{5.6}$$

where

$$\eta=\frac{r}{l_{ch}}. \tag{5.7}$$

According to self-similarity concepts, f and g are generally two distinct functions that preserve their own shape with respect to the x coordinate. Consequently, f and $\eta f'$ in Eq. (5.6) must be constant. Since $u=$ const, we obtain the following equations:

$$\frac{l_{ch}}{U_s^2}\frac{dU_s}{dx} = \text{const}, \tag{5.8}$$

$$\frac{1}{U_s}\frac{dl_{ch}}{dx} = \text{const}. \tag{5.9}$$

For the general solution of Eqs. (5.8) and (5.9), we use $l \sim x^n$ and $U_s \sim x^{n-1}$ but the solution of Eq. (5.6) requires an extra equation other than these two. From the basics knowledge of fluid dynamics, the momentum integral can be used as an auxiliary equation here, which is written as

$$\rho \int_{-\infty}^{\infty} U(U-u)2\pi r\,dr = -\rho\pi u^2\theta_M^2. \tag{5.10}$$

In this equation, θ_M is called the *momentum thickness*. Using Eq. (5.4) in Eq. (5.10) yields

$$\int_{-\infty}^{\infty} f\eta\,d\eta - \frac{U_s}{u}\int_{-\infty}^{\infty} f^2\eta\,d\eta = \frac{u}{2l_{ch}^2 U_s}\theta_M^2. \tag{5.11}$$

In free-shear wakes, $\frac{U_s}{u}$ approaches zero at infinity; hence the second term in Eq. (5.11) can be neglected resulting in

$$2U_s l_{ch}^2 \int_{-\infty}^{\infty} f\eta\,d\eta = u\theta_M^2. \tag{5.12}$$

Since θ_M is constant, we have

$$U_s l_{ch}^2 = \text{const}. \tag{5.13}$$

From Eqs. (5.8) and (5.13), we conclude that

$$l_{ch} \sim x^{\frac{1}{3}}, \quad U_s \sim x^{\frac{-2}{3}}. \tag{5.14}$$

Using Eqs. (5.14) in Eq. (5.6) results in

$$\lambda(\eta^2 f' + 2\eta f) = g + \eta g', \tag{5.15}$$

where λ is an unknown constant parameter.

Solving Eq. (5.15) requires a relation between f and g. In the present model, the eddy viscosity concept is used to obtain a proper relation:

$$-\frac{1}{r}\frac{\partial(r\overline{uv})}{\partial r} = \nu_T \frac{1}{r}\frac{\partial}{\partial r}\left(r\frac{\partial(U-u)}{\partial r}\right), \tag{5.16}$$

in which v_T is called the eddy viscosity. Applying Eqs. (5.4) and (5.5) to this equation results in

$$-\frac{v_T}{U_s l_{ch}}(f' + \eta f'') = g + \eta g'. \tag{5.17}$$

Applying Eq. (5.17) in Eq. (5.15) yields

$$-\frac{v_T}{U_s l_{ch}}(f' + \eta f'') = \lambda(\eta^2 f' + 2\eta f). \tag{5.18}$$

The term $-\frac{v_T}{U_s l_{ch}}$ is a constant parameter; hence Eq. (5.18) can be rewritten as

$$(\epsilon \eta^2 f + \eta f')' = 0. \tag{5.19}$$

Since l_{ch} and U_s are not known, we have no known value for ϵ. The general solution for Eq. (5.19) is in the form of

$$f(\eta) = \exp(-\epsilon \frac{\eta^2}{2}). \tag{5.20}$$

The parameter ϵ is unknown and should be obtained. This parameter depends on the initial wake shape or the wake that is formed just behind the turbine. This parameter depends on U_s and l_{ch} that themselves depend on the initial shape of the wake. Therefore, to solve Eq. (5.20), we need to obtain the profile of the wake just behind the wind turbine.

5.3 Similarity solution for a single wind turbine

The similarity solution suggests that the initial wake just behind the wind turbine will decay exponentially according to Eq. (5.17). Moreover, the wake expansion is proportional to the cubic root of distance, or $r \sim x^{\frac{1}{3}}$, and the velocity deficit behaves as $U_s \sim x^{-\frac{2}{3}}$. In obtaining these values, it is supposed that the eddy viscosity, v_T, is constant, which is not a valid statement. Therefore, for actual simulations, the error that arises from this assumption must be compensated. There are some models handling this issue, and, according to one of them, a correction term is multiplied by the decay function. Therefore, the velocity at any distance, x, and radius, r, is

$$u(r, x) = u_\infty - U_s(c_0 \eta^2 + 1) \exp(c_1 \eta^2), \tag{5.21}$$

where η is defined by Eq. (5.7).

5.3.1 Initial wake profile just behind the wind turbine

It is quite clear that the shape differs for each turbine, and we have to use numerical methods to obtain its profile. For the present model, as explained, BEM

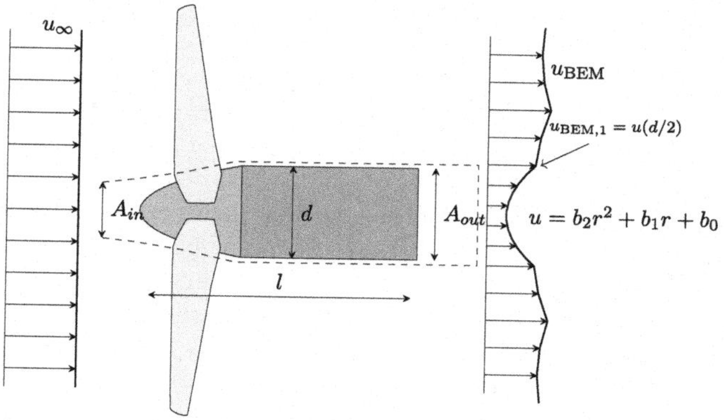

FIGURE 5.1 Streamlines around a nacelle.

is used to obtain the wake profile just behind the blades and a two- or third-order polynomial for the nacelle.

Fig. 5.1 shows the concept of the second-order profile for the wake generated by the nacelle. In some cases, we can choose a third- or higher-order profile as well. But in this case, proper boundary conditions must be imposed so that the shape of the profile becomes physically consistent. Experience shows that mathematically, a second-order profile can be obtained for all the situations, but higher-order ones may not be obtained in some cases. In the following chapter, we use a third-order profile for further discussions, but the same procedure can be applied to any desired profile.

Although nacelles have different shapes and sizes, a bullet-like blunt cylinder is reasonable for most cases. The third-order wake profile is assumed as:

$$u(r) = b_3 r^3 + b_2 r^2 + b_1 r + b_0, \qquad r \in [0, d/2], \qquad (5.22)$$

where r is the radial coordinate starting from the nacelle axis up to the nacelle diameter, or $r = d/2$. The dashed lines in the figure show the streamlines that pass over the nacelle. From the fundamental laws of fluid dynamics, we know that no mass crosses the streamlines. This fact will help us find the unknown coefficients.

Since the wake profile is assumed to be third-order, we need to find four coefficients. As we will see, for continuity, A_{in} becomes the fifth unknown that should be determined as well. Hence, we have five unknowns, and we need five relations to obtain all of them. The physical equations, as well as the boundary conditions, that are used here for obtaining these unknowns are:

1. **Continuity**

   According to Fig. 5.1, the mass flow that enters A_{in} must leave at A_{out}. This implies the principle of conservation of mass because no mass crosses the streamlines, and we have no mass source or sink. The continuity yields

$$\rho u A_{in} = \rho \int_0^{d/2} u(r) 2\pi r \, dr. \tag{5.23}$$

2. **Conservation of momentum**

The conservation of momentum dictates that the drag force exerted on the blunt cylinder is the difference between the momenta of the inlet and outlet. The inlet momentum is the momentum of the fluid at A_{in} and the outlet momentum is that at A_{out}. Thus, writing the conservation of momentum, we get the following equation:

$$D = \rho u^2 A_{in} - \rho \int_0^{d/2} u^2(r) 2\pi r \, dr. \tag{5.24}$$

On the other hand, the drag of a blunt cylinder can be obtained using

$$D = \frac{1}{2} c_D \rho A_n u^2. \tag{5.25}$$

Equating Eqs. (5.24) and (5.25) gives the second relation for our unknowns. However, as we can see, the momentum equation introduces two new unknown parameters, which are the drag coefficient, c_D, and a reference surface, A_n. We can obtain this value from the work documented in (Hoerner, 1965). Fig. 5.2 shows the drag coefficient, c_D, of a blunt cylinder with an aspect ratio l/d. For this data, the reference area is

$$A_n = ld = l \times d. \tag{5.26}$$

Using the data of Fig. 5.2 and Eq. (5.21), for any nacelle, the left-hand side of Eq. (5.25) becomes a known parameter.

3. **No jumps at the blade root**

The third assumption is that the velocity of the nacelle wake profile at $r = \pm d/2$ must be equal to the blade velocity profile obtained from BEM. This assumption implies jumpless velocity, which is physically quite understandable. Mathematically, it means

$$u(d/2) = u_{BEM,1} = u_\infty(1 - a_1), \tag{5.27}$$

where $u_{BEM,1}$ is the BEM algorithm's velocity for the first section (root section, or $r = d/2$), and a_1 is the induction factor at the same radius.

4. **Assumption of axisymmetric wake**

The profile must be axisymmetric with respect to the nacelle axis. This is a geometrical constraint that mathematically means

$$\left. \frac{du}{dr} \right|_{r=0} = 0. \tag{5.28}$$

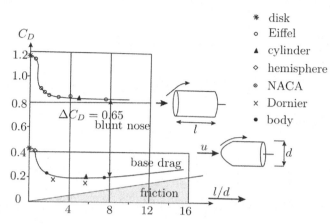

FIGURE 5.2 Drag of cylinder and blunt cylinder (Hoerner, 1965).

This assumption simply results in

$$(3b_3r^2 + 2b_2r + b_1)|_{r=0} = 0, \qquad (5.29)$$

or

$$b_1 = 0. \qquad (5.30)$$

5. **Point of inflection**

For a second-order profile, we do not need any further assumption. But for a third-order polynomial, we have to specify another one. Here we assume that the wake profile has its minimum value at the axis, or $r = 0$. By this assumption, we guarantee that the wake will not have any other maximum from $r = 0$ to $r = d/2$. The assumption of an inflection point on the axis mathematically is written as

$$\frac{d^2u}{dr^2}\bigg|_{r=0} = 0. \qquad (5.31)$$

Applying Eq. (5.17) to Eq. (5.25) gives

$$(6b_3r + 2b_2)|_{r=0} = 0, \qquad (5.32)$$

which yields

$$b_2 = 0. \qquad (5.33)$$

By these assumptions, we have five proper equations that are used to obtain the five unknowns.

5.3.2 Calculation of wake expansion radius and velocity deficit at axis line

The wake expansion radius is proportional to $x^{\frac{1}{3}}$, but for calculations we need an exact relation. Before proceeding, we define a far distance for wake propagation that is not influenced directly by blade rotations. Experimental observations show that for most cases, the far distance starts at $x = 2D$, or two times the diameter of the turbine. By this definition, the wake expansion radius is defined as

$$r(x) = r_1 (\frac{x}{4r_0})^{\frac{1}{3}}, \tag{5.34}$$

where r_0 is the radius of the turbine at $x = 0$ and r_1 is the wake distance at the beginning of the far distance, or $x = 4r_0$. At this point, from the concept of an ideal wind turbine, we can obtain a relation for r_1. When analyzing the ideal turbine in Chapter 2 as shown in Fig. 2.9, the relation of wind velocity at turbine section and a far location is expressed by Eqs. (2.47) and (2.48). Applying continuity requirement to these equations, the radius of wake at the far distance or r_1 can be calculated as

$$r_1 = r_0 \sqrt{\frac{1-a}{1-2a}}. \tag{5.35}$$

Therefore, the wake expansion radius is expressed as

$$r(x) = r_0 \sqrt{\frac{1-a}{1-2a}} (\frac{x}{4r_0})^{\frac{1}{3}}. \tag{5.36}$$

The similarity solution (Eq. (5.14)) suggests that the velocity deficit at the center of the wake is proportional to $x^{-\frac{2}{3}}$. Therefore we can write

$$U_s = \tilde{U} (\frac{x}{4r_0})^{-\frac{2}{3}}. \tag{5.37}$$

Again, to find the proportionality factor, $\tilde{U}$, we make use of the condition at far distance, or $x = 4r_0$, where according to (2.48) the velocity is equal to $u(4r_0) = u(1 - 2a)$. Hence if we subtract both sides from u, we get

$$u - u(4r_0) = u - u(1 - 2a). \tag{5.38}$$

The left-hand side is U_s at the far distance. Comparing Eqs. (5.37) and (5.38) (at $x = 4r_0$), we conclude that

$$\tilde{U} = 2au, \tag{5.39}$$

and therefore we find the following equation for U_s:

$$U_s = 2au (\frac{x}{4r_0})^{-\frac{2}{3}}. \tag{5.40}$$

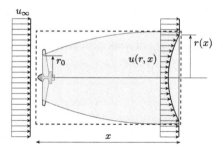

FIGURE 5.3 Control volume for obtaining analytical profile.

Eq. (5.40) shows that U_s is a function of the overall induction factor, or a. Therefore, the overall induction factor of the turbine must be calculated for each turbine. It was fully discussed in the previous chapter that the overall induction factor is a function of turbine geometrical properties and operational conditions. In the present chapter, we use Ghadirian's model to obtain the overall induction factor according to Eq. (4.37) using the BEM method.

5.3.2.1 Velocity profile at far distances

When applying Eq. (5.21) for the wake profile, we see that all the parameters are obtained, except for c_0 and c_1. These two parameters must be calculated using continuity and boundary conditions.

The air mass that flows through the wind turbine can be found by integrating the velocity profile shown in Fig. 5.1. As it is clear from the figure, a part of the profile is a third-order polynomial, and a part of it is obtained using the BEM method. As discussed before, the choice of the third-order profile is quite arbitrary, and any desirable profile could be possible.

For such a profile, we can write

$$\dot{m}_{\text{num}} = \int \rho V dA = \int_0^{r_0} \rho V (2\pi r dr), \qquad (5.41)$$

where $\dot{m}_{\text{num}}$ is the mass flow rate obtained from numerical calculation. Additionally, V is the velocity profile that is a combination of nacelle profile from $r = 0$ to $r = d/2$ and BEM profile from $r = d/2$ to $r = r_b$. For the control volume shown in Fig. 5.3, the continuity yields

$$\dot{m}_{\text{num}} + \rho \pi (r^2(x) - r_0^2)u = \int_0^{r(x)} \rho U(r, x) 2\pi r dr, \qquad (5.42)$$

where we can define the analytical mass flow rate by

$$\dot{m}_{\text{analytic}} = \int \rho U(r, x) dA = \int_0^{r_1, x} \rho U(r, x)(2\pi r dr). \qquad (5.43)$$

Equating the numerical and analytical mass flow rates gives one equation for obtaining c_0 and c_1. The other equation can be found using a proper boundary condition. We know that at $r = r(x)$, the velocity reaches the undisturbed far-field velocity. Hence we have

$$r = r(x) \Rightarrow U(r(x), x) = u. \tag{5.44}$$

Eq. (5.44) is the second equation required for finding c_0 and c_1.

5.3.2.2   Analytical wind velocity just behind the wind turbine

The wind profile just behind the wind turbine was obtained using the BEM method for blade section and a third-order polynomial for the nacelle. This profile is shown in Fig. 5.3. However, if we want to search for an analytical profile, we can make use of the self-similarity solution the same way we obtained the wind profile far from the turbine. In other words, we can use Eq. (5.21) for the wake profile just behind the wind turbine and find c_0 and c_1 for $x = 0$. Just note that at $x = 0$, the wake radius and velocity deficit cannot be obtained using Eqs. (5.39) and (5.40), since these equations are obtained for far distances. Instead, we know that at $x = 0$, the following equations are valid:

$$x = 0 \Rightarrow r(x) = r_0 \tag{5.45}$$

and

$$x = 0 \Rightarrow U_s = u - U_c, \tag{5.46}$$

where U_c is the wind velocity at the center line which can be easily obtained using the nacelle profile at $r = 0$. Thus, from Eq. (5.22), we get

$$U_c = b_0. \tag{5.47}$$

Consequently, for near wake at $x = 0$, instead of Eqs. (5.42) and (5.44), we use the following equations:

$$\dot{m}_{num} = \int_0^{r_0} \rho U(r, 0) 2\pi r \, dr \tag{5.48}$$

and

$$U(0, 0) = u - b_0. \tag{5.49}$$

These equations can be used to obtain unknown parameters, c_0 and c_1. It should be noted that this profile slightly differs from the numerical one but gives a reasonable estimate.

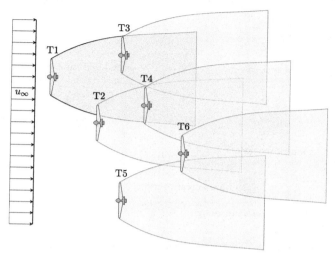

FIGURE 5.4 Effect of wake on downstream turbines.

5.4 Wake interaction

In a wind farm, the wakes of different turbines interact with each other and generate a complex turbulent region resulting in a low-speed region downstream of the interacted wakes. According to the placement of the turbines, Fig. 5.4 illustrates different expected scenarios.

- The first scenario applies to the turbines that are not affected by their upstream turbines. For example, T1, T2, and T5 in the figure are facing the undisturbed wind velocity.
- The second scenario applies to the turbines that are entirely located inside the wake of a single turbine. For instance, T3 is located just inside the wake of T1. From the previous discussions, we can find the wake velocity profile of T1 at the location of T3. Then we have to find an average wind velocity for T3 because the BEM method requires a single value for upstream wind speed. Unfortunately, steady BEM does not accept a variable profile for calculation. Now the question is how to obtain a single mean velocity out of a velocity profile. This question will be answered later.
- The third scenario applies to those turbines that are located in the interacted wakes of different turbines. In the shown wind farm, T4 depicts such a scenario. This turbine is located in an area that is affected by both T1 and T2. For simplicity, in the present model, we assume that the effect of T1 is considered on the calculation of the mean velocity for T2; thus, T2 contains the effect of T1. Consequently, we assume that T4 is entirely located inside the wake of T2.

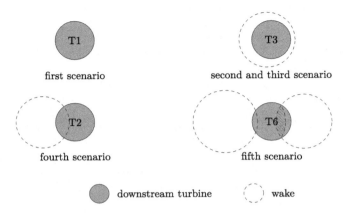

FIGURE 5.5 Different interaction scenarios.

- The fourth scenario applies to the turbines located where different wind velocity reaches the rotating surfaces. In the figure, T2 illustrates such a situation. As shown in the figure, a part of T2 is located at the wake of T1, and another part faces the upstream undisturbed wind speed.
- The last scenario applies to the turbines that are located where different wind velocity and a wind interaction reaches the rotating surfaces. Such a situation happens for T6. As shown in the figure, a part of T6 is located in the wake of T4, a part in the wake of T2, a part in the wake of T5, and finally a part in the region where T2's wake interacts with T5's wake. This situation seems to be very complicated, but in fact, the situation is the same as the fourth scenario.

The five different scenarios are illustrated in a different way by Fig. 5.5. The turbine names on the figure are the same as those shown in Fig. 5.4. Investigating this figure shows that the mean velocity for all the different scenarios can be obtained using a single algorithm. The steps of the algorithm are as follows:

Step-1: Obtain how many different profiles are reaching the rotor's swept area.
Step-2: Obtain the subareas on which each profile interacts.
Step-3: On each subarea, calculate the adequate velocity profiles.
Step-4: Perform integration on mass or momentum basis (discussed later).
Step-5: Calculate the mean velocity for the turbine.

Example 5.2. For the different scenarios shown in Fig. 5.4, apply the abovementioned algorithm.

Answer. To solve this problem, we assume that the turbines shown in Fig. 5.5 are the same as those in Fig. 5.4. According to the figures, the algorithm for different scenarios is as follows:

Step-1: From Fig. 5.5, it is clear that T1 and T3 are located in one single subdomain, T2 faces two different subareas, and the last turbine, T6, is divided into four subareas.

Step-2: The figure graphically shows the subareas. But for each subarea, the coordinates of the turbine and wake must be used to obtain the interacting area.

Step-3: T1 faces u_∞, hence we can easily obtain the mean velocity, which is $\bar{u} = u_\infty$. T3 is located in the wake of an upstream turbine (T1 according to Fig. 5.4). From the geometrical positions, the location of the wake and its distance from T1 is known. Hence the velocity profile is quite known. The left subarea of T2 is located in the wake of T1, and the right subdomain faces u_∞. The situation is a little bit more complicated for T6 because it faces different incoming wind speeds. The leftmost subarea is located in the wake of T4; hence its profile can be obtained using T4's wake. The subarea next to is under the influence of T2's wake. The enclosed area is influenced by the wakes of T4 and T5. Hence for this area, both wakes should be combined. Finally, the rightmost subarea is under the influence of T5. For this area, the T5's wake profile can be obtained using the present chapter's method.

Step-4: Depending on the selected method, on each subarea, the velocity profile should be integrated either on a mass or momentum basis. Having the velocity profiles and the geometry of the area, it would not be a big deal.

Step-5: Having the integral of Step-4, the mean velocity can be easily calculated. This method will be discussed further on.

Such scenarios can be discussed for any other configuration. □

The mean velocity can be obtained from either a mass or momentum basis. The mass basis uses continuity to obtain a mean velocity. In other words, it states that

$$\rho \int \rho u(r) dA = \rho \bar{u} \pi (\frac{d}{2})^2, \tag{5.50}$$

where $u(r)$ is the velocity profile for each subarea and dA refers to the differential area for the subarea on which $u(r)$ is applied. For the cases where a turbine encounters different wakes, the integral at the left-hand side should be divided into proper subareas. The left-hand side of Eq. (5.50) represents the mass flow rate passing the turbine if the velocity were uniform, or u_m. Eq. (5.50) yields

$$\bar{u} = \frac{\int \rho u(r) dA}{\pi (\frac{d}{2})^2}. \tag{5.51}$$

This equation is the same as (4.19).

According to the conservation of momentum, the mean velocity states that the momentum that the nonuniform profiles exerted on the turbine must be equal to the momentum that the uniform velocity exerts on it. Equating these two values gives a proper relation for calculating the uniform wind velocity on the

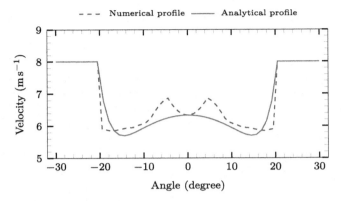

FIGURE 5.6 Wake profile just behind the wind turbine.

wind turbine. This fact is consistent with Eq. (4.21). Thus we have

$$\bar{u} = \sqrt{\frac{1}{A} \int u^2(r) dA}. \tag{5.52}$$

5.5 Simulation of a wind farm

The algorithms discussed in this chapter are best understood via some samples. The power curve of NTK 500/41 turbine is shown in Fig. C.11 and its other characteristics are given in Section C.11. This turbine is a stall-regulated turbine with the rated speed of $14 \, \mathrm{m\,s^{-1}}$ and the furling, or cut-off, speed of $25 \, \mathrm{m\,s^{-1}}$. This turbine is simulated using the BEM method, and its results are given in Chapter 3. Therefore, we use the results of the simulation, and all the following discussions are based on the simulation of the present turbine.

5.5.1 Wind profile just behind a wind turbine

The BEM method is applied to simulate NTK 500/41 turbine at $u_\infty = 8 \, \mathrm{m\,s^{-1}}$. As mentioned, Eq. (5.21) gives an analytical profile for the wind wake just behind the wind turbine by putting $x = 0$. The first step is to calculate the wake profile using the BEM method to find numerical wake for the blade sections and then use (5.22) to obtain a third-order profile for the nacelle wake. The result is shown by dashed lines in Fig. 5.6. Then using Eqs. (5.45) to (5.49), an analytical profile is obtained. This profile is shown in Fig. 5.6 by the solid line. As it can be seen, the analytical profile has a slightly different shape but, at the same time, shows a good prediction of the wind turbine wake. The advantage of the analytical profile is that it can be used for obtaining the wind turbine wake using the self-similarity solution, as discussed in the present chapter. Moreover, the analytical profile has a known function that may be used for any future cal-

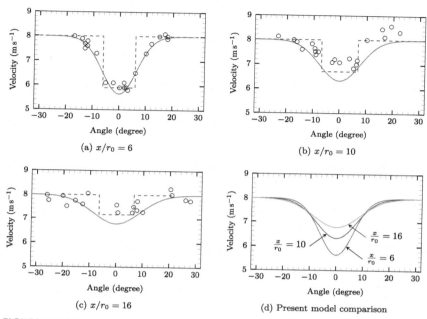

FIGURE 5.7 Wake profile comparison of the present model, —; Jensen's model (Jensen, 1983), ---; and experiment (Vermeulen et al., 1979), o.

culations such as calculation of momentum, thrust, or any other aerodynamic calculations.

5.5.2 Wake profile at far distances behind the wind turbine

After finding the analytical profile just behind the wind turbine, the mass flow rate is found by integrating the velocity profile, using Eq. (5.43). Having the analytical flow rate, and using Eqs. (5.42) and (5.44), the unknown parameters of Eq. (5.20), i.e., c_0 and c_1 can be found. These parameters give the wake distribution at far distances from the wind turbine. The results are shown in Fig. 5.7 where the results of the present model (solid lines) are compared with the experimental tests (Vermeulen et al., 1979) (circular symbols) and the wake profile calculated by Jensen's model (Jensen, 1983). As the figures indicate, the present model gives a better prediction than Jensen's model, which provides a step, or hat-shape, profile, while the current model provides a bell-shape prediction. The experimental results also agree better with the bell-shape profile.

Fig. 5.7d compares the wake profile for different distances. As it can be seen, the wake profiles become flatter as the distance increases. This fact is in agreement with the real tests. The value of the wake profiles are also validated by Figs. 5.7a–5.7c. Fig. 5.7d also shows that self-similarity is a good assumption while the profiles are self-similar.

5.6 Simulation of wind farms

The simulation of a wind farms depends on different parameters. The first thing is haveing an accurate model. Many researchers have tried to investigate the wake of wind turbines by actual field tests (Chu and Chiang, 2014; Chamorro and Porté-Agel, 2009; Christopher et al., 2016; Frandsen et al., 2006; Hashemi Tari et al., 2016; Kermani et al., 2013; Krogstad and Adaramola, 2012; Qing'an et al., 2016). Others have tried numerical schemes (Bianchini et al., 2021; Crasto et al., 2012; Crespo and Hernańdez, 1996; Espana et al., 2011; Ghadirian et al., 2014; Javaheri and Canadillas, 2013; Sørensen et al., 1998; Wang et al., 2016). Since in the numerecal schemes the turbulent modling has a great influence on the results, the effect of turbulent models is very important (Castellani and Vignaroli, 2013; Castellani et al., 2015; Ivanell et al., 2010, 2009; Yakhot et al., 1992). The turbulent models have been tested from differnt points of veiw (Cabezón et al., 2011; Naderi and Torabi, 2017; Nedjari et al., 2017; van der Laan et al., 2015; Wu and Porté-Agel, 2011, 2015), and still is onder consideration.

In addition to the wake model and the effect of turbulence modeling, wind turbine models are also important (Hassanzade and Naughton, 2016; Hosseini et al., 2019; Mikkelsen, 2003; Moradtabrizi et al., 2016; Newman, 1983). The more accurate the model, the more realistic the result will be. The turbine models can be integrated into the wind farm simulator (Burton et al., 2001; Johnson, 1985; Højstrup, 1983; Katic et al., 1986; Mosetti et al., 1994). The final goal of a wind farm simulation is obtaining the net annual energy.

In a wind farm, the turbines are located in the wake of their upstream turbines. Hence, not all the turbines experience the same wind speed. Therefore, to have a more realistic estimate of the total gained energy, each wind turbine should be analyzed independently and the net gained energy must be obtained by summing the energy of all the wind turbines. To illustrate the method, some simple wind farms are analyzed here.

5.6.1 Four wind turbines in a row

Fig. 5.8 shows a simple wind turbine farm consisting of four identical turbines. The distance between each turbine is 6 times the diameter D_0. The upstream wind velocity is u, and the wind blows from the west to the east. In this situation, the first turbine operates with the undisturbed upstream wind, while turbine 2 is located in the wake of turbine 1. Similarly, turbine 3 is located in the wake of turbine 2, and finally, turbine 4 is located in the wake of turbine 3. Naturally, the turbines do not generate the same amount of power since they do not experience the same wind velocity. The left-most turbine generates the highest power and the right-most one generates the least.

As we discussed for Fig. 5.4, we neglect the hierarchial wake distribution to the downstream turbines. For example, since the effect of turbine 1 is con-

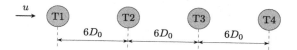

FIGURE 5.8 A wind farm with 4 turbines in a row.

sidered on turbine 2, we assume that its effect is included for the calculation of wind speed at the location of turbine 3. Thus the wind speed at the location of turbine 3 is calculated by just using the wake of turbine 2. The same argument is true for turbine 4.

The procedure for finding the wind velocity for each turbine is as follows:

1. Calculate wind turbine 1 using the BEM method. By this method, the power of turbine 1 is obtained. Also, the required parameters for the calculation of its wake become available.
2. Calculate the wake of wind turbine 1 at the location of wind turbine 2.
3. Obtain a uniform wind velocity using Eq. (5.51).
4. Repeat the above steps for the other turbines.

As repeatedly stated, for each wind turbine, we do not consider the combined effect of wakes. For example, to calculate the wind velocity for turbine 3, we do not use the results of turbine 1 since its effect has been already considered on turbine 2.

The algorithm of the above procedure is shown in Fig. 5.9.

The mean velocity of each turbine differs when the upstream wind speed changes. The variation of this parameter is shown in Fig. 5.10 where the prediction obtained by the present model is compared with the prediction of Jensen's model. For all the turbines, the present model predicts a higher velocity. It is worth mentioning that as the wind velocity increases, the difference between the present model and the Jensen's model increases. For example, for the 4th turbine at $u = 18\,\mathrm{m\,s^{-1}}$, the Jensen's model predicts a mean velocity about $u_m = 7\,\mathrm{m\,s^{-1}}$ while the present model gives $u_m = 13\,\mathrm{m\,s^{-1}}$. This means that these two models predict the mean velocity with almost 100% difference. As discussed in the previous chapter, the Jensen's model cannot give good predictions when the wind speed increases since it uses a value for the induced velocity, which is valid for low-speed velocities (around $u = 8\,\mathrm{m\,s^{-1}}$). Therefore, the present model is superior for high wind speeds.

Fig. 5.11 shows the net power of the whole farm with respect to the upstream wind velocity. It is clear that for the entire range of the wind velocity, the Jensen's model predicts less power. Since, according to Fig. 5.7, the present model shows more consistency with the real data, we can conclude that the results of the present model are more reliable. Consequently, if the Jensen's model is used for optimization purposes, the obtained layout will not necessarily be the optimum one.

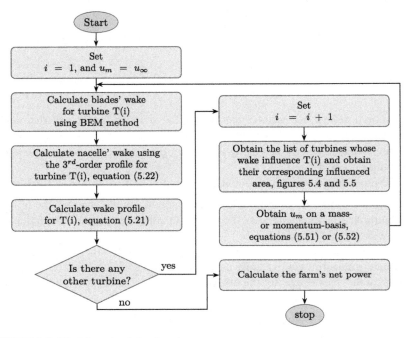

FIGURE 5.9 Wind farm simulation flowchart.

Example 5.3. Assume that the results of the present model are more accurate. Plot the error aroused by Jensen's model with respect to upstream wind velocity and discuss the results.

Answer. The resulting error is calculated using the following equation:

$$Error = \left| \frac{P_{\text{Present}} - P_{\text{Jensen}}}{P_{\text{Present}}} \right| \times 100\%.$$

These calculations are plotted in Fig. 5.12. It shows that in all the velocity ranges, Jensen's model has an error around 30%. □

The above example shows the importance of an accurate model. Unfortunately, due to the turbulent nature of wind and wake, an accurate analytical model is very hard to achieve. All the available models (including the present one) contain lots of simplifications; hence they have many sources of inaccuracy. But, from an engineering point of view, the models based on turbulence models can yield a better estimate and are more accurate. No need to reemphasize that the numerical models based on turbulence modeling can give a more realistic prediction, but they cannot be incorporated in optimization loops since they require lots of computational time.

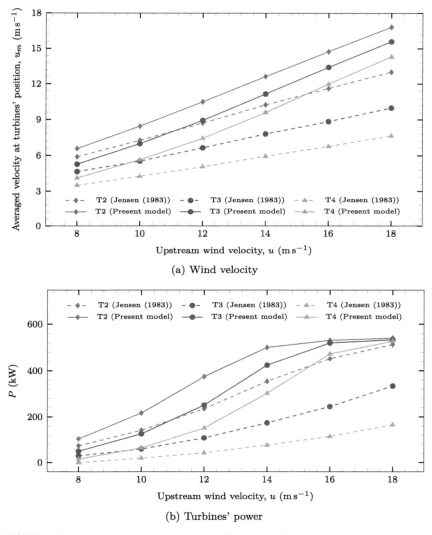

(a) Wind velocity

(b) Turbines' power

FIGURE 5.10 Results of farm simulation using different methods.

5.6.2 A 4 × 4 wind farm

The second wind farm consists of 16 identical turbines, which are arranged in a 4 × 4 matrix form as shown in Fig. 5.13. The wind blows from the west to the east, and as it is indicated, we consider its angle as zero. The vertical and horizontal distances of each turbine are the same and are equal to x_D times the turbine diameter, or D_0. Each turbine is labeled with two indices that indicate its distance from the origin. For example, T_{00} means that this turbine is located

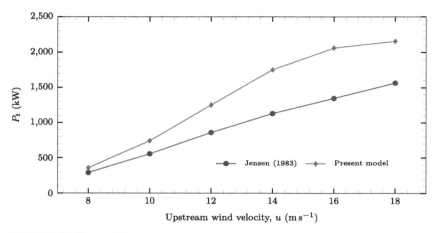

FIGURE 5.11 Net wind farm power comparison.

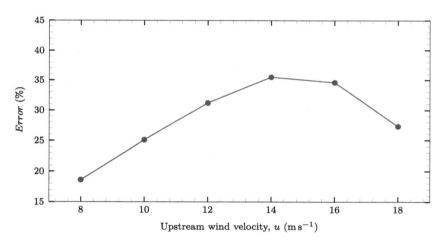

FIGURE 5.12 Error of calculation.

at $x = 0$ and $y = 0$, while T_{23} means that the turbine is located at $x = 2D_0$ and $y = 3D_0$. For convenience, the vertical and horizontal axes are indicated by x_D.

When the wind direction changes, the upstream wind turbines of each turbine vary. It is better to use two different coordinates to have a better sense and simpler calculations. One rigid coordinate system is designated by x and y, and one local coordinate system is designated by x' and y'. The x' axis of the x'–y' coordinate system is always parallel to the wind direction, as is shown in Fig. 5.14. The angle between these two coordinate systems is called δ. Note that by this definition, δ is not equal to the standard wind angle. As the standard

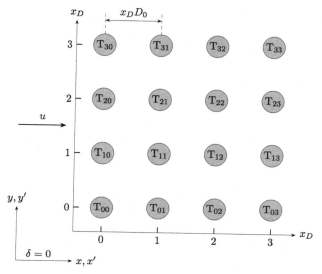

FIGURE 5.13 Wind blowing with $\delta = 0°$.

indicates, the wind angle is zero for the north wind (blowing from the north to the south) and also is measured clockwise. The relation between the standard wind angle and the convention shown in Fig. 5.14 is

$$\delta = -\delta_w - 90, \qquad (5.53)$$

where δ is the direction angle considering from the left to the right (as is conventional in fluid dynamics) and δ_w is the standard wind angle. The wind angle shown in this figure is considered to be negative since the local coordinate system is rotated clockwise.

The local coordinate system is important because it is in the wind direction and also determines the order in which the turbines must be specified. Fig. 5.15 demonstrates the situation for two different wind angles, namely $\delta = -30°$ in Fig. 5.15a and $\delta = -45°$ in Fig. 5.15b. The numbers on the turbine's location indicate the simulation priority. For example, in Fig. 5.15a the first turbine that is going to be simulated is turbine T_{30} which is indicated by number 1. The second to be simulated is T_{20}, and so forth. As it is clear for these two figures, the order by which the turbines are going to be simulated is determined from their horizontal distance in the local $x'-y'$ coordinate system. Or better said, the x' coordinate determines the priority. The turbines that have the same x' coordinate also have the same level of priority. For example, in Fig. 5.15b, turbines 4, 5, and 6 have the same priority, and there is no difference which one is going to be simulated first.

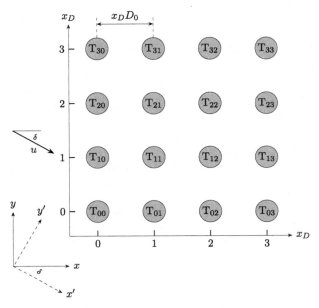

FIGURE 5.14 Wind blowing with angle δ.

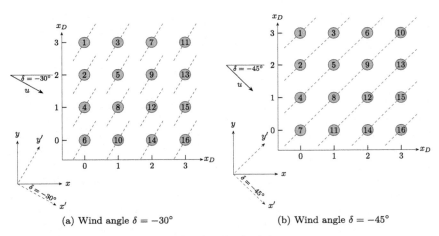

(a) Wind angle $\delta = -30°$ (b) Wind angle $\delta = -45°$

FIGURE 5.15 Determining the order of turbines for simulation.

Now that the farm geometry is known, it is going to be simulated and studied. For this purpose, we focus on variation of different parameters, including:

1. Change in wind direction or variation of δ,
2. Change in wind speed, or u,
3. Change in turbines' positions, or variation of x_D.

5.6.2.1 Wind blowing with $\delta = 0°$

When the wind blows from the west to the east (also known as the west wind), the global and local coordinate systems are the same and $\delta = 0$. In this situation, the priority of simulation is for the turbines located in the same column. Also, it is well known that by increasing the turbine distances, the wind velocity at downstream turbines becomes higher, and the gained energy is increased.

Fig. 5.16 shows the simulation results for the current situation. As we can see from Fig. 5.16a, increasing the distances from $x_D = 5$ to $x_D = 10$ results in a higher net power. The increase would be much noticeable as the wind speed decreases. For example, at $u = 8\,\mathrm{m\,s}^{-1}$, the net farm power increases from 1.3 to 1.6 MW which shows a nearly 20% increase in power. But the increase in power at $u = 18\,\mathrm{m\,s}^{-1}$ is from 8.62 to 8.66 MW which means about 0.4%. To have a better overview, it would be better to plot the percentage of power increase with respect to $x_D = 5$ for each wind speed. The plot is shown in Fig. 5.16b in which the net power at x_D is chosen as the reference point for each wind speed. Hence, the power at x_D would be zero, and the curves show the increase in farm power on a percentage scale. From this figure, we see that the turbine distances have more effect on lower wind velocities. The effect is quite meaningful since the wind turbine generates more resistance in low-speed winds. In other words, the induced velocity decreases when the wind velocity increases. This physical phenomenon was studied in the previous chapter. Consequently, Fig. 5.16b shows that if the wind rose and other statistical data of a particular place predict higher possibility in the lower wind ranges then the increase in distances between the turbines is suggested.

Although the increase in power at moderate wind speeds is quite noticeable, it should be noted that the economic factors and geometrical features are more important here. If we have plenty of space to install the turbines, then longer distances are more favorable. But we can neglect the energy increase and install the turbines using shorter distances in case of insufficient space.

5.6.2.2 Investigation of wind blowing angle, δ

In this section, the effect of wind direction is studied. When the direction of the wind changes, many different parameters change. The change in these parameters highly affects the net farm's power. For example, the distance between the turbines change. Also, each turbine is located inside the wake of different upstream turbines. Hence, the net farm's power should be recalculated and cannot be related to the previous results.

Hopefully, the present approach can consider the situation by presenting the local coordinate system. For considering the effect of wind rotation, we need to adjust two parameters, namely (a) the new distance and (b) the priority of calculation. For the local coordinate system, we can simply set

$$x' = x \cos(\delta) + y \sin(\delta), \tag{5.54}$$
$$y' = y \cos(\delta) - x \sin(\delta). \tag{5.55}$$

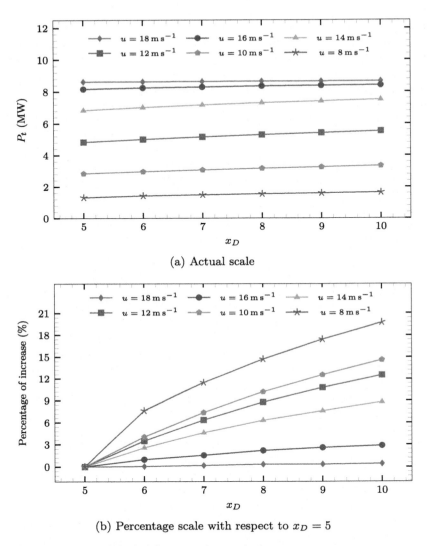

(a) Actual scale

(b) Percentage scale with respect to $x_D = 5$

FIGURE 5.16 Effect of physical distance on the net gained power.

As discussed before, the priority order is determined by the value of x'. For two different directions, these priorities are shown in Fig. 5.15.

The results of the simulation are plotted in Fig. 5.17. The figure shows that the net power of the farm is higher at $\delta = -30°$ although, according to Fig. 5.15, the distance between the turbines is lower compared to $\delta = -45°$. This happened because the wake pattern differs in both situations. For example, consider

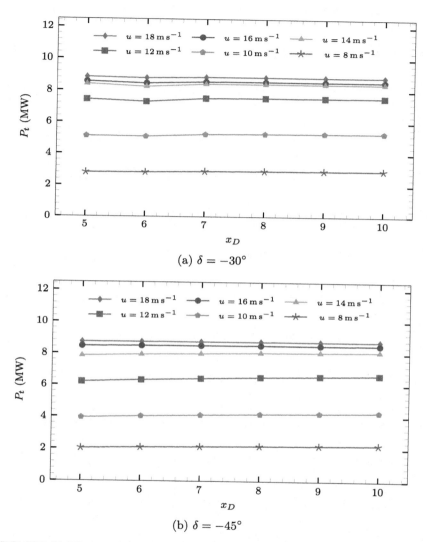

FIGURE 5.17 Effect of wind direction on net farm's power.

turbine 2 in two cases when $\delta = -30°$ and $\delta = -45°$. In case 1, it is much closer to turbine 1, but as Fig. 5.15a shows, it is not influenced by its wake. Hence, for case 1, it is more likely for turbine 2 to be operated by u_∞ instead of being affected by turbine 1's wake. However, in case 2, when the turbines are close to each other, turbine 2 may be affected by turbine 1's wake and produce less power. Such a statement can be applied to other turbines, too. Consequently, the

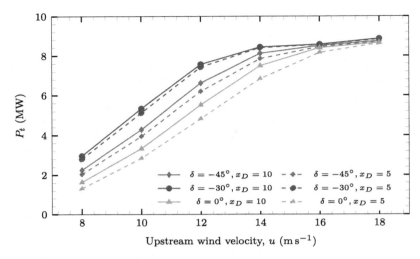

FIGURE 5.18 Comparison of farm's net power for different wind direction.

wake arrangement is such that the net power of the farm becomes higher when $\delta = -30°$.

In comparison to $\delta = -45°$, more turbines are located in the wake of upstream turbines when $\delta = 0°$. Hence, we expect that the net gained power is lower when we have the west wind. To have a better understanding, the net power for different wind directions is plotted in Fig. 5.18. The solid lines are for the situation where $x_D = 10$, and the dashed lines are for $x_D = 5$. This figure shows that higher energy is obtained when $\delta = -30°$, followed by $\delta = -45°$, and the least net power is when $\delta = -0°$. Therefore, such an arrangement is suitable for the location when the majority of the winds blow from the northwest to the southeast. If the wind rose of a particular place shows a different situation, the arrangement should be reconsidered.

It is worth noting that for $\delta = -30°$, the dimensionless distance, or x_D, has minimum effect on the net power because, at this angle, the turbines are located far enough from each other. Hence, there is enough room for the wakes to be recovered before reaching the downstream turbines. The situation is worse for $\delta = -45°$ and the worst case is when $\delta = 0°$.

5.7 The gained energy of a wind farm

Calculation of the net gained energy of a wind farm is important since, in a real wind farm, not only does the wind speed change, but its direction also is varying. In fact, in a real wind farm, we have a different probability for wind speed for each direction. This results in a bivariate probability density function (PDF) which depends on wind speed and direction. Therefore, the simple Weibull PDF discussed in Chapter 2 is not sufficient for the calculation of the net energy of the farm, and the wind rose diagrams must also be used.

There are many different ways to obtain the bivariate PDF, and many researchers have used other methods. One of the easiest methods is the method of bins. In this method, the wind direction is divided into several bins. Then, for each direction bin, a separate Weibull PDF is obtained out of the meteorological data. Then for each bin, the annual energy is calculated according to Eq. (2.69) and in the same way as in Example 2.22. Then the net energy of the farm is the sum of the energy of all the individual turbines. Just note that since each bin has its own probability, the summation must be performed according to its probability. Eq. (5.56) is the proper equation that represents the net annual gained energy of a wind farm:

$$E_{\text{farm}} = \sum_{i=1}^{\text{All turbines}} E_i. \tag{5.56}$$

Example 5.4. For the city of Torbat-e-Jaam, obtain the proper bins and corresponding Weibull PDF parameters.

Answer. Torbat-e-Jaam is a city in the eastern Iran. The wind rose diagram of the city is shown in Fig. 2.7. As can be seen, in that area, the major wind direction is NNE (meaning that the wind blows from NNE to SSW). As the rose diagram suggests, the wind direction can be divided into 16 bins shown in the same figure. The angular width of each bin then becomes

$$\Delta\theta = \frac{360}{16} = 22.5°.$$

Since the weather data are available with the accuracy of $1°$, the width of each bin is assumed to be $\Delta\theta = 23°$. The meteorological data available in Satba site (SATBA, 2020) gives the wind velocity at different heights, including 10, 30, and 40 meters. In addition to the wind speed, it contains the wind direction. Therefore, the data can be sorted according to the direction column, and statistical data can be found for each bin. The obtained data are given in Table 5.1. For the calculation of wind angle, the standard convention is used. That is the 0-angle indicates the north wind and the positive direction is clockwise. It is completely clear from the data shown in the table.

These data are obtained for the height of 30 meters and the columns k and c are the shape and scale parameters of the Weibull PDF. ☐

Note that there are two different probabilities defined here. The first is the probability of wind blowing in a specified direction, and the second defines the distribution of wind speed in the same direction. For better distinguishability, we can call them *directional* and *velocity* probabilities. In the previous example, the former is shown in the table, and the latter is defined by its parameters, i.e., k and c.

TABLE 5.1 Directional statistical wind data of Torbat-e-Jaam.

Bin name	Angle	From angle	To angle	Probability	k	c
N	0	−11	11	0.15199	1.87	5.53
NNE	22.5	12	34	0.32738	2.53	9.88
NE	45	35	56	0.09006	2.25	8.21
ENE	67.5	57	79	0.03086	1.69	4.15
E	90	80	101	0.03681	2.04	4.41
ESE	112.5	102	124	0.04151	2.23	4.49
SE	135	125	146	0.04175	2.19	4.26
SSE	157.5	147	169	0.04922	1.89	4.13
S	180	170	191	0.03745	1.38	4.03
SSW	202.5	192	214	0.02656	1.08	3.62
SW	225	215	236	0.01825	1.30	2.97
WSW	247.5	237	259	0.02025	1.25	2.78
W	270	260	281	0.01980	1.23	3.00
WNW	292.5	282	304	0.02484	1.49	3.40
NW	315	305	326	0.03231	1.64	3.76
NNW	337.5	327	348	0.05090	1.78	4.18

Example 5.5. If a single wind turbine, as that discussed in Example 2.22, is located in Torbat-e-Jaam, calculate its annual net energy.

Answer. To calculate the net produced energy in a real situation, the net directional energy of the turbine must be calculated for all 16 directions. In other words, Example 2.22 must be solved for all the 16 directions. In each direction, all the parameters are the same except for the Weibull parameters, k and c. To be more clear, these calculations are given for some selected directions in Table 5.2.

Fig. 5.19 summarizes all the calculated data. The two curves show directional produced energy without considering the wind rose diagram or the probability of direction of wind, and the share of produced energy in each direction, E_{share}, which is obtained by multiplication of E to the direction probability. Hence, the net yearly produced energy is the sum of E_{share}s for all the 16 directions:

$$E_{net} = \sum_{i=1}^{16} E_{share,i} = 16265705.8 \, \text{kWh} = 16.2657058 \, \text{GWh}. \qquad \square$$

Example 5.6. Calculate the capacity factor of the turbine of Example 5.5.

Answer. The capacity factor is defined by Eq. (2.70). The theoretical energy is defined as the product of its rated capacity and the total number of hours of a

TABLE 5.2 Results of Example 5.5.

N	Velocity	3	4	5	6	...	22	23	24
$k = 1.87$	Power	0	100	500	1000	...	8000	8000	8000
$c = 5.53$	Probability	0.144	0.147	0.135	0.113	...	2×10^{-6}	6×10^{-7}	2×10^{-7}
	Energy	0	129492	592659	992156	...	142	46	14
	Net Energy in direction N = 9806018 Wh								
	Directional Probability = 0.152								
NNE	Velocity	3	4	5	6	...	22	23	24
$k = 2.53$	Power	0	100	500	1000	...	8000	8000	8000
$c = 9.88$	Probability	0.039	0.058	0.075	0.089	...	4×10^{-4}	2×10^{-4}	8×10^{-5}
	Energy	0	50811	330946	7879507	...	31224	13560	5518
	Net Energy in direction NNE = 33298683 Wh								
	Directional Probability = 0.327								
NE	Velocity	3	4	5	6	...	22	23	24
$k = 2.25$	Power	0	100	500	1000	...	8000	8000	8000
$c = 8.21$	Probability	0.070	0.091	0.106	0.113	...	9×10^{-5}	3×10^{-5}	1×10^{-5}
	Energy	0	80142	465372	990012	...	6739	2710	1030
	Net Energy in direction NE = 23658776 Wh								
	Directional Probability = 0.090								

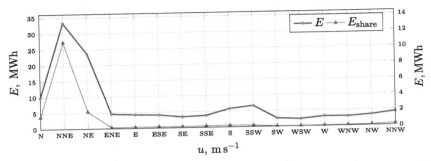

FIGURE 5.19 Directional produced energy (E) and its share to yearly net gained energy (E_{share}).

year. In other words, it means the produced energy of the turbine as if it works 24 hours a day throughout the year:

$$E_{\text{theo}} = 8760 \times P_R = 8760 \times 8000 = 70.08 \, \text{GWh}.$$

Consequently, the capacity factor of CF of the turbine becomes

$$CF = \frac{E_{\text{act}}}{E_{\text{theo}}} = \frac{16.2657058}{70.08} = 0.2321.$$

It means that the turbine can be considered as a $P = 8000 \times 0.2321 = 1856.8 \, \text{W}$ turbine that works 24 hours a day, 365 days a year. □

Example 5.7. For wind farm of Section 5.6.1, and weather data of Torbat-e-Jaam, calculate the annual gained energy.

Answer. The procedure is quite straightforward. As expected:

1. We divide the wind rose chart into 16 directions.
2. For each direction, the wind angle is mapped into the direction angle consistent with what is shown in Fig. 5.15.
3. For the specified direction, the energy output of the farm is calculated using the procedure of Example 5.5 and the Weibull PDF parameters given in Table 5.1.
4. We calculate the net gained energy according to the directional probability of each specific direction.

The result of this algorithm shows that the net yearly energy of the farm is

$$E_{\text{net}} = 14077436.954 \, \text{kWh} = 14.077 \, \text{GWh}.$$

For this farm, we can also calculate the capacity factor. Since the rated capacity of each turbine is 500 kW, then the theoretical energy is

$$E_{\text{theo}} = 16 \times 8760 \times 500 = 70.08 \, \text{GWh}.$$

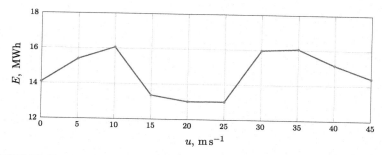

FIGURE 5.20 Yearly net produced energy of the wind farm with respect to rotation angle.

Therefore, CF of this farm becomes

$$CF = \frac{14.077}{70.08} \sim 0.2008 \sim 20.08\%.$$

The above examples show that the calculation method of a wind farm is quite the same as the procedure of calculation of a single turbine. The only difference is that it requires extra calculations for obtaining the mean wind velocity for each turbine. Other than that, the rest of the procedure remains the same. The role of wind rose chart and directional PDF concepts are used in the same as for a single wind turbine.

It is clear that the position of the wind turbines in a farm affects the net output energy of a farm. This fact can be further studied through the following example.

Example 5.8. Assume that the positions of wind turbines in the wind farm of Example 5.7 are rotated around the origin (i.e., around the point (0, 0)). Calculate the change of energy output by this rotation and discuss the results.

Answer. The farm is calculated using different rotational angles. The result is shown in Fig. 5.20 and, as it can be seen, the rotation of the farm has a great influence on the net gained energy. Therefore, this problem can be put into optimization loops to investigate the best rotation angle. □

Example 5.8 gives us the idea of optimization. It shows that the arrangement of the turbines in a farm is important since, in each region, the wind rose diagram and weather data will influence the net energy output. It is worth noting that what we did in Example 5.8 was to rotate the whole farm by a constant angle. In a real farm design, the position of each individual turbine can be selected as a design variable, and the best positions of all the turbines can be obtained individually.

5.8 Summary

In the present chapter, a new wake model is presented based on self-similarity solution. Since the model is based on the turbulent flow concepts, it has a very good accuracy compared to some other simple methods. However, the present model can be considered as semi-analytical since it requires some numerical analysis to obtain specific parameters. For the present time, these parameters are taken from the BEM method, which provides a fast algorithm for simulation. Therefore, the present approach is relatively fast and suitable for wind farm modeling and also optimization.

Some notes for the present modeling are worth mentioning:

- All the analytical models, including the present one, require many simplifications. Hence, the results will not quite agree well with field data. However, based on the accuracy of the model, the error may become acceptable. Many different scientists and engineers are trying to give relations or correlations to improve the models.
- The analytical models discussed in this book do not consider the effect of terrain and other geographical features. Hence, their results agree better when the field is flat or has less terrain roughness. For instance, these models will give good accuracy for off-shore farms or the farms that are installed in deserts.
- The analytical models are pretty good for optimization algorithms. These models are quite fast and can be easily integrated into optimization algorithms.
- Since the analytical models inherently include some errors, it would be better to have a numerical simulation of the final field. This fact will be discussed in the next chapters. With the power of currently available simulation applications and also different turbulence models, the numerical simulations will give very accurate results, and then the results can be compared with the results of analytical models.

The method was combined with a wind rose diagram to calculate the realistic yearly net gained energy of a farm. This method was successfully applied to some worked examples, and the procedure was studied.

5.9 Problems

1. For the wind farm shown in Fig. 5.8, plot the net power generation with respect to the distances between the turbines.
2. For the wind farm shown in Fig. 5.8, plot the net power generation with respect to the upstream wind velocity.
3. For the wind farm shown in Fig. 5.8, plot the net power generation with respect to the wind direction.

4. If the turbine model changes, the net power of the wind farm changes, too. Study this fact by simulating Mapna MWT2.5-103-I (see Section C.21) instead of Nordtank NTK 500/41 in the wind farm shown in Fig. 5.8.
5. Considering Mapna MWT2.5-103-I, plot the variation of net gained power of the farm shown in Fig. 5.8 with respect to the wind velocity change from 5 to 25 m s^{-1}. Note that Mapna MWT2.5-103-I is a pitch-regulated turbine and will give a constant power when the wind velocity is over its rated velocity.
6. Repeat the same study as requested in Problem 3 using Mapna MWT2.5-103-I.
7. For the wind farm shown in Fig. 5.13, plot the net power with respect to the wind direction for Nordtank NTK 500/41 at its rated wind speed. Which direction gives the highest power?
8. In the present chapter, wind farm in Fig. 5.13 is studied for wind velocity in the range of 10 to 18 m s^{-1}. Extend the range from 5 to 25 m s^{-1}.
9. Repeat Problem 7 using Mapna MWT2.5-103-I.
10. Repeat Problem 8 using Mapna MWT2.5-103-I.

References

Bianchini, Alessandro, Balduzzi, Francesco, Gentiluomo, Domenico, Ferrara, Giovanni, Ferrari, Lorenzo, 2021. Comparative analysis of different numerical techniques to analyze the wake of a wind turbine. In: Proc. of the ASME Turbo Expo 2017.

Burton, Tony, Sharpe, David, Jenkins, Nick, Bossanyi, Ervin, 2001. Wind Energy Handbook, vol. 2. Wiley Online Library.

Cabezón, D., Migoya, E., Crespo, A., 2011. Comparison of turbulence models for the computational fluid dynamics simulation of wind turbine wakes in the atmospheric boundary layer. Wind Energy 14 (7), 909–921. https://doi.org/10.1002/we.516.

Castellani, F., Vignaroli, A., 2013. An application of the actuator disc model for wind turbine wakes calculations. Applied Energy 101, 432–440. https://doi.org/10.1016/j.apenergy.2012.04.039.

Castellani, Francesco, Astolfi, Davide, Piccioni, Emanuele, Terzi, Ludovico, 2015. Numerical and experimental methods for wake flow analysis in complex terrain. Journal of Physics: Conference Series 625 (1), 012042.

Chamorro, Leonardo P., Porté-Agel, Fernando, 2009. A wind-tunnel investigation of wind-turbine wakes: Boundary-layer turbulence effects. Boundary-Layer Meteorology 132 (1), 129–149. https://doi.org/10.1007/s10546-009-9380-8.

Christopher, L. Kelley, Maniaci, David C., Resors, Brian R., 2016. Scaled aerodynamic wind turbine design for wake similarity. In: 34th Wind Energy Symposium, AIAA SciTech Forum.

Chu, Chia-Ren, Chiang, Pei-Hung, 2014. Turbulence effects on the wake flow and power production of a horizontal-axis wind turbine. Journal of Wind Engineering and Industrial Aerodynamics 124, 82–89.

Crasto, G., Gravdahl, A.R., Castellani, F., Piccioni, E., 2012. Wake modeling with the actuator disc concept. Energy Procedia 24, 385–392. https://doi.org/10.1016/j.egypro.2012.06.122.

Crespo, A., Hernańdez, J., 1996. Turbulence characteristics in wind-turbine wakes. Journal of Wind Engineering and Industrial Aerodynamics 61 (1), 71–85. https://doi.org/10.1016/0167-6105(95)00033-X.

Davidson, Lars, 1997. An Introduction to Turbulence Models.

Espana, G., Aubrun, S., Loyer, S., Devinant, P., 2011. Spatial study of the wake meandering using modelled wind turbines in a wind tunnel. Wind Energy 14 (7), 923–937. https://doi.org/10.1002/we.515.

Frandsen, Sten, Barthelmie, Rebecca, Pryor, Sara, Rathmann, Ole, Larsen, S., Højstrup, J., Thøgersen, Morten, 2006. Analytical modelling of wind speed deficit in large offshore wind farms. Wind Energy 9 (1), 39–53.

Torabi, F., Hamedi, R., 2015. An analytical model for prediction of wind velocity profile and wind turbine wake. Renewable Energy 75, 945–955.

Ghadirian, A., Dehghan, M., Torabi, F., 2014. Considering induction factor using BEM method in wind farm layout optimization. Journal of Wind Engineering and Industrial Aerodynamics 129, 31–39.

Hamedi, Razieh, Javaheri, Alireza, Dehghan, Omid, Torabi, Farshad, 2015. A semi-analytical model for velocity profile at wind turbine wake using blade element momentum. Energy Equipment and Systems 3 (1), 13–24.

Hashemi Tari, Pooyan, Siddiqui, Kamran, Hangan, Horia, 2016. Flow characterization in the near-wake region of a horizontal axis wind turbine. Wind Energy 19 (7). https://doi.org/10.1002/we.1895.

Hassanzade, Arash, Naughton, Jonathan, 2016. Design and analysis of small wind turbine blades with wakes similar to those of industrial scale turbines. In: APS Division of Fluid Dynamics.

Hoerner, Sighard F., 1965. Fluid Dynamic Drag: Practical Information on Aerodynamic Drag and Hydrodynamic Resistance, 2nd edition. ISBN 978-9991194448.

Højstrup, J., 1983. Nibe wake. Part I. Internal technical report. Risø National Laboratory.

Hosseini, Radmarz, Roohi, Reza, Ahmadi, Goodarz, 2019. Parametric study of a novel oscillatory wind turbine. Energy Equipment and Systems 7 (4), 377–387.

Ivanell, Stefan, Mikkelsen, Robert, Sørensen, Jens N., Henningson, Dan, 2010. Stability analysis of the tip vortices of a wind turbine. Wind Energy 13 (8), 705–715. https://doi.org/10.1002/we.391.

Ivanell, Stefan, Sørensen, Jens N., Mikkelsen, Robert, Henningson, Dan, 2009. Analysis of numerically generated wake structures. Wind Energy 12 (1), 63–80. https://doi.org/10.1002/we.285.

Javaheri, A., Canadillas, B., 2013. Wake modeling of an offshore wind farm using OpenFOAM. DEWI Magazin.

Jensen, Niels Otto, 1983. A note on wind generator interaction.

Johnson, Gary L., 1985. Wind Energy Systems. Citeseer.

Katic, I., Højstrup, J., Jensen, N.O., 1986. A Simple Model for Cluster Efficiency. European Wind Energy Association, Rome, Italy.

Kermani, N.A., Andersen, S.J., Sørensen, J.N., Shen, W.Z., 2013. Analysis of turbulent wake behind a wind turbine. In: International Conference on Aerodynamics of Offshore Wind Energy Systems and Wakes.

Krogstad, P., Adaramola, Muyiwa S., 2012. Performance and near wake measurements of a model horizontal axis wind turbine. Wind Energy 15 (5), 743–756. https://doi.org/10.1002/we.502.

Launder, B.B.E., Spalding, D.D.B., 1972. Lectures in Mathematical Models of Turbulence. Academic Press. ISBN 9780124380509.

Mikkelsen, Robert, 2003. Actuator Disc Methods Applied to Wind Turbines. PhD thesis. Department of Mechanical Engineering, Technical University of Denmark.

Moradtabrizi, Hamid, Bagheri, Edris, Nejat, Amir, Kaviani, Hamid, 2016. Aerodynamic optimization of a 5 megawatt wind turbine blade. Energy Equipment and Systems 4 (2), 133–145.

Mosetti, G., Poloni, C., Diviacco, B., 1994. Optimization of wind turbine positioning in large wind farms by means of a genetic algorithm. Journal of Wind Engineering and Industrial Aerodynamics 51 (1), 105–116. https://doi.org/10.1016/0167-6105(94)90080-9.

Naderi, Shayan, Torabi, Farschad, 2017. Numerical investigation of wake behind a HAWT using modified actuator disc method. Energy Conversion and Management 148, 1346–1357. https://doi.org/10.1016/j.enconman.2017.07.003.

Nedjari, Hafida Daaou, Guerri, Ouahiba, Saighi, Mohamed, 2017. CFD wind turbines wake assessment in complex topography. Energy Conversion and Management 138, 224–236. https://doi.org/10.1016/j.enconman.2017.01.070.

Newman, B.G., 1983. Actuator-disc theory for vertical-axis wind turbines. Journal of Wind Engineering and Industrial Aerodynamics 15 (1), 347–355. https://doi.org/10.1016/0167-6105(83)90204-0.

Paofsky, H.A., Dutton, J.A., 1984. Atmospheric Turbulence Models and Methods for Engineering Applications. John Wiley & Sons.

Pope, Stephen B., 2000. Turbulent Flows. Cambridge University Press.

Qing'an, Li, Murata, J., Endo, M., Maeda, T., Kamada, Y., 2016. Experimental and numerical investigation of the effect of turbulent inflow on a horizontal axis wind turbine (Part II: Wake characteristics). Energy 113, 1304–1315. https://doi.org/10.1016/j.energy.2016.08.018.

SATBA, 2020. Map of wind and solar energy of Iran. http://www.satba.gov.ir/en/regions. (Accessed 20 April 2020).

Sørensen, J.N., Shen, W.Z., Munduate, X., 1998. Analysis of wake states by a full-field actuator disc model. Wind Energy 1 (2), 73–88. https://doi.org/10.1002/(SICI)1099-1824(199812)1:2<73::AID-WE12>3.0.CO;2-L.

van der Laan, M. Paul, Sørensen, Niels N., Réthoré, Pierre-Elouan, Mann, Jakob, Kelly, Mark C., Troldborg, Niels, Hansen, Kurt S., Murcia, Juan P., 2015. The k-ε-f_P model applied to wind farms. Wind Energy 18 (12), 2065–2084. https://doi.org/10.1002/we.1804.

Vermeulen, P.E.J., Builtjes, P., Dekker, J., Lammerts van Bueren, G., 1979. An experimental study of the wake behind a full scale vertical-axis wind turbine.

Wang, Longyan, Tan, Andy C.C., Cholette, Michael, Gu, Yuantong, 2016. Comparison of the effectiveness of analytical wake models for wind farm with constant and variable hub heights. Energy Conversion and Management 124, 189–202. https://doi.org/10.1016/j.enconman.2016.07.017.

Wu, Yu-Ting, Porté-Agel, Fernando, 2011. Large-eddy simulation of wind-turbine wakes: evaluation of turbine parametrisations. Boundary-Layer Meteorology 138 (3), 345–366. https://doi.org/10.1007/s10546-010-9569-x.

Wu, Yu-Ting, Porté-Agel, Fernando, 2015. Modeling turbine wakes and power losses within a wind farm using LES: an application to the Horns Rev offshore wind farm. Renewable Energy 75, 945–955. https://doi.org/10.1016/j.renene.2014.06.019. http://www.sciencedirect.com/science/article/pii/S0960148114003590.

Yakhot, V., Orszag, S.A., Thangam, S., Gatski, T.B., Speziale, C.G., 1992. Development of turbulence models for shear flows by a double expansion technique. Physics of Fluids A: Fluid Dynamics 4 (7), 1510–1520. https://doi.org/10.1063/1.858424.

Chapter 6

Numerical simulation of a wind turbine

In wind farms, studying the interaction between downstream turbines and the wake of upstream ones in order to predict output power has a significant role. In addition, by forming a wake, the turbulence intensity of the flow is increased, and waked turbines are exposed to more turbulent flows. Hence, destructive forces on downstream turbines increase, which leads to more failure. Accordingly, providing a model for studying the behavior of wind farms can be helpful both technically and economically to optimize the performance of the whole set.

Many studies have been done on the simulation of wind turbine wake. The methods of these studies are divided into three categories, including analytical, numerical, and experimental methods. Each of these methods has its own advantages and drawbacks. The analytical methods have been covered in previous chapters. In the present chapter, the numerical methods are considered. In all the analytic models, an approximate solution is obtained for the problem by simplifying the governing equations of the fluid flow. In order to obtain more precise results, Navier–Stokes equations should be solved in a computational domain.

Hopefully, thanks to the advances in numerical methods and computer resources, a wind turbine or even a wind farm can be modeled and simulated using available commercial or open-source applications. These applications solve the Navier–Stokes equations by means of advanced numerical schemes such as finite volume method (FVM), finite element method (FEM), and other numerical techniques. Some of these programs are commercial, including FLUENT (ANSYS Inc., 2021), PHONICS (CHAM, 2021), WAsP (Danish Technical University, 2021), and COMSOL (COMSOL Inc., 2021), and some others are open source, including OpenFOAM (OpenFOAM, 2021) and FAST (NREL, 2021). The above applications are mostly general-purpose CFD codes that are able to solve the Navier–Stokes equations for any applications. But some of them are specially prepared for wind turbines. For example, WAsP, which is developed by DTU, uses a potential flow model to predict the flow of wind over terrain at a site. Similarly, FAST is a CFD toolkit that is developed based on OpenFOAM by NREL to simulate wind turbines.

OpenFOAM is popular among CFD experts due to its availability and ease of usage. Especially, the software can be used for the simulation of wind turbines as well as wind farms. However, as with any CFD software, a complete wind farm simulation with full detail is very computationally expensive. Therefore, many scientists try to give simplified models that are less costly and, at the same time, keep good accuracy. OpenFOAM can be used for the simulation of wind farms since it is a general-purpose CFD package. Using this program, we can simulate a wind farm with as many details as we need. Obviously, more accurate simulation requires more computations, hence costs more. Thus, depending on our needs, accuracy, simulation time, and cost, we can generate as complex a model as needed. For example, we can model a time-dependent simulation in which a full wind turbine is modeled with all its details. Also, the terrain can be modeled if proper information is available. On the other hand, if simplified models are enough for our simulation, we can make any simplification that is required. This package also includes different turbulence models which can be used for better accuracy. Consequently, OpenFOAM is a complete package for the simulation of wind turbines and wind farms.

6.1 Basic fluid dynamics concepts

When dealing with fluid mechanics, it is crucial to recognize the type of flow. Flows can be classified by different features (White, 2011). In general, a given flow is:

- Steady or unsteady,
- Viscous or inviscid,
- Compressible or incompressible, and
- Gas or liquid.

We need to select one from each pair before starting the analysis. For example, we assume that the flow is that of a steady, incompressible, viscous gas. When studying a viscous flow, it may be laminar or turbulent. Whether we are using simulation software or developing our own CFD codes, it is very important to have a good understanding of these features. Briefly, we have to have enough information about these concepts:

Transient or steady flow Steady-state assumption means that the fluid properties are constant at different times and do not change with time. Being steady-state does not mean that the properties do not change in space. In fact, any parameter has a specific space distribution that does not change with respect to time. In mathematical terms, $\frac{\partial}{\partial t} = 0$ in all the governing equations of the fluid flow.

Viscosity In addition to thermodynamic variables such as temperature, density, and pressure, some variables are important in fluid mechanics. One of these variables is viscosity, that is, a connection between local stresses and the strain rate of the fluid element. When a fluid is under shear stress, its motion is proportional to the inverse of the coefficient of viscosity, μ.

Compressibility Compressibility means that the density of the fluid changes and is not constant. The changes may be due to space or time. The criterion for checking the compressibility is the magnitude of Mach number:

$$\text{Ma} = \frac{V}{C},\tag{6.1}$$

in which V is the flow velocity, and C is the sound speed in the fluid that is about $340\,\text{m}\,\text{s}^{-1}$ in ABL. Mach numbers larger than 0.3 corresponds to compressible flow, and smaller values mean that the flow is incompressible. Note that incompressible flow does not actually mean that the density of the fluid is really constant, rather it means that we can neglect its variation. Since the operating range of the installed wind turbines is below $100\,\text{m}\,\text{s}^{-1}$, the flow regime around them is considered incompressible ($\delta\rho = 0$).

6.1.1 Governing equations of fluid flow

In general, the fluid flow must conserve mass, momentum, and energy under corresponding boundary conditions. These equations relate to the flow variables and their rate of change in time and space. Since in incompressible flows the density is constant, just the equations for conservation of mass and momentum must be solved, and if the variation of temperature is considered, also energy equation must be added (Sayma, 2009).

6.1.1.1 Conservation of mass (continuity)

The continuity equation applies to all kinds of flows, including compressible, incompressible, Newtonian, and non-Newtonian flows. It says that matter is conserved in a flow, meaning that the difference between the entering and exiting flow is equal to the mass rate. With assumption of incompressible flow, the mass conservation law is expressed using

$$\frac{\partial u}{\partial x} + \frac{\partial v}{\partial y} + \frac{\partial w}{\partial z} = 0,\tag{6.2}$$

in which u, v, and w represent the velocity components in the x, y, and z directions (Sayma, 2009).

6.1.1.2 Newton's second law (Navier–Stokes equations)

By exerting an external force to a specified system, it accelerates proportional to the exerted force in its direction. In a fluid motion, the viscosity resists the motion by exerting an opposite force. The net sum of viscous and external forces accelerates the flow. Eqs. (6.3) to (6.5) express the mathematical form of this law. These equations are known as the Navier–Stokes equations:

$$\rho\left[\frac{\partial u}{\partial t} + u\frac{\partial u}{\partial x} + v\frac{\partial u}{\partial y} + w\frac{\partial u}{\partial z}\right] = \rho g_x - \frac{\partial p}{\partial x} + \mu\left[\frac{\partial^2 u}{\partial x^2} + \frac{\partial^2 u}{\partial y^2} + \frac{\partial^2 u}{\partial z^2}\right] + f_x,\tag{6.3}$$

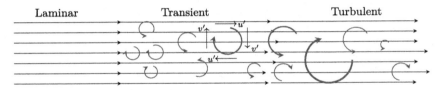

Laminar   Transient   Turbulent

FIGURE 6.1 Production and motion of eddies.

$$\rho\left[\frac{\partial v}{\partial t} + u\frac{\partial v}{\partial x} + v\frac{\partial v}{\partial y} + w\frac{\partial v}{\partial z}\right] = \rho g_y - \frac{\partial p}{\partial y} + \mu\left[\frac{\partial^2 v}{\partial x^2} + \frac{\partial^2 v}{\partial y^2} + \frac{\partial^2 v}{\partial z^2}\right] + f_y,$$
(6.4)

$$\rho\left[\frac{\partial w}{\partial t} + u\frac{\partial w}{\partial x} + v\frac{\partial w}{\partial y} + w\frac{\partial w}{\partial z}\right] = \rho g_z - \frac{\partial p}{\partial z} + \mu\left[\frac{\partial^2 w}{\partial x^2} + \frac{\partial^2 w}{\partial y^2} + \frac{\partial^2 w}{\partial z^2}\right] + f_z,$$
(6.5)

in which μ and p denote the viscosity and pressure, respectively (White, 2011).

6.1.2 Turbulence

Regarding the nature of the flow around wind turbines, they should be considered turbulent. As its name implies, this flow has an unorganized and random behavior. In such flows, because of the intense mixing processes, apart from close areas to the wall, the shape of the flow layers cannot be recognized, and the fluid particles do not pass a particular path, as shown in Fig. 6.1. In other words, turbulent flow is a type of flow in which the fluid is under fluctuations and severe mixing processes; this behavior is in contrast to laminar flow behavior in which the fluid moves in certain layers and paths. In a turbulent flow, the velocity at each point constantly fluctuates and changes, both in value and direction (Pope, 2000). In brief, a turbulent flow has the following characteristics (Davidson, 2015):

Irregularity It has chaotic behavior and consists of eddies, including large eddies whose size are in the order of flow geometry and small scale eddies that dissipate by viscosity.

Diffusivity In turbulent flows, the diffusion of properties and momentum increase. Accordingly, the molecular diffusion becomes negligible.

Large Reynolds numbers The criterion for separating the laminar and turbulent flow in fluid mechanics is based on Reynolds number. This limit is 2300 for internal flow and 100000 for the boundary layer, and larger Reynolds numbers correspond to turbulent flow.

Three-dimensional Turbulence is always a three-dimensional concept.

Dissipative In a turbulent flow, in a cascade process, the large eddies extract their energy from the mean flow and transform it to the small-scale eddies. This energy then is dissipated by viscosity and converted to internal energy.

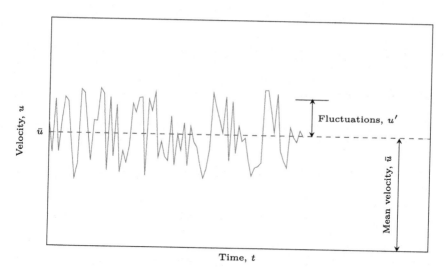

FIGURE 6.2 Instantaneous and average velocities.

In turbulent flows, the velocity can be divided into an average ($\bar{u}$) and an oscillation part (u'). The overall direction of the flow is determined by the mean flow, and the fluctuations result in some deterioration in the fluid movement. This fact is illustrated in Fig. 6.1 and in another illustration of Fig. 6.2. This variation of speed exists in each point of the domain. By averaging the velocity over an appropriate time interval, that is not too small and not too large, the mean value for velocity is obtained using

$$\bar{u} = \frac{1}{\Delta t} \int_{t}^{t+\Delta t} u \, dt, \tag{6.6}$$

in which

$$u = \bar{u} + u'. \tag{6.7}$$

This averaging can be used for all of the flow variables, such as pressure and viscosity. It is the way for reducing computational cost in a numerical simulation.

Several models are presented to solve the turbulence equations. The k–ϵ family is a popular turbulence model, and among the k–ϵ models, the standard k–ϵ proposed in (Launder and Spalding, 1972) is used frequently. Turbulence kinetic energy and dissipation equations are the two governing equations in this model.

Renormalization group (RNG) k–ϵ proposed by Yakhot et al. (1992) is another turbulence model that is a modification of the standard k–ϵ. In this model, the dissipation equation is modified in order to take into account smaller scales of motion.

6.1.3 Wake

When a turbine absorbs wind kinetic energy, the wind energy decreases, and this phenomenon appears as a speed deficit. In this situation, some vortices are formed at blade edges, and the turbulence intensity of the flow increases significantly. The produced turbulence and vortices are named wake that has a specific shape downstream of the turbine. There are two kinds of vortices behind the blade, namely the vortices produced near the root and tip vortices. By advancing downstream, the lost energy will recover while the wake interacts with the free stream. The root vortices are destroyed earlier because of the hub induction, and tip vortices move downstream, so the structure of the wake depends on tip vortices.

The wake recovery highly depends on freestream turbulence intensity. In high turbulence flows, the lost kinetic energy recovers faster because of the high momentum diffusion that is known as turbulence diffusion. This recovery also can be reinforced when some turbines are operating close to each other, and the wakes of multiple turbines interact with each other. In this situation, the speed deficit behind the first turbine is more than the speed deficit behind the waked turbines.

6.2 Different types of modeling

As mentioned above, CFD codes can simulate wind turbines with as many details as we want. The numerical simulation of the Navier–Stokes equations (Eqs. (6.3) to (6.5)) is not an easy task. One of the most challenging parts is the way the pressure is to be solved for. In order to solve the equations of continuity and momentum simultaneously, several methods have been developed. In this research, the semiimplicit method for pressure-linked equations (SIMPLE) is utilized, which was proposed by Patankar and Spalding in 1971 (Patankar and Spalding, 1972). In this approach, the momentum equation is first solved, assuming that the pressure value is known (using initial guesses). The obtained velocity field does not satisfy the continuity equation since the pressure field is not real. Accordingly, a pressure and velocity correction must be conducted. For a two-dimensional velocity field, the correction equation for pressure field is

$$p = p^* + p_c, \tag{6.8}$$

in which p^* is the assumed initial p in the first iteration, and in the succeeding iterations, it is the pressure obtained from the previous iteration. Also, p_c is the pressure correction.

For the velocity field, the correction equations are:

$$
\begin{aligned}
u &= u^* + f_u(p_c), \\
v &= v^* + f_v(p_c), \\
w &= w^* + f_w(p_c),
\end{aligned}
\tag{6.9}
$$

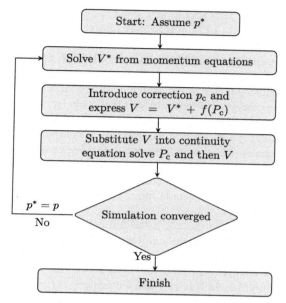

FIGURE 6.3 Flow chart of the SIMPLE algorithm.

where u^*, v^*, and w^* are the velocities in the x, y, and z directions that are calculated by old pressure value using momentum equation where p_c is the pressure correction. The value of p_c can be calculated by substituting u, v, and w into continuity equation. Then, the corrected pressure is used in the momentum equation to compute the final velocity field in each iteration. At last, by using the under relaxation factor for each variable, its value for subsequent iteration can be found. Fig. 6.3 shows the described algorithm.

6.2.1 Full rotor

Full rotor modeling means that a wind turbine is modeled as is. The whole tower, nacelle, blades, rotor, etc., are modeled, and unsteady flow is simulated using a proper solver. In wind turbine modeling, the fluid regime is subsonic and incompressible since the rotational velocity of the rotor is chosen such that the highest velocity (at the blades' tip) becomes less than $100 \, \text{m s}^{-1}$. Knowing that the sound speed is about $330 \, \text{m s}^{-1}$, we conclude that the maximum Mach number is less than $M < 100/330 < 0.3$. Hence, the assumption of incompressible flow is quite meaningful.

For such a fluid regime, FVM solvers are capable of giving accurate results. Different solvers are developed, among them SIMPLE, SIMPLER, and SIMPLEC are commonly used in many different applications. The unsteady SIMPLE solver is applicable for the simulation of wind turbines. In addition to

FVM, FEM solvers are also able to give good results for incompressible flow. Thus, applications that use FEM as their native solver can be used for such simulations.

As an FVM software, OpenFOAM includes SIMPLE algorithm and its different variants. This package also contains different turbulence models that are important for capturing proper wake, which in turn influence the wind farm layout. One of the advantages of OpenFOAM is that it can be easily run on parallel architectures, which is very important from both speed and memory points of view. Since the full rotor simulation requires lots of computational mesh, a single personal computer (PC) may not be able to provide enough memory for the simulation of a whole wind farm. However, PCs can be connected via a network to make a Beowulf cluster in which each PC in the network is responsible for the simulation of a part of the whole farm. Then they communicate with each other through the network.

One of the most precise models is the full rotor simulation that is used in many studies such as (AbdelSalam and Ramalingam, 2014) and (Abdulqadir et al., 2017). It is worth mentioning that the computational cost of this simulation is high because of resolving the boundary layer around the blades since, for full rotor simulation, the whole Navier–Stokes equations (i.e., Eqs. (6.2) to (6.5)) must be solved. In addition, the equations of turbulence models should also be added to the system of equations.

6.2.2 Actuator line

The full rotor simulation requires lots of computational resources. Even a single turbine can hardly be simulated on a single PC due to the numerous computational grids and lots of computations since the flow is unsteady and requires remeshing of the turbine.

To overcome the problem of high computational cost, different models have been developed. One of these methods is called *actuator line method,* or simply ALM. In this method, instead of simulating the whole rotor, its effect is simulated inside the computational domain. In other words, in this method, the blade is divided into some elements as it was in BEM. Then, the aerodynamic forces of each element are calculated. This can be done by the BEM algorithm or any other method. After that, these forces are inserted into momentum equation via the body force terms, f_x, f_y and f_z in Eqs. (6.3) to (6.5).

ALM needs unsteady state simulation (Ivanell et al., 2009, 2010) and is computationally expensive, but not as expensive as a full rotor simulation. In this method, individual tip vortices are simulated using the AL approach. It is preferred when near wake characteristics are studying, but when the case is a wind farm including a large number of turbines, AL imposes a high computational cost. Recently, a new method named *virtual blade model* (VBM) was introduced by Bianchini et al. (2017, 2021) that simulates the wind turbine by body forces, which is not computationally complex in comparison to the AL method. The BEM method is also used in VBM in order to compute the loads.

6.2.3 Actuator disc

As a simple and low-cost way, the *actuator disc* method (ADM) can be introduced. In this investigation, the wind turbine is considered as a porous, permeable disc that absorbs wind momentum. Neglecting the boundary layer region around the blade leads to a low cost but efficient way. ADM follows the same concept as ALM, but in this time the whole extracted momentum of the wind turbine is expressed as a source term in the momentum equation.

Newman (1983) simulated a wind turbine using the concept of ADM and examined the effect of different turbine parameters on the power coefficient. Sørensen et al. (1998) have done a two-dimensional simulation by solving Navier–Stokes and ADM equations simultaneously. Mikkelsen (2003) studied the turbine performance in different operational conditions, such as different yaw and cone angles. The obtained results show that the AD method, despite its simplifications, can model the interaction between the turbine and its surroundings. Some researchers used ADM simulation as the input for full rotor simulation in order to decrease the simulation time (Nedjari et al., 2017; Sturage et al., 2015).

In contrast to ALM, the rotor rotation is not simulated in ADM, but it does not mean that the flow cannot be transient. In fact, the solver can be set to transient mode and unsteady flow with any suitable turbulence model. Hence, an unsteady turbulent wake can be modeled. The results show that ADM is capable of predicting reasonable results in capturing turbine wakes and can be used for farm simulation, although it is not as accurate as ALM.

To better understand the method presented in this book, it is needful to provide a summary of the concept of induction factor and forces applied to the turbine and the fluid flow. The concept of ADM is fully discussed in Chapter 2. In Fig. 2.9, the concept of actuator disc is shown and its formulation is given in Section 2.3.3. According to the one-dimensional momentum theory, an ideal wind turbine, as a permeable disc, is a device that receives mechanical energy through the reduction of the kinetic energy of wind. The term "ideal" means that there is no rotary component inside the wake behind the turbine and the disc has no friction. Thus, the turbine acts as an obstacle that reduces the wind velocity from the value u_1 in infinity to the value u_2 on the turbine surface and u_3 after the turbine, as shown in Fig. 2.9.

The results state that the trust coefficient in theory is given as

$$c_T = \frac{\frac{1}{2}\rho A_2 (u_1 + u_4)(u_1 - u_4)}{\frac{1}{2}\rho u_1^2 A} = 4a(1-a). \quad (6.10)$$

Also, it was found that the flow velocity after the turbine is

$$u_2 = u_3 = (1 - 2a)u_1. \quad (6.11)$$

6.3 Development of actuator disc method using OpenFOAM

OpenFOAM uses a simple model for the implementation of ADM. To implement the model, first, the geometry must be set up. Then the momentum equation is modified such that the effect of the wind turbine is included inside the model through body forces. Then some necessary solver settings must be adjusted. Finally, the problem is solved using a proper solver, and the results can be viewed using ParaView software. In the proposed model, a uniform source term is applied all over the disc, and we are not able to apply variable load along with the blades. This imposes some error in turbine calculations because, in the real world, the momentum source varies along the turbine's blade.

Although many research groups use the method and it gives good results, the procedure can be further improved since, as mentioned, the load distribution is not uniform all across the turbine swept area. To overcome the problem, OpenFOAM proposed a nonuniform AD model in which the source term induced by the turbine is modeled as a fourth-order polynomial. Having the proper load distribution, this model gives a better estimate compared to a uniform distribution. These two models will be discussed shortly in the succeeding subsections.

Although the nonuniform distribution gives more accurate results, the point is that the load distribution of wind turbines is neither a constant value nor necessarily of a polynomial shape. As repeatedly mentioned in the previous chapters, the induction factor is not uniform along the blades; thus, imposing a constant value or predefined shapes is not a universal solution. Some researchers proposed that OpenFOAM becomes coupled with another solver such as BEM. In this way, the BEM solver calculates the radial thrust, $T(r)$, and sends it back to OpenFOAM. Then OpenFOAM marches an iteration and gives the new conditions to BEM to modify the load distribution. This way, a more realistic simulation takes place.

In the following subsections, the method is discussed in more detail. Note that the implementation of the original method is shipped with OpenFOAM package through its standard tutorials and is called *turbineSiting* (the current version of OpenFOAM is V8 in June, 2021).

6.3.1 Geometry and mesh generation using BlockMesh

For simulation of a single wind turbine using ADM, the whole domain is modeled without any details of the actual wind turbine. Therefore, all the details of the wind turbine are neglected, which results in a very simple domain and mesh generation. Fig. 6.4 shows such a domain in which a simple disc models the turbine. As shown in the figure, the turbine geometry, nacelle, hub, and turbine tower are not modeled. Moreover, the whole domain is a simple rectangular shape with the boundary conditions that are indicated in the figure.

As can be seen, the wind flows perpendicular to the front, or y–z plane, or the wind direction is aligned with the x axis. Therefore, on the front surface, the velocity inlet is the proper boundary condition. However, it is worth mentioning

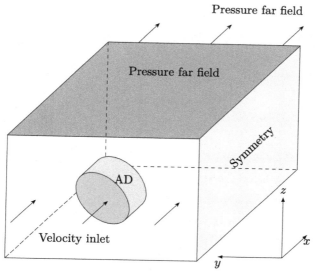

FIGURE 6.4 Proper domain for ADM.

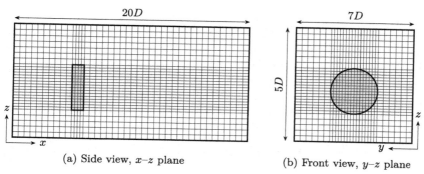

(a) Side view, x–z plane

(b) Front view, y–z plane

FIGURE 6.5 Mesh generation for ADM.

that a nonuniform wind speed can be set for this surface to better model the ABL.

The upper and the rear surfaces are set as pressure far-field, since they are far from the turbine or the AD to ensure that the turbine does not affect the boundaries. The left and right surfaces (at x–z planes) are set to be symmetrical. This will result in stress-free boundary conditions in the y direction. Finally, the lower surface is considered a wall because it simulates the ground. Note that the geographical features can be modeled if their information is available.

Since the aerodynamic of wind turbines take place in a subsonic regime, the flow has an upstream influence. Hence, as mention above, the domain must be set up such that the outlet boundaries are not affected by the turbine. One of

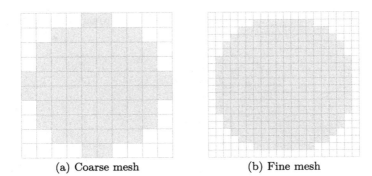

(a) Coarse mesh   (b) Fine mesh

FIGURE 6.6   Effect of clustering on AD meshing.

the solutions is to choose a very large field for simulation. Fig. 6.5 shows such a domain from two different views. For instance, Fig. 6.5a shows a side view of the domain shown in Fig. 6.4. The domain length in the x direction is about 20 times the turbine diameter, as illustrated by the figure. Fig. 6.5b also shows that the domain in the y and z directions are extended about 7 times the turbine diameter.

The important point is that these values are just some typical values that are used by some researchers. In practice, the values must be selected by trial and error so that the results become geometry independent. Hence, the domain dimensions must be determined according to the specific problem. The turbine size, wind speed, and operational conditions affect the size.

Fig. 6.5 also demonstrates the meshing strategy that is suitable for ADM. Since the source terms are to be applied only on the volumes that compose the AD, it is better to cluster the computational mesh around the disc. In this way, a lesser numerical mesh is generated while the best accuracy is obtained. The front view also shows that the clustering gives a better accuracy where the AD is located.

The effect of clustering can be better understood by studying Fig. 6.6. As the figure illustrates, the coarser the mesh, the less accurate the turbine. By increasing the number of mesh points, the AD shape becomes more similar to an actual circle. But it should be noted that the higher accuracy results in higher computational cost. Consequently, as in any other numerical simulation, the clustering must be studied. The best mesh size must be identified so that the results become mesh independent with the least computation.

6.3.2   Source term definition

To simulate an AD, the effect of the turbine must be inserted in momentum equations. In OpenFOAM, this is done by means of Eq. (6.12) where the whole wind turbine is modeled as a single disc with a uniform load. It means that on all the turbine surface, a uniform load is distributed such that

$$T = \frac{1}{2} c_T \rho A u_1^2 = 2 \rho A u_1^2 a (1 - a), \tag{6.12}$$

in which the total induction factor, a, is constant across the disc and its value is defined by

$$a = \frac{c_P}{c_T}, \tag{6.13}$$

where c_P and c_T are the power and thrust coefficients, respectively. These values can be extracted from the power and thrust curves at an operating wind speed.

As discussed before, OpenFOAM also contains a radial distribution for the source term, which is in the form of

$$T(r) = \frac{T}{C_0 + \frac{C_1 R^2}{2} + \frac{C_2 R^4}{3}} \left(C_0 + C_1 r^2 + C_2 r^4 \right). \tag{6.14}$$

As can be seen, the denominator of the above fraction is a constant since it is a function of the blade's length, or R. Also, the radial function gives a specific value at $r = 0$, which is C_0. But as we know, the turbine's blades do not generate power up to a specific radius (the blade radius is measured from the hub axis). Therefore, this model exerts greater force than a real turbine. Javaheri and Canadillas (2013b) worked on this and gave the following relation for correcting this issue:

$$T(r) = \begin{cases} 0 & r < r_{\text{Hub}}, \\ T \times \frac{V_{\text{cell}}}{V_{\text{eff.}}} & r > r_{\text{Hub}}. \end{cases} \tag{6.15}$$

In this equation, V_{cell} is the volume of the computational cell and $V_{\text{eff.}}$ is the turbine's effective volume, which is the actual swept volume of the turbine, or as can be seen in Fig. 6.7, it is the volume of the actuator disc minus the volume of the hub, which has a radius of r_{Hub}.

6.3.3 Turbulence modeling

Turbulence modeling affects the turbines' wake. Hence, choosing a proper turbulence model is essential. Unfortunately, there is no classical method for selecting an appropriate model. Thus, it is a trial-and-error process and requires experience.

Fortunately, OpenFOAM comes with a vast library containing lots of different turbulence models. Some of the models are tabulated in Table 6.1. These models are suitable for incompressible flow, which is the case in wind turbines. Therefore, the user is able to simulate the turbine or wind farm with different turbulence models and compare the results.

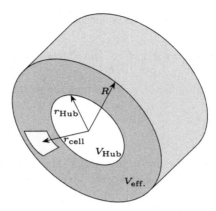

FIGURE 6.7 Illustrations of the parameters of Eq. (6.15) (Javaheri and Canadillas, 2013b).

TABLE 6.1 Different available turbulence models in OpenFOAM.

Turbulence model	Specification
Laminar	Dummy turbulence model for laminar flow
kEpsilon	Standard high-Re k–ε model
kOmega	Standard high-Re k–ω model
kOmegaSST	k–ω-SST model
RNGkEpsilon	RNG k–ε model
NonlinearKEShih	Nonlinear Shih k–ε model
LienCubicKE	Lien cubic k–ε model
qZeta	q–ζ model
LaunderSharmaKE	Launder–Sharma low-Re k–ε model
LamBremhorstKE	Lam–Bremhorst low-Re k–ε model
LienCubicKELowRe	Lien cubic low-Re k–ε model
LRR	Launder–Reece–Rodi RSTM model
LaunderGibsonRSTM	Launder–Gibson RSTM with wall-reflection terms
realizableKE	Realizable k–ε model
SpalartAllmaras	Spalart–Allmaras 1-equation mixing-length model

6.3.4 Solver selection and settings

OpenFOAM comes with lots of different solvers, one of which is simpleFOAM. As the name indicates, this solver is an implementation of the SIMPLE algorithm (see Fig. 6.3 which is suitable for incompressible flows).

When using simpleFOAM, note that the solver is developed for steady-state simulation and cannot capture the transient effects. For transient simulation, other solvers such as pisoFOAM must be used. The details of the software are well explained in (Behrens, 2009).

One of the advantages of OpenFOAM is that many advanced techniques are implemented in the solver. Also, each equation can be solved using a different numerical scheme to gain the best performance out of the solver. By default, the pressure field is solved for using the intrinsic multigrid solvers, as GAMG (generalized (or geometric) algebraic multigrid), and other variables are solved for using smoothSolver, which is an iterative method for symmetric and asymmetric matrices. These values can be changed according to the needs of the problem.

The solver selection requires lots of experience; however, to begin with, the default values are quite suitable for simulation. The solver can be run using parallel processing without any modification to the original problem. It is quite sufficient to determine the number of processors to the solver. All the required processes are performed internally, and the program runs on the specified parallel architecture.

6.4 Modified actuator disc

The ADM mentioned in Section 6.3 is applied to the numerical domain by adding a constant or radius-dependent force in the momentum equation. To do this, the source terms should be evaluated in a separate solver, and the results are set as inputs to OpenFOAM. The point is that the data being processed may not be accurate since they are not dynamically coupled with OpenFOAM. This fact will be much more critical in the simulation of wind farms, where the turbines affect their downstream turbines, and the wind velocity for each turbine is not known before calculation. Since the velocity is not known, the static input data for wind turbine simulation produces an error in simulation. In the present chapter, a new coupled method is introduced, and its application to wind farm simulation is postponed to the next chapter.

6.4.1 Radial load distribution

To have a more realistic model, OpenFOAM should be coupled with a solver that is able to calculate the induction factors of the wind turbine. For instance, the BEM method is one of the best choices since it is very simple to implement and very fast, which is very important from the computational cost perspective. Another important advantage of BEM is that it can calculate a nonuniform distribution of induction factors along the blade. Hence, the same distribution can be implemented in OpenFOAM to obtain a better model.

Fig. 5.3 shows the radial distribution of wind velocity obtained by BEM method. By the methods of the present book, as discussed in Chapter 5, the wake of the hub can also be found. Therefore, the induction factor distribution can be easily obtained, which in turn can be used to calculate the radial distribution of thrust according to

$$T_{el} = \frac{1}{2} \rho \, A_{el} \, u_1^2 \, 4a_{el} \, (1 - a_{el}). \tag{6.16}$$

In the above equation, T_{el} is the thrust on each element, a_{el} is the induction factor in each element, and A_{el} is the area of each element. The total thrust on the disc can be calculated as follows:

$$T_t = \sum_{n=1}^{N} T_{el,n}, \tag{6.17}$$

where T_t represents the total thrust on the disc and N is the number of elements.

For coupling Eq. (6.16) with OpenFOAM, a simple programming code in any available language must be developed. Here, a C++ code is written for calculation of Eq. (6.16) and coupling the results with OpenFOAM. The method is quite clear and straightforward.

All the above explanations are done in an algorithm that is written in the C++ programming language. In order to couple it with OpenFOAM and AD method, the code is simultaneously solved with Reynolds averaged Navier–Stokes equations in the computational domain and updated in each iteration. It is important to use fine grids in order to consider load variation in the spanwise direction. Such a grid distribution is shown in Fig. 6.6b. This disc is located in the main domain as is illustrated by Fig. 6.5a. For the volumes that are located in the AD location, an extra source term is calculated using Eq. (6.16) but not for the other volumes.

6.4.2 Modified source term

The AD configuration is performed using utilities in OpenFOAM in two main steps. First, the geometry of the AD must be defined, which can be done by using the *topoSet* dictionary in the *system* folder. The *topoSet* helps us to collect faces, cells, or points based on specific rules. The required entries are described below:

- Type: By this entry, the type of cell group is selected; *pointSet*, *faceSet*, *cellSet*, *faceZoneSet*, and *cellZoneSet* are the options. Using the *cellSet* option, a group of cells is constructed in this study.
- Action: The action that must be performed on a set is defined here using one of the following options: *clear*, *invert*, *remove*, *new*, *add*, *delete*, *subset*, or *pointSet*. The *new* option creates a new zone.
- Source: This option specifies the shape of the cell group that can be a box, a cylinder, or a plane. By using the *cylinderToCell* option, a cylinder with a specified diameter and one cell length is created as an AD to simulate a wind turbine.

After generating the AD, the source is added to the momentum equation. The configurations are done on $fvOption$ dictionary in $system$ folder. The primary entries in this dictionary are based on classical AD that exerts uniform load on the wind flow. These are $type, active, fields, selectionMode, cellSet, discDir, C_P, C_T, dicArea$, and $upstreamPoint$.

The modified AD is based on the BEM method, and input data such as turbine characteristics and operational parameters must be entered. They are entered in the $fvOption$ dictionary, and the main code reads the entries from this dictionary in each iteration. If a change is applied to input values, it will be updated in the next iteration. The following items are added to modify the old $fvOption$ dictionary:

- Radius of the hub in order to separate hub from the blades
- Radial divisions
- Twist distribution in the spanwise direction
- Chord distribution in the spanwise direction
- Rotor radius
- Angular velocity

6.5 Simulation example

In this section, wake characteristics behind a single HAWT are investigated using the modified AD. To do this, again Nordtank 500/41 is selected (see Appendix C.11) since its characteristics are known, and this turbine was simulated using the BEM method in the previous chapters. The experimental data that were represented by Højstrup (1983) are used in order to evaluate the accuracy of the presented methodology. Therefore, we can verify the numerical results with experimental data and also compare the simulation with the analytical model developed in the previous chapter.

6.5.1 Numerical model

Each element in the BEM code corresponds to a cell on the disc in the numerical model. The AD surface after mesh generation is shown in Fig. 6.6b. It is located in a domain and solved simultaneously using the simpleFOAM solver. As we know, simpleFOAM is a steady-state solver for incompressible flows. The equations were discretized by a finite volume method using the upwind scheme, and the iteration stopped when the residuals reached 10^{-5}.

Since turbulence modeling affects the results, some of the available turbulence models are solved in the present study, and the results are compared. These models are all categorized as Reynolds averaged Navier–Stokes, or simply as RANS models, and are available in the OpenFOAM package.

TABLE 6.2 Domain size in three directions.

Direction	Dimension
Axial	23D
Lateral	7D
Vertical	5D

6.5.2 Computational domain

Wind speed has a significant influence on the characteristics of the air. In low wind speeds, where the Mach number is less than 0.3, the air can be considered an incompressible viscous flow. In subsonic flow, the information regarding the presence of the turbine moves upstream with speed equal to the sound. This fact is better known as *upstream influence,* meaning that any obstacle in a moving fluid influences its upstream regions. For an accurate simulation, this effect must be considered; hence, the domain must be large enough so that the effect of numerical boundaries become minimum based on the accuracy of the solver. In other words, it is necessary to select the appropriate input and output distances in order to make the boundaries independent of the turbine effect. Therefore, different distances were checked, and the inlet distance was selected $3D$ upstream, and outlet distance was selected $20D$ downstream the turbine, in which D represents the rotor diameter. This requirement has also been checked for other patches.

Domain sensitivity has been investigated in order to determine the effect of domain size on the simulation results. Table 6.2 represents the selected domain size in three directions and it is illustrated in Fig. 6.8. In this figure, M_1, M_2, and M_3 are three masts installed by Højstrup (1983) in order to measure the wind speed at different distances downstream the wind turbine.

6.5.3 Boundary conditions

The proper domain of solution is shown in Fig. 6.8. As was discussed, the turbine is located at $3D$ from the inlet boundary and $20D$ away from the outlet. The wind blows in the x direction and has an ABL shape. It means that it follows the logarithmic profile, which is an internal function in OpenFOAM.

The solver uses a velocity inlet type at the inlet and calculates the velocity with the ABL model. Thus, a nonuniform wind profile is applied to the inlet boundary. At the inlet, other variables except for the velocity follow a zero-gradient condition. It means that the flow has no stress when flowing to the solution domain.

Since the outlet boundary is chosen very far from the wind turbine, the turbine effect would be minimal at these boundaries. Hence, the best choice for these patches is zero gradients for all the variables. This condition, in some ref-

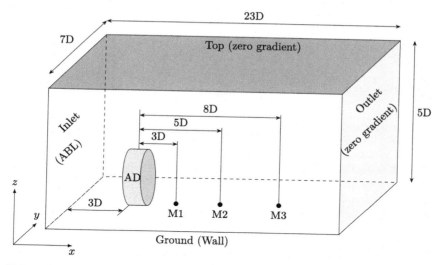

FIGURE 6.8 Domain outline.

erences, is known as a stress-free boundary, pressure outlet, or simply as an outlet.

The boundary condition for the ground patch is no known as the wall on which no-slip boundary condition is applied. Moreover, to decrease the number of the cells adjacent to the wall, wall function is used.

Finally, for the top, front, and back patches, the slip boundary condition is applied. This selection is physically natural since these patches are actually parallel to the wind direction.

6.5.4 Grid sensitivity

In CFD basic science, it is proved that the size of the computational mesh directly affects the final results. Therefore, this parameter induces a numerical error, which is not desirable. Also, it is proved that by decreasing grid size, the error can be ignored. Hence, the computational mesh must be as fine as possible. However, using an ultrafine mesh means that we have increased the computational cost. Therefore, a balance must be achieved between accuracy and computational costs.

A grid sensitivity study must be done to ensure that the results are not altered by numerical errors aroused by mesh size. For this purpose, first, the computational domain must be solved using the initial mesh size. Then it must be recalculated using a finer mesh. If the results are the same, then we have got a mesh-independent solution. However, if the results differ, then we have to continue to use a finer grid. The process continues until the solution becomes grid or mesh independent.

TABLE 6.3 Grid resolution and number of cells.

Case	Axial direction	Lateral direction	Vertical direction	Total
Grid1	177	60	69	750000
Grid2	190	69	80	1000000
Grid3	210	80	95	1596000

For the present example, a mesh-independent test in three directions was carried out to achieve an efficient grid resolution and minimize the computational time. The effect of number of cells has been determined on the final results. The specifications of the grid sizes are summarized in Table 6.3 in which the axial, lateral, and vertical directions are corresponding to the x, y, and z, respectively (see Fig. 6.8).

The velocity profile behind the turbine for three different mesh sizes is plotted in Fig. 6.9. These plots are in three different directions, and as it can be seen, the axial distance along the x axis has the most sensitivity to the grid size. According to this figure, Grid2 has similar results in comparison with Grid3 and has fewer cells that leads to less computational time. For this reason, it was decided to use Grid2 for further analysis.

It was mentioned that, for a better resolution, the generated grid is clustered around the disc. A cross-section at the disc location is shown in Fig. 6.10. This section is in the y–z plane.

6.5.5 Results

Fig. 6.11 illustrates the values of induction factor that are obtained using BEM theory in the spanwise direction. It can be seen that as radius increases from 4.5 to 10 m, a increases moderately. Then from 10 to 20 m, it is approximately constant, again at the blade tip, a sharp increase in a is visible. This variable behavior of a is not considered in classical AD. From this variable behavior, one can conclude that as the radius increases, the blade effect on downstream velocity becomes stronger, and a larger velocity deficit occurs at higher radii, almost in the range of $0.7 < \frac{r}{R} < 0.9$ (Hashemi Tari et al., 2016).

Different methods are used to simulate Nordtank 500/41 wind turbine, and the results are shown and compared in Fig. 6.13. These methods are (a) classical ADM, (b) the present modified ADM, (c) the measured experimental data, and (d) the analytical model discussed in Chapter 5. It is clear that the present modified model has a better prediction in all the different distances than the classical method. Other methods show more accurate results in some distances than the present modified ADM but deviate at different distances. In general, the present approach behaves more naturally than the other ones. Hence the modified ADM is proved to have more accuracy than the classical ADM. The analytical model also has great accuracy, especially in near wake regions; however, in far away regions, the modified ADM shows a better accuracy.

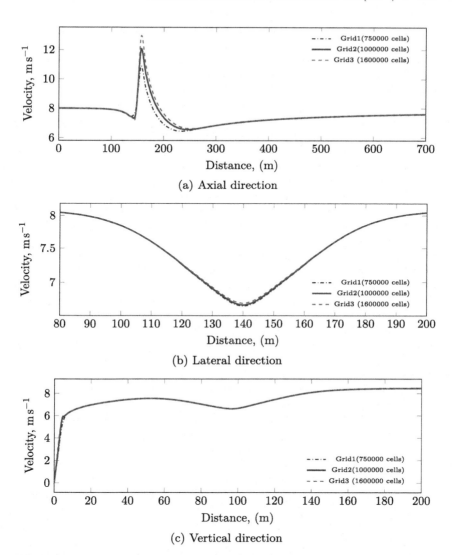

FIGURE 6.9 Grid analysis in different directions.

The present numerical method indicates that the concept of self-similarity that was used to obtain the analytical profile was an acceptable assumption. Comparing the wake profile at different distances downstream, the turbine shows a self-similar behavior. The wake shape has a conserved bell-shape profile with different stretching in the wind direction. Therefore, the present simulation is another confirmation for the accuracy of the obtained analytical profile in Chapter 5.

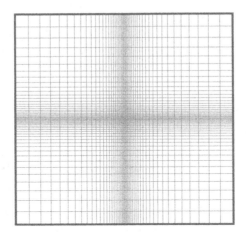

FIGURE 6.10 Inlet mesh resolution.

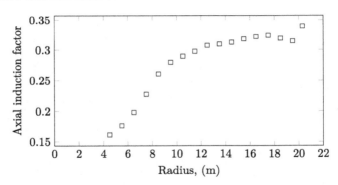

FIGURE 6.11 Radial distribution of induction factor in the spanwise direction at $u_1 = 8\,\mathrm{m\,s}^{-1}$.

Classical ADM is incapable of estimating the exact value of velocity deficit behind the turbine at the area near the hub centerline. It underestimates the velocity deficit in all downstream distances. In closer distances downstream of the wind turbine, the classical ADM gives unrealistic results because it neglects radial load distribution.

It is evident in Fig. 6.12, the modified ADM has a better performance than classical ADM in predicting velocity deficit almost in near wake and wake center. At $3D$ downstream, blades have a dominant effect on the velocity profile. As we further go downstream, the effect of the blades becomes weaker and atmospheric effects become dominant.

The analytical model that is represented in Chapter 5, follows the experimental data in the near wake very well. Comparison between this model and the modified ADM shows a significant difference in far wake, almost in wake center. Neglecting turbulence effects makes some inaccuracies in far wake because

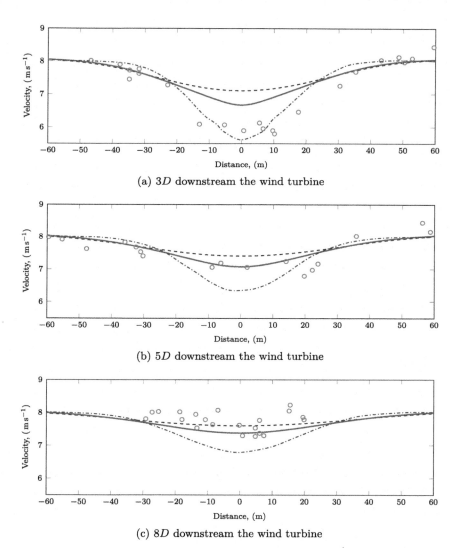

(a) $3D$ downstream the wind turbine

(b) $5D$ downstream the wind turbine

(c) $8D$ downstream the wind turbine

FIGURE 6.12 Comparison between different wake models at $u_1 = 8\,\mathrm{m\,s}^{-1}$ (classical ADM, - - -; modified ADM, —; experiment (Højstrup, 1983), o; analytical model of Chapter 5, - · - · -).

in this region atmospheric forces are dominant and turbulence effects are more important than load distribution on the blades.

It is good to note that all the different methods have predicted a bell-shaped wake, which is very consistent with the measured data and the self-similarity solution. However, the measured data, as it is shown in the figure, shows a chaotic behavior in the wake center. It is because in the real world, the turbine wake is very complex, and turbine vortexes strongly affect the wake region. But in the

TABLE 6.4 k–ε coefficients for different methods.

	C_μ	C_{ε_1}	C_{ε_2}	σ_κ	σ_ε
Industrial	0.09	1.44	1.92	1	1.3
ABL	0.033	1.176	1.92	1	1.3

numerical methods, we have neglected the blades' rotation; hence, its effect was not considered. If a more realistic simulation is required, the problem should be solved using an unsteady full-rotor simulation.

After determining the important role of turbulence effect on wake modeling, a brief comparison is performed between three different turbulence models. These models are the (a) standard k–ε which is the most commonly used two-equation turbulence model with industrial coefficients (Javaheri and Canadillas, 2013a; Sørensen, 1995), (b) standard k–ε with coefficients that is optimized for atmospheric boundary layer (ABL) (Javaheri and Canadillas, 2013a; Paofsky and Dutton, 1984), and (c) RNG k–ε turbulence model that is proved to give realistic results in the near wake. The coefficients of the two different k–ε models are given in Table 6.4.

All the above turbulence models are available in the OpenFOAM package, and for comparison, all of them are used. The results are shown in Fig. 6.13. The standard k–ε model shows the best performance in predicting the far wake profile. However, for near wakes, it underestimates the wake deficit.

In the literature, full rotor simulation (AbdelSalam and Ramalingam, 2014) and experimental measurement (Krogstad and Adaramola, 2012) show that there is a double pick shape for velocity profile in the near wake. As it can be seen, the RNG model can capture the velocity deficit, and double pick shape in the same way AbdelSalam and Ramalingam (2014) predicted. Unlike the standard k–ε model, it takes into account the smaller eddy motions. This feature makes it capable of producing good estimates of wake characteristics in the near wake. Therefore, RNG k–ε is appropriate for regions close to the wind turbine body. However, as it can be seen, for far away wakes, the model overestimates the velocity deficit compared to the other methods.

In the near wake, the ABL k–ε has a better prediction than the industrial k–ε; however, for far wakes, the industrial k–ε is superior. For far distances downstream the turbine, the industrial k–ε gives the best prediction compared to the other methods and its results are more consistent with the experimental data and have negligible errors.

Fig. 6.14 illustrates a top view of wake recovery behind the wind turbine. As the wake recovery becomes faster, higher energy production by downstream turbines is achievable. It can be seen that using the modified AD leads to a faster wake recovery than the classical one in the wake center. It is worth mentioning that the modified AD makes more wake expansion in far wake.

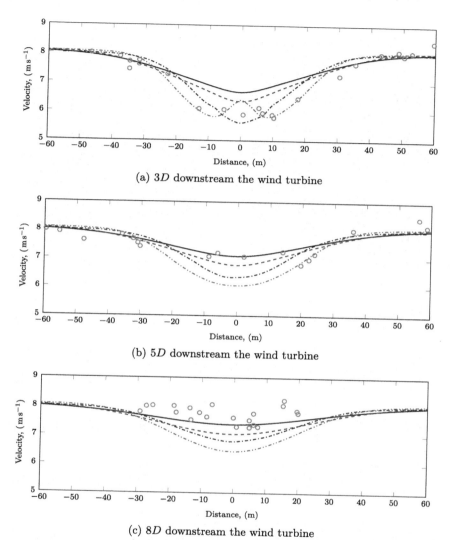

(a) $3D$ downstream the wind turbine

(b) $5D$ downstream the wind turbine

(c) $8D$ downstream the wind turbine

FIGURE 6.13 Comparison between different turbulence models at $u_1 = 8\,\mathrm{m\,s}^{-1}$ (standard $k\text{--}\varepsilon$, —; atmospheric $k\text{--}\varepsilon$, - - -; RNG $k\text{--}\varepsilon$, - · · · ·; experiment (Højstrup, 1983), o; Analytical model of Chapter 5, - · - · -).

6.6 Summary

In the present chapter, it is shown that wind turbines can be modeled using simple ADM. The model creates huge simplifications in numerical calculations. A simple wind turbine can be simulated in a couple of seconds. This is a great

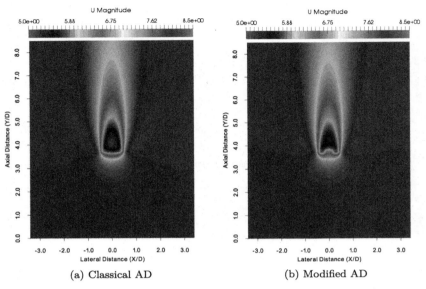

(a) Classical AD   (b) Modified AD

FIGURE 6.14 Top view of wake recovery just behind the wind turbine at $u_1 = 8\,\mathrm{m\,s}^{-1}$.

fact because the low computational time enables the method to be incorporated in optimization loops.

In addition to computational time, the ADM requires minimum effort for mesh generation. A full-rotor algorithm requires detailed modeling of the wind turbine. The blades, nacelle, tower, and other details of the wind turbine are necessary to be modeled. The detailed geometry will affect the mesh generation and make the problem a very time-consuming task. In this regard, the ADM does not require any detailed modeling and by creating simple geometries.

It should be noted that for improving the accuracy of the model, some simple modifications can be applied to the model. Some simple static geometries can be modeled inside the model. For instance, we can model the tower since it is not very hard to draw, and also, its boundary conditions are quite simple to implement. Other features such as nacelle, etc., can be modeled in the same manner.

As explained in the present chapter, the classical ADM is quite simple and conventionally used by many researchers. But it is shown that the present modified ADM gives better results while it is not very hard to implement. The present model requires a BEM solver coupled with OpenFoam, which can be done by any programming or scripting language. We used the C++ programming language, but other famous languages such as Python are quite suitable for this purpose.

Finally, the ADM model is quite suitable for the simulation of wind farms since many turbines are operating simultaneously in a wind farm. Modeling an actual farm with a full-rotor scheme requires lots of computational meshes, which requires lots of computational resources and time. However, ADM can be easily applied to a wind farm without too much increase in the computational mesh. A full wind farm can be simulated on a single personal computer with moderate amount of RAM, and it takes a couple of minutes for full simulation.

6.7 Problems

1. Use modified ADM method to simulate NTK 500/41 wind turbine for different wind speeds and compare the wake shapes in different downstream distances.
2. Compare the results of the previous problem with the analytical model.
3. Using modified ADM to simulate Mapna MWT2.5-103-I turbine and compare the results with analytical data.
4. To investigate the effect of turbulent modelings, rerun the above problems with different turbulence models.
5. For Problem 3, model the tower with a simple cylinder with the specifications shown in Section C.21. Compare the results with Problem 3.
6. Pitch-regulated turbines can also be modeled using the present approach. Simulate Mapna turbine with the present CFD tool and compare the results with the analytical models.
7. Apply different turbulence modelings to the previous problem and compare the results.
8. Apply the present model to large wind turbines such as SeaTitan, described in Section C.25.

References

AbdelSalam, Ali M., Ramalingam, Velraj, 2014. Wake prediction of horizontal-axis wind turbine using full-rotor modeling. Journal of Wind Engineering and Industrial Aerodynamics 124, 7–19.

Abdulqadir, Sherwan A., Iacovides, Hector, Nasser, Adel, 2017. The physical modelling and aerodynamics of turbulent flows around horizontal axis wind turbines. Energy 119, 767–799. https://doi.org/10.1016/j.energy.2016.11.060.

ANSYS Inc., 2021. Ansys fluent, fluid simulation software. https://www.ansys.com/products/fluids/ansys-fluent/. (Accessed 21 April 2021).

Behrens, Tim, 2009. OpenFOAM's basic solvers for linear systems of equations. Chalmers, Department of Applied Mechanics 18 (02).

Bianchini, Alessandro, Balduzzi, Francesco, Gentiluomo, Domenico, Ferrara, Giovanni, Ferrari, Lorenzo, 2017. Potential of the virtual blade model in the analysis of wind turbine wakes using wind tunnel blind tests. Energy Procedia 126, 573–580. https://doi.org/10.1016/j.egypro.2017.08.212.

Bianchini, Alessandro, Balduzzi, Francesco, Gentiluomo, Domenico, Ferrara, Giovanni, Ferrari, Lorenzo, 2021. Comparative analysis of different numerical techniques to analyze the wake of a wind turbine. In: Proc. of the ASME Turbo Expo 2017.

Comsol Inc., 2021. Comsol Multiphysics®. https://www.comsol.com/. (Accessed 21 April 2021).

Concentration, Heat & Momentum Limited (CHAM), 2021. PHOENICS. Concentration, Heat and Momentum Limited (CHAM) Engineering-Software Company. http://www.cham.co.uk/. (Accessed 23 April 2021).

Danish Technical University, 2021. WAsP. https://www.wasp.dk/wasp/. (Accessed 21 April 2021).

Davidson, Lars, 2015. An Introduction to Turbulence Models.

Hashemi Tari, Pooyan, Siddiqui, Kamran, Hangan, Horia, 2016. Flow characterization in the near-wake region of a horizontal axis wind turbine. Wind Energy 19 (7). https://doi.org/10.1002/we. 1895.

Højstrup, J., 1983. Nibe wake. Part 1. Internal technical report. Risø National Laboratory.

Ivanell, Stefan, Sørensen, Jens N., Mikkelsen, Robert, Henningson, Dan, 2009. Analysis of numerically generated wake structures. Wind Energy: An International Journal for Progress and Applications in Wind Power Conversion Technology 12 (1), 63–80.

Ivanell, Stefan, Mikkelsen, Robert, Sørensen, Jens N., Henningson, Dan, 2010. Stability analysis of the tip vortices of a wind turbine. Wind Energy 13 (8), 705–715.

Javaheri, A., Canadillas, B., 2013a. Wake modeling of an offshore wind farm using OpenFOAM. DEWI Magazin.

Javaheri, Alireza, Canadillas, Beatriz, 2013b. Wake modelling of an offshore wind farm using Open-FOAM. In: WE KnOW WInD, p. 118.

Krogstad, P., Adaramola, Muyiwa S., 2012. Performance and near wake measurements of a model horizontal axis wind turbine. Wind Energy 15 (5), 743–756. https://doi.org/10.1002/we.502.

Launder, B.B.E., Spalding, D.D.B., 1972. Lectures in Mathematical Models of Turbulence. Academic Press. ISBN 9780124380509.

Mikkelsen, Robert, 2003. Actuator Disc Methods Applied to Wind Turbines. PhD thesis. Department of Mechanical Engineering, Technical University of Denmark.

Nedjari, Hafida Daaou, Guerri, Ouahiba, Saighi, Mohamed, 2017. CFD wind turbines wake assessment in complex topography. Energy Conversion and Management 138, 224–236. https://doi.org/10.1016/j.enconman.2017.01.070.

Newman, B.G., 1983. Actuator-disc theory for vertical-axis wind turbines. Journal of Wind Engineering and Industrial Aerodynamics 15 (1), 347–355. https://doi.org/10.1016/0167-6105(83) 90204-0.

NREL, 2021. Physics-based engineering tool for simulating the coupled dynamic response of wind turbines. https://www.nrel.gov/wind/nwtc/fast.html/. (Accessed 21 April 2021).

OpenFOAM, 2021. OpenFOAM: the open source CFD toolbox. https://www.openfoam.com/. (Accessed 19 April 2021).

Paofsky, H.A., Dutton, J.A., 1984. Atmospheric Turbulence Models and Methods for Engineering Applications. John Wiley & Sons.

Patankar, S.V., Spalding, D.B., 1972. A calculation procedure for heat, mass and momentum transfer in three-dimensional parabolic flows. International Journal of Heat and Mass Transfer 15 (10), 1787–1806. https://doi.org/10.1016/0017-9310(72)90054-3.

Pope, Stephen B., 2000. Turbulent Flows. Cambridge University Press.

Sayma, Abdulnaser, 2009. Computational Fluid Dynamics. BookBoon. ISBN 9788776814304.

Sørensen, J.N., Shen, W.Z., Munduate, X., 1998. Analysis of wake states by a full-field actuator disc model. Wind Energy 1 (2), 73–88. https://doi.org/10.1002/(SICI)1099-1824(199812)1:2<73:: AID-WE12>3.0.CO;2-L.

Sørensen, N.N., 1995. General purpose flow solver applied to flow over hills. PhD thesis. Technical University of Denmark.

Sturage, D., Sobotta, D., Howell, R., While, A., Lou, J., 2015. A hybrid actuator disc – full rotor CFD methodology for modelling the effects of wind turbine wake interactions on performance. Renewable Energy, 525–537. https://doi.org/10.1016/j.renene.2015.02.053.

White, F.M., 2011. Fluid Mechanics. McGraw-Hill Series in Mechanical Engineering. McGraw Hill. ISBN 9780073529349.

Yakhot, V., Orszag, S.A., Thangam, S., Gatski, T.B., Speziale, C.G., 1992. Development of turbulence models for shear flows by a double expansion technique. Physics of Fluids A: Fluid Dynamics 4 (7), 1510–1520. https://doi.org/10.1063/1.858424.

Chapter 7

Numerical simulation of a wind farm

Wind farms are modeled using different methods. As explained different simple methods (Ghadirian et al., 2014; Hamedi et al., 2015; Torabi and Hamedi, 2015) have been used in which the analytical models are used. The models can becom as complicated as the one used by some researchers (Castellani and Vignaroli, 2013; Crasto et al., 2012; Javaheri and Canadillas, 2013). These models are based on the fundamentals of fluid mechanics. Since the wind turbine's wake is turbulent, it is very important to take care of turbulent modeling (Pope, 2000; Chu and Chiang, 2014; Nedjari et al., 2017).

The modified ADM developed in the previous chapter can be easily applied to simulate a wind farm. But for simulating a complete farm, some additional issues should be considered. First of all, it is clear that each turbine experiences a different wind speed in a wind farm. Hence, the controlling program that manages the interaction between the BEM solver and OpenFOAM must be modified. The controlling code must obtain the wind velocity for each individual turbine from OpenFOAM and pass it to the BEM solver. Secondly, the wind direction becomes important in the simulation of a wind farm. Hence, the program must manage the wind direction and make simulations according to the wind rose diagram of the specific area. Finally, the program must be able to automatically calculate the position of the turbines as the wind direction changes and specifies the proper volumes as to the AD of each individual turbine.

Other than the above-mentioned changes, there is no other major difference between the simulation of a single wind turbine and the whole wind farm. Fortunately, all the steps required for the simulation of a single wind turbine are the same for the simulation of a wind farm. Also, all the numerical methods and turbulence models can be applied to the simulation of a wind farm in the same way we did in the previous chapter.

In the present chapter, first, we discuss the differences between the simulation of a single wind turbine and a wind farm. Then the method is applied to a specific wind farm to show the capability of the present modified ADM and OpenFOAM for simulation of a wind farm. Finally, the results are given in different charts and figures, and each of them is discussed.

Fundamentals of Wind Farm Aerodynamic Layout Design. https://doi.org/10.1016/B978-0-12-823016-9.00013-9

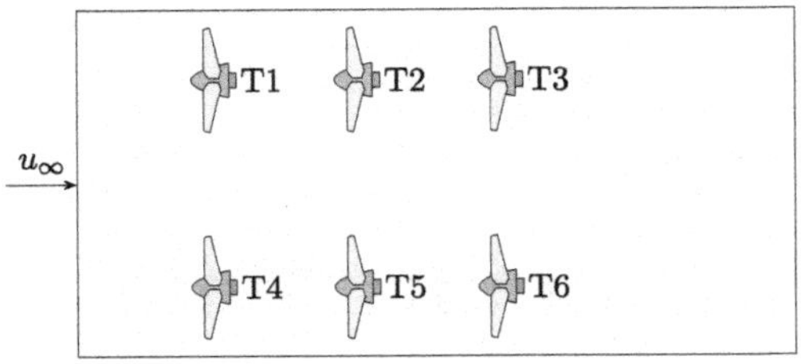

FIGURE 7.1 A wind farm in wind direction equal to 270°.

7.1 Wind farm layout generation

Two major differences between simulation of a single wind turbine and a wind farm are (a) the layout generation and (b) calculation of the upstream wind velocity. The wind direction does not affect anything for a single turbine because it is assumed that the wind turbines rotate towards the wind. But for a wind farm the wind direction directly affects the numerical mesh generation and the farm layout must be corrected to account for the wind direction.

7.1.1 Automatic layout generation

There are lots of different strategies that can handle the wind direction. Therefore, the procedure explained here is just one of the solutions, and there may be other applied methods in the literature.

To illustrate the current approach, consider a wind farm as shown in Fig. 7.1. In this farm, six different turbines are located in a matrix form. A rectangular domain is generated for its simulation, which works as the computational domain. In this situation, the wind direction is 270° according to the wind-direction standard.

The situation shown in Fig. 7.1 can be handled in the same way as we did for a single turbine. The only difference is that we need to determine six different cylindrical patches instead of each turbine in this problem. Then, for each patch, or AD, proper source terms must be determined and applied to OpenFOAM. The method is quite straightforward, as explained in the previous chapter.

Now consider the situation when the north wind is blowing as shown in Fig. 7.2. According to the standard, the north wind's angle is zero (or we can consider it as 360°). As the figure illustrates, the wind turbines have rotated toward the wind direction. Now for this situation, there are two different strategies:

1. The farm can be modeled as shown in Fig. 7.2a. Therefore, it should be noted that the simulation domain must be recalculated since it must be extended in the wind direction.
2. A better approach is to use the layout shown in Fig. 7.2b. In this situation, the simulation domain does not change, but the location of the turbines does.

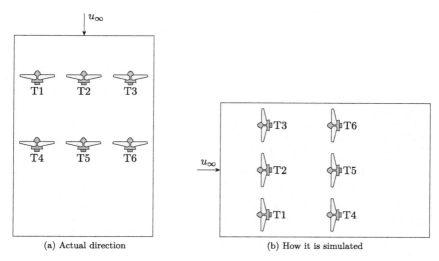

(a) Actual direction   (b) How it is simulated

FIGURE 7.2 A wind farm in wind direction equal to $0°$.

The second approach is much more convenient in the simulation of the wind farms since the simulation domain does not change with the wind direction. Then for each wind direction, the new location of the wind turbines must be obtained. This is a very simple task by using a simple rotation matrix as

$$R = \begin{bmatrix} \cos\theta & -\sin\theta \\ \sin\theta & \cos\theta \end{bmatrix}, \tag{7.1}$$

where θ is the angle that converts the actual situation to the simulated domain. For example, $\theta = -90°$ for Fig. 7.2.

Example 7.1. For the wind farm of Fig. 7.1, discuss the situation when the wind blows at an angle of $300°$.

Answer. The situation is shown in Fig. 7.3 where the wind blows from the northwest with an angle of $330°$. Fig. 7.3a shows the actual direction, and, as it can be seen, the turbines are rotated towards the wind. Again, the farm can be modeled as is, but a better strategy is to rotate the whole farm so that it looks like in Fig. 7.3b. This can be done by simply setting $\theta = -60°$ in Eq. (7.1). $\square$

The present strategy in layout modeling greatly reduces the computational cost and makes it easy to handle any wind farm with any wind direction. The only remarkable thing here is that intermediate scripting or programming language is required to recalculate the new locations of the wind turbines in a new domain. The selection of the intermediate language makes no difference in the final results.

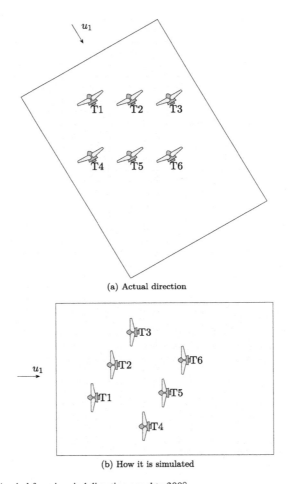

FIGURE 7.3 A wind farm in wind direction equal to 300°.

7.1.2 Upstream velocity

In order to use the modified ADM obtained in the previous chapter for simulating a wind farm, a reference speed at the inlet of each turbine is required. For a single turbine, the undisturbed wind velocity at infinity is chosen as the reference speed. This speed is set in the BEM algorithm to find the spanwise induction factor. However, in a wind farm, the turbines are located in the wake of each other. Therefore, each turbine experiences a different wind velocity. Thus, except for the first turbine (or row of turbines) located in an undisturbed region, the far-field velocity cannot be used in the BEM code. Now the question is how the wind velocity (or reference wind speed) must be calculated for the turbines that are located in the wake of upstream turbines.

The upstream velocity is used for two reasons:

1. First, it is used as an input for the BEM algorithm in order to calculate the induction factor and forces applied on the turbine, and
2. Second, it is used as an inlet velocity for the ADM to which the momentum deficit is applied on.

As shown in Fig. 7.4, in a wind farm, different turbines experience different wind velocities. Since the upstream velocity is not known for each turbine, we cannot choose a single point for each turbine as its reference point for calculating the upstream wind velocity. Therefore, we need to use an iterative method to find a single wind velocity, which is considered the upstream wind velocity for each turbine. The proposed algorithm is as follows:

1. An initial value, or guess, is selected for the upstream wind velocity.
2. Based on the initial value, the BEM method is applied, and the induction factors are calculated accordingly.
3. Having the induction factors, the farm is simulated using OpenFOAM.
4. By using the spanwise induction factors, the total induction factor is obtained using Eq. (4.37).
5. From the CFD model, the average velocity is calculated at the rotor plane, which is called u_d.
6. From the concept of the induction factor, the upstream velocity is calculated using

$$U_{up} = \frac{U_d}{1 - a_{total}}. \tag{7.2}$$

7. Then we choose U_{up} and return back to Step 1 and repeat the procedure until U_{up} does not change during the succeeding iterations.

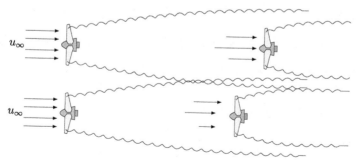

FIGURE 7.4 Variable profile of velocity upstream the turbines.

7.2 Simulation example: Horns Rev offshore wind farm

In the present study, the Horns Rev offshore wind farm is simulated using the modified ADM in OpenFOAM. The wind turbines located in the wind farm are

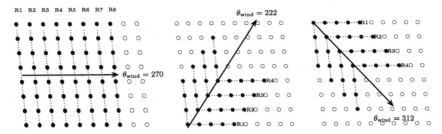

FIGURE 7.5 Layout of the investigated wind farm and different wind angles.

Vestas-V80 with 2 MW rated power. The characteristics of the turbines are summarized in Section C.17. There are 80 turbines in the wind farm, but, in order to compare the represented model with experimental data published by Wu and Porté-Agel (2015) and older models, 64 turbines are simulated here according to Fig. 7.5. This farm has been investigated by many researchers (Barthelmie et al., 2009; Naderi and Torabi, 2017).

The simulations have been performed for three main wind angles, namely 270°, 222°, and 312°, according to figure Fig. 7.5 and four deviation angles, including ±1°, ±5°, ±10°, and ±15°. In the square layout, the minimum space between wind turbines is $7D$. Changing the wind angle, θ_{wind}, leads to a change in the space between wind turbines, and, as a result, the power production alters significantly.

All of the simulations are performed for $8\,\text{m}\,\text{s}^{-1}$ wind speed at the chosen height. The friction velocity $v^* = 0.442$ and aerodynamic roughness length $z_0 = 0.05$ are obtained from the wind log-law. In this situation, the turbulence intensity at hub height is about 7.7%, which is consistent with experimental measurement. The experimental data were collected for $8\,\text{m}\,\text{s}^{-1}$ inlet velocities by Wu and Porté-Agel (2015). Because of the turbulence and stochastic behavior of the flow, the produced power of the turbines in the same row is different. The normalized power is calculated using the averaged produced power of the first row of turbines and averaged power in each row.

7.2.1 Numerical settings

Since the simulation is in steady-state, the simpleFOAM solver is the proper choice. Two different turbulence models are used in this simulation, namely the standard k–ϵ and RNG k–ϵ.

7.2.2 Boundary conditions

The specified boundary conditions are listed in Table 7.1. The ABL is set with $8\,\text{m}\,\text{s}^{-1}$ wind speed at hub height. At the ground patch, the wall function is used

TABLE 7.1 Specified boundary conditions.

Patch	Type
Ground	Wall (Fixed value)
Inlet	Atmospheric boundary layer inlet velocity
Outlet	Zero gradient
Sides	Slip
Top	Slip

TABLE 7.2 Computational domain size.

Direction	Size
Axial distance between the first turbine and inlet boundary	$5D$
Lateral distance between wind turbine and side boundary	$5D$
Axial distance between the last turbine and outlet boundary	$25D$
Vertical distance	$7D$

to decrease the number of cells. Also, the slip condition used for sides and top patches means that the normal component of a vector variable is zero, and the tangential component remains constant through the boundary.

7.2.3 Computational domain and mesh study

As discussed in the previous chapter, the size of the computational domain must be chosen sufficiently large so that the simulation results are no longer dependent on the size of the domain and blockage. Thus for the present simulation, the distances from wind turbines to the boundaries are selected according to Table 7.2.

To examine the effect of mesh resolution on the simulation results, four cases (labeled Mesh 1 to Mesh 4) have been built using four different structured hexahedral mesh sizes according to Table 7.3. Two parameters, including power production in each row and velocity profile between rows 3 and 4, are selected to evaluate the mesh size effect. This evaluation is carried on by calculating the coefficient of determination (COD) (Bianchini et al., 2021, 2017), which is a statistical factor between each case and the finest case (Mesh 4), and the minimum acceptable COD is selected as 0.99. By studying Fig. 7.6, it can be seen that Mesh 3 passed this criterion for both parameters; hence, it is selected for further analysis. In all cases, 65% of the cells are located between the ground and $z = 250$ m according to Fig. 7.7.

TABLE 7.3 Different mesh resolutions.

Case	ΔX (m)	ΔY (m)	Number of cells in the z direction	Total number of cells (million cells)
Mesh 1	15.2	33	50	6.1
Mesh 2	14.2	28	55	8.4
Mesh 3	13.1	24	60	10.8
Mesh 4	11.2	20.4	60	14.4

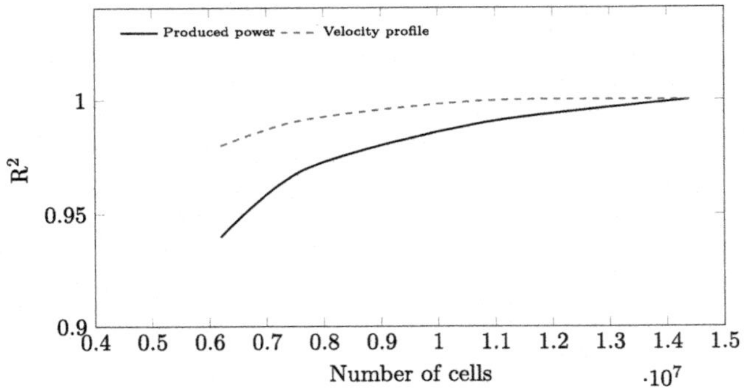

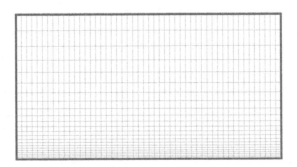

FIGURE 7.6 R^2 obtained by different mesh resolutions.

FIGURE 7.7 Grid resolution at inlet patch.

7.2.4 Results

7.2.4.1 Induction factor and load distribution

As mentioned before, there is a radial load distribution on the blades. This distribution is obtained using the BEM method in different operational conditions. The dependency of the induction factor on inlet velocity is shown in Fig. 7.8. As the inlet velocity increases, induction factors decrease. This trend leads to a lower thrust coefficient at higher inlet velocities.

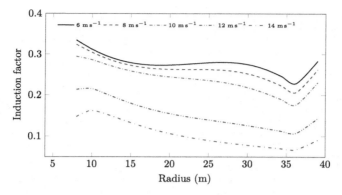

FIGURE 7.8 Induction factor distribution along the Vestas-V80 blades at different inlet velocities.

Fig. 7.8 also illustrates another important fact that the spanwise variation of the induction factor cannot be ignored. As the figure shows, the induction factor has a considerable variation from root to tip such that in some cases the difference is more than 50%. Therefore, considering a constant coefficient as is used in the conventional ADM will not give appropriate results. Note that setting a variable induction factor as is the way in the modified ADM not only generates a more accurate simulation but also causes a complex rotational wake which is more consistent with the real cases.

7.2.4.2 Velocity contours

The results of the numerical simulation of the wind farm are shown in Fig. 7.8. As can be seen, the simulation has been done for four different wind angles. Also, as was discussed before, the simulation domain has remained the same, but the turbines were rotated according to the wind angle. The wind direction is from the bottom to the top in the present simulation, as the velocity contours confirm. Hence, the boundary conditions of simulation are:

- The bottom surface is the velocity inlet where the wind comes in. Therefore, ABL is applied on this surface.
- The upper surface is the outlet, and zero gradients are set.
- The left and right surfaces are side surfaces, and the slip condition is set (zero normal velocity).

One of the advantages of the present method on simulation of wind farms is that the boundary conditions do not change when the wind direction changes. This will make the simulation much easier. As it can be seen in Fig. 7.9, the position of the turbines is changed, which can be automatically found by using a simple scripting code.

A comparison of Figs. 7.9a and 7.9b shows that the wind direction has a very important effect on wake generation and the way the turbines are located in

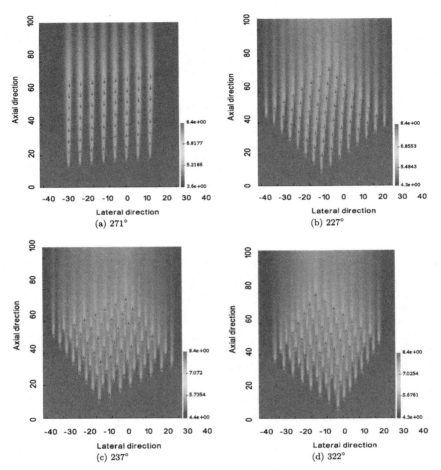

FIGURE 7.9 Top view of velocity contours at $V_0 = 8\ \text{ms}^{-1}$. Distances are normalized by rotor diameter.

the wake of each other. The same conclusion is made by comparing Figs. 7.9c and 7.9d. For example, when the wind blows with the angle of 237°, the turbines' axial distance becomes larger than when the wind direction is 270°. Hence, we can estimate that the generated power at the angle of 237° must be greater than at the angle of 271°. In order to justify the above deduction, the quantity of the generated power will be discussed here.

7.2.4.3   Turbulence models

Two different turbulence models, including the standard k–ε and RNG k–ε, are examined in order to investigate their influence on the final results and to be able to select the appropriate one. Figs. 7.10 to 7.12 illustrate a comparison between

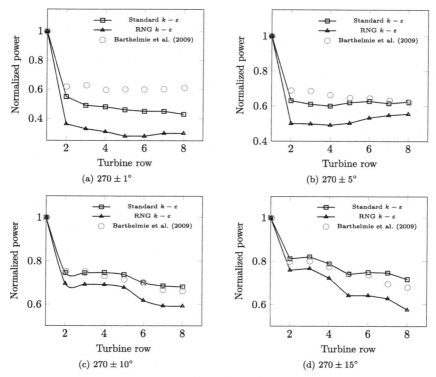

FIGURE 7.10 Comparison between different turbulence models in 270° wind angle and four wind sectors.

these models for three main wind angles. For 270° wind angle and low deviation angles, the standard $k–\varepsilon$ has better performance and lower errors. As the deviation angle increases, both models predict more accurate power production compared to the experimental data. For 222° ± 5° inflow angle, both turbulence models predict a decreasing trend, but for 222° ± 10° and 222° ± 15° inflow angles, the performance of the standard $k–\varepsilon$ model came closer to the experimental data, while the RNG $k–\varepsilon$ turbulence model still had a significant error.

By increasing the wind angle to 312° and the separation distance to 10D, the results changed according to Fig. 7.12. Both turbulence models could predict the trend of power generation, however, the standard $k–\varepsilon$ gave values which agreed better with the field data.

In this case, as can be deduced from Fig. 7.9, the distance between the turbines is larger than for two previous wind angles. Therefore, if the turbine axial distances increase, the standard $k–\varepsilon$ is a better choice. This result confirms those of the previous chapter where a single turbine was modeled by different turbulence models. As discussed before, the RNG $k–\varepsilon$ model is suitable for near-wake

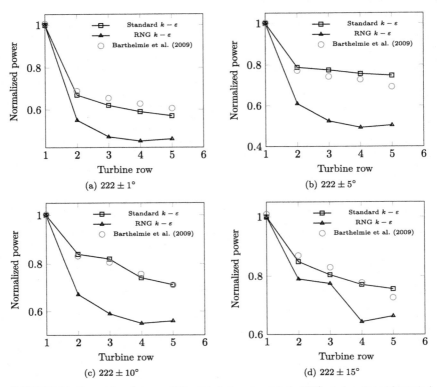

FIGURE 7.11 Comparison between different turbulence models for 222° wind angle and four wind sectors.

regions but gives less accurate results for far wakes. For a wind farm, such a conclusion can also be made: comparing the results obtained by the RNG and standard $k-\varepsilon$ models in Figs. 7.10 to 7.12 indicates that the RNG $k-\varepsilon$ model gives better performance when the distance between the turbines is short, and when the axial distance increases the standard $k-\varepsilon$ model gives better predictions.

7.3 Summary

The simulation of a wind farm using ADM or modified ADM is very simple and straightforward. There are two major differences between the simulation of a single wind turbine and the whole farm:

1. Grid generation as the wind direction changes, and
2. Numerical calculation of the upstream wind speed which is needed in BEM or any other algorithm that is used for calculation of the spanwise induction factor.

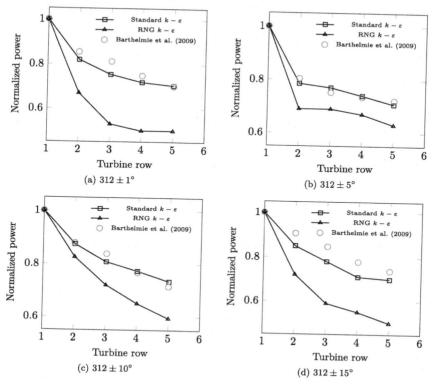

FIGURE 7.12 Comparison between different turbulence models for 312° wind angle and four wind sectors.

To overcome the problem of the wind direction, the strategy of the present chapter was adopted for the numerical simulation. Although many different strategies can be selected, the presented method greatly simplifies the boundary conditions. By the present method, all the boundary conditions remain unchanged during the simulations, and only the turbine locations must be changed. This is a very simple task since we have to apply a simple rotation matrix to the position of each wind turbine. Other tasks, including mesh generation and domain decomposition, are done by OpenFOAM facilities without requiring any human work. Hence, by a simple scripting code, all the procedures can be done automatically.

It should be noted that in some numerical articles, the researchers preferred to use other strategies. In other words, instead of using the present mesh generation strategy, they have generated a static computational mesh, and, to impose the wind direction, they have changed the boundary conditions. This method also works fine but has a smaller convergence rate. Hence in our studies, the first strategy is chosen as the best choice.

In the present modeling, the modified ADM is used. The induction factor was calculated using the methods described in Chapter 5 which was a combination of the BEM method (for the induction factor of the blades), and a second- or third-order polynomial for the calculation of nacelle's wake. This code is dynamically coupled with CFD code. It means that between any two succeeding CFD simulations, the BEM method is run on each individual turbine to update its induction factor.

The coupling between BEM and CFD can be done through other techniques such as neural network or machine learning algorithms. To do this, wind turbines can be modeled using BEM for some selected wind speeds. Then the results are inserted into a neural network or any other similar algorithm for training. Next, during the CFD computations, the trained algorithm generates the results and can be used instead of BEM. This technique can also influence the computational cost while the same accuracy is achieved.

As observed in the present chapter, turbulence modeling has a great role in numerical results. Hence, different turbulence models must be checked to see which one gives the best answers. Fortunately, OpenFOAM comes with lots of different models which can be used in the simulation without any difficulty. The best turbulence model is that which has the best agreement with the field data.

Finally, a full wind farm calculation can be made using the wind rose diagram of a specific region. The method is similar to the analytical method of Chapter 5. To do this, the farm must be simulated for different wind angles with different possibilities, and the energy of the farm is the sum of the whole simulation based on the wind rose diagram. The only difference between the numerical and analytical approaches is the way the farm power is calculated, and all the other concepts are the same.

7.4 Problems

1. OpenFOAM supports different RAS turbulence models, including RNGkEpsilon, SSG, ShihQuadraticKE, SpalartAllmaras, kEpsilon, and kOmega. These models are all suitable for incompressible solver and can be run using simpleFoam. For Horns Rev offshore wind farm, simulate the farm using different methods and compare the results.
2. Some useful results may be obtained using large-eddy simulation (LES) instead of RAS models. Simulate Horns Rev offshore wind farm using LES and compare the results with field data and also RAS models.
3. Obtain the yearly energy of Horns Rev offshore wind farm using the actual data at the same location. Use the wind rose diagram of the region for a realistic calculations.

References

Barthelmie, R.J., Hansen, K., Frandsen, S.T., Rathmann, O., Schepers, J.G., Schlez, W., Phillips, J., Rados, K., Zervos, A., Politis, E.S., Chaviaropoulos, P.K., 2009. Modelling and measuring flow and wind turbine wakes in large wind farms offshore. Wind Energy 12 (5), 431–444. https://doi.org/10.1002/we.348.

Bianchini, Alessandro, Balduzzi, Francesco, Gentiluomo, Domenico, Ferrara, Giovanni, Ferrari, Lorenzo, 2017. Potential of the virtual blade model in the analysis of wind turbine wakes using wind tunnel blind tests. Energy Procedia 126, 573–580. https://doi.org/10.1016/j.egypro.2017.08.212.

Bianchini, Alessandro, Balduzzi, Francesco, Gentiluomo, Domenico, Ferrara, Giovanni, Ferrari, Lorenzo, 2021. Comparative analysis of different numerical techniques to analyze the wake of a wind turbine. In: Proc. of the ASME Turbo Expo 2017.

Castellani, F., Vignaroli, A., 2013. An application of the actuator disc model for wind turbine wakes calculations. Applied Energy 101, 432–440. https://doi.org/10.1016/j.apenergy.2012.04.039.

Chu, Chia-Ren, Chiang, Pei-Hung, 2014. Turbulence effects on the wake flow and power production of a horizontal-axis wind turbine. Journal of Wind Engineering and Industrial Aerodynamics 124, 82–89.

Crasto, G., Gravdahl, A.R., Castellani, F., Piccioni, E., 2012. Wake modeling with the actuator disc concept. Energy Procedia 24, 385–392. https://doi.org/10.1016/j.egypro.2012.06.122.

Ghadirian, A., Dehghan, M., Torabi, F., 2014. Considering induction factor using BEM method in wind farm layout optimization. Journal of Wind Engineering and Industrial Aerodynamics 129, 31–39.

Hamedi, Razieh, Javaheri, Alireza, Dehghan, Omid, Torabi, Farshad, 2015. A semi-analytical model for velocity profile at wind turbine wake using blade element momentum. Energy Equipment and Systems 3 (1), 13–24.

Javaheri, A., Canadillas, B., 2013. Wake modeling of an offshore wind farm using OpenFOAM. DEWI Magazin.

Naderi, Shayan, Torabi, Farschad, 2017. Numerical investigation of wake behind a HAWT using modified actuator disc method. Energy Conversion and Management 148, 1346–1357. https://doi.org/10.1016/j.enconman.2017.07.003.

Nedjari, Hafida Daaou, Guerri, Ouahiba, Saighi, Mohamed, 2017. CFD wind turbines wake assessment in complex topography. Energy Conversion and Management 138, 224–236. https://doi.org/10.1016/j.enconman.2017.01.070.

Pope, Stephen B., 2000. Turbulent Flows. Cambridge University Press.

Torabi, F., Hamedi, R., 2015. An analytical model for prediction of wind velocity profile and wind turbine wake. Renewable Energy 75, 945–955.

Wu, Yu-Ting, Porté-Agel, Fernando, 2015. Modeling turbine wakes and power losses within a wind farm using LES: an application to the Horns Rev offshore wind farm. Renewable Energy 75 (Suppl. C), 945–955. https://doi.org/10.1016/j.renene.2014.06.019.

Chapter 8

Optimization for wind farm layout design

The ultimate goal of a wind farm is to produce maximum energy. Therefore, the farm layout should be designed such that the maximum energy is gained during a year. Since the wind rose diagram of each location is unique, we cannot give a general guideline for the placement of turbines of all the farms. It means that for each individual field, a specific layout must be designed.

On the other hand, the relation between turbine generation and wind speed is nonlinear, meaning that the calculation of the maximum gained energy of a farm is not an easy task. Moreover, the wind velocity itself is not known and must be treated via statistical procedures.

The above statements reveal that the only way to obtain the best layout is via optimization algorithms. The optimization algorithms are able to deal with nonlinear, multidimensional, and complex systems. Maximizing the gained energy of a farm by means of an optimization algorithm is relatively straightforward. These algorithms have been used in many different scientific and industrial fields and have been proved to work well in obtaining the maxima (Mosetti et al., 1994). In the present chapter, the optimization of wind farms using optimization algorithms is explained.

8.1 Optimization algorithms

Many different optimization algorithms can be used for the wind farm layout design. Since the annual gained energy of a farm as a function of wind turbines is not differentiable, classical search methods are not suitable. For such problems, the best practices are stochastic algorithms such as genetic algorithm (GA), particle swarm optimization (PSO), crow search algorithm (CSA), whale optimization algorithm (WOA), and teaching–learning-based optimization algorithm (TLBO). These methods are evolutionary algorithms in which an initial population is located in the field. Then through evolutionary algorithms, the population moves toward the minimum or maximum of that field. The details of these algorithms are explained in Appendix E. For a better understanding of the algorithms, two of these methods, namely GA and PSO, are implemented in C++ language and are given in Appendix F.

The optimization algorithms usually search for the minimum of a function. Hence, if we are looking for the maximum, we can define a new function whose

Fundamentals of Wind Farm Aerodynamic Layout Design. https://doi.org/10.1016/B978-0-12-823016-9.00014-0

minimum is the maximum of the desired function. For example, the maximum of the function

$$f(x) = -x^2 \qquad (8.1)$$

is the same as the minimum of the function

$$g(x) = x^2. \qquad (8.2)$$

Therefore, from now on, we talk about the minimization of a function as a general statement. Obviously, it also covers the maximization of any function.

In applying and using optimization algorithms, it should be kept in mind that all the stochastic algorithms may find a local minimum instead of the absolute minimum. Therefore, if we use different algorithms, we see that we find different solutions. Even if we use the same algorithm, we get other answers in different algorithm runs. Obtaining different minimums may have two meanings:

1. The function has really many absolute minima. For example, the function $\sin(x)$ has infinitely many absolute minima, which all are equivalently acceptable. Any of the minimum values is the answer for such a situation, and we can choose whichever we want.
2. The function has an absolute minimum but different local ones. In this case, we can check and compare the results of different algorithms or rerun the same optimization algorithms with different population size or generations.

8.2 Cost function and constraints

Any optimization algorithm requires three different parts:

1. Objective function, which is the function that is going to be minimized;
2. Design variables, or parameters, that are to be changed so that the objective function becomes minimum;
3. Constraints, that is, the relations between the design variables that should be satisfied.

The design variables in wind farm optimization are the positions of the turbines since we want to find the best place for the turbines in the field. Hence, the x and y positions of each turbine act as design parameters.

Example 8.1. For the wind farm given in Section D.1, obtain the design variables.

Answer. The wind farm given in Section D.1 has four turbines, namely T1, T2, T3, and T4. Each turbine has two variables corresponding to x and y of its center. Therefore, the problem has 8 design variables. ☐

Defining a proper cost function is crucial in any optimization. Usually, the cost function in a wind farm is known and we want to maximize the annual

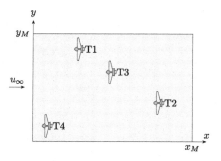

FIGURE 8.1 Illustration of optimization constraints.

gained energy. This function ensures that the farm is economical. Therefore, the optimization code must find the best positions for the turbines to maximize the total annual energy.

After defining the cost function, proper constraints should also be defined. The constraints of the optimization are applied to both the design variables and the cost function. For example, when we have a field with limited geographical dimensions, the field boundaries are the limits of our design variables. In addition to the constraints of the design variables, other regulations must also be defined. For example, the distance between the turbines must be defined as a physical constraint. As another example, we can name the cases where geographical features such as rivers, hills, etc., limit the positions of the turbines.

Example 8.2. For the wind farm of Example 8.1, discuss the constraints of design variables.

Answer. Since the design variables are the x and y coordinates of each turbine, they are bounded by the boundaries of the field. An example of such a situation is shown in Fig. 8.1. Therefore, for such a problem, we have:

$$0 \leq x \leq x_M, \qquad 0 \leq y \leq y_M.$$

Applying these constraints makes the optimization algorithm find the positions of the turbines within the field. □

Example 8.3. Discuss other geometrical constraints of the optimization problem.

Answer. In addition to the limits of the design variables, we have to define other constraints so that the solution becomes physically possible. One of the proper constraints is the distance between the turbines. If the optimization process is not aware of the distance between the turbines, it may obtain unrealistic solutions. Such a solution is shown in Fig. 8.2a. All the four turbines are located in a vertical column and are facing the upstream velocity in such a situation. It is

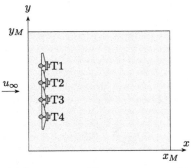

(a) Optimization without proper constraint

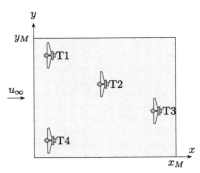

(b) Optimization with proper constraint

FIGURE 8.2 The role of constraint in optimization.

quite clear that the situation reaches the maximum energy since the turbines are not located in the wake of each other (note that for the present problem, it is assumed that the wind always blows from the west).

Obviously, the solution is not physically possible. The turbines cannot be located in such a way. There should be a minimum distance between the turbines to make the layout possible. The minimum distance between the turbines should be given to the optimization algorithm as an additional constraint. Fig. 8.2b shows the situation when the additional constraint is imposed on the code. As illustrated, the turbines are positioned properly. □

In some cases, the constraints cannot be fully satisfied. The optimization algorithm then tries to obtain a solution in which the constraints are satisfied as much as possible. In other words, it finds a solution that is the best possible answer.

Example 8.4. In studying Example 8.3, obtain the solution if

$$x_M = 100, \qquad y_M = 100,$$

and the turbines distance defined to be 200 m is a constraint to the code.

Answer. As we can see, the constraint of the problem is stricter than the boundaries of the field. Optimization algorithm never violates the boundaries of the design variables. Therefore, it tries to locate the turbines as far as possible. Fig. 8.3 shows such a situation. □

As discussed before, an optimization process may reach many different solutions. In many cases, the problem has many physical solutions. Hence, all the solutions are acceptable. But in many other situations, the optimization algorithm might not be able to find the real *best answer*. Therefore, we need to repeat the optimization process again and again to obtain different layouts and

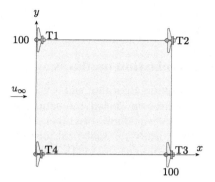

FIGURE 8.3 Satisfying the constraint of Example 8.4.

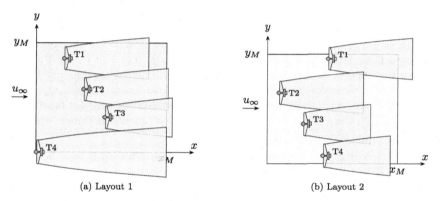

(a) Layout 1   (b) Layout 2

FIGURE 8.4 Different optimized scenarios.

compare the objective functions. We can also try different optimization methods to compare their performance and become sure that the absolute maximum energy of the farm is reachable.

Example 8.5. Discuss different optimization scenarios for Example 8.1.

Answer. According to Example 8.1 where the wind strictly blows from the west, we can find different scenarios, for all of which the final gained energy is the same. Two possible cases are shown in Fig. 8.4 where the turbines are not located in any wake. Therefore, all turbines are experiencing the upstream wind and produce their maximum possible power. □

In the above examples, it is assumed that the wind always blows from the west. But in the real world, the wind blows from all directions according to the wind rose diagram. Therefore, achieving a perfect design as shown in Fig. 8.4 is not possible. However, the concept of optimization, the method, and the results are the same. It means that the optimizing algorithm tries to find the best position of the wind turbine, so that the net gained energy of the field becomes maximized.

8.3 Coupling of optimization methods and wake models

The coupling of an optimization algorithm and a wake model is quite simple. The optimization code requires an objective function which can be the wind farm solver. Theoretically, any wind farm solver that calculates the annual farm's energy, can be used. But practically, the analytical solvers (Frandsen et al., 2006; Ghadirian et al., 2014; Hamedi et al., 2015; Hosseini et al., 2019; Torabi and Hamedi, 2015), are superior to the numerical ones (Javaheri and Canadillas, 2013; Naderi and Torabi, 2017). This is because for optimization purposes, the objective function (or the solver) is executed many times. Therefore, if the numerical solvers are used, the total execution time would be very large. In such cases, supercomputers or parallel clusters may be used.

The wind farm solvers use the wake models to calculate the annual gained energy of the farm. Therefore, as mentioned above, they can be used as the objective function. However, the additional constraints (other than the limits of the design variables) are mixed with the objective function. Typically, such restrictions are added to the objective function by means of a weight function.

In the following pseudocode, a typical objective function is shown. The parameters of the code are as follows:

objective is the main objective of the optimization. This value must be minimized at the end of the optimization process.

minDistance is the distance constraint. In this code, the minimum distance is set to be 200 m.

actualDistance is used to calculate the actual distance between the turbines. This value is compared with the minDistance to check if the constraint is violated or not.

energy is the annual gained energy of the farm which is obtained by windFarmEnergy() program.

windFarmEnergy() is the main program that calculates the wind farm energy. This function is responsible for linking the optimization method to the wind farm calculations. Using such a configuration makes the program coupling as simple as possible.

nTurbines is the number of wind turbines.

xTurbines holds the x coordinate of the center of each wind turbine.

yTurbines holds the y coordinate of the center of each wind turbine.

```
1  double objFunc()
2  {
3      double objective = 0;        // The final objective
4
5      double minDistance = 200;    // The distance constraint
6      double actualDistance = 0;   // A variable to store the minimum distance
```

```
7
8     energy = windFarmEnergy();  // Calculates the annual gained energy of the farm
9
10    // Calculates the distance between the turbines
11    for(int i=0; i<nTurbines-1; i++)
12    {
13        for(int j=i+1; j<nTurbines; j++)
14        {
15            actualDistance += max(0.0,minDistance-sqrt((pow((xTurbines[i]-
        xTurbines[j]),2)+(pow((yTurbines[i]-yTurbines[j]),2)))));
16        }
17    }
18
19    // The objective is a combination of the annual energy and the constraint
20    objective = -energy + 1000000*(actualDistance);
21
22    return objective ;
23 }
```

In the listed code, at line 20, we see how the energy of the wind farm is coupled with the minimum distance. First of all, the energy of the farm is multiplied by a minus sign since the code is going to minimize the objective function. Therefore, we minimize -energy, which means +energy is being maximized.

Secondly, the minimum distance between the turbines multiplied by 10^7 is added to the objective function. It means that the minimum distance of the turbines reduces the farm's energy. By this definition, the closer the turbines are located, their enrgy becomes less. Therefore, the optimization routine tries to find the best solution with the maximum energy in a year while keeping note of the minimum distance between the turbines.

As is clear from the coupling method, the optimization code does not guarantee the exact minimum distance to be applied. If the boundaries of the field are limited such that the minimum distance constraint cannot be achieved, the code finds the largest distance possible. Hence, as we can see, the additional constraints may be violated. The situation was explained by Example 8.4 and illustrated by Fig. 8.3.

8.4 Some worked examples

In this section, we try to use the optimization method for optimizing some wind farms. For the present studies, we use the wind farms from Sections D.1 and D.2 to illustrate the current discussion about wind farm layout design. For the present worked examples, we have used the PSO algorithm. But the same procedure can be carried on by means of other algorithms such as those discussed in Appendix E.

For all the following examples, we assume that NTK 500/41 wind turbines are going to be used. Hence, all the simulations are based on this type of turbine. The diameter of the turbines is 41 m, and we have chosen the minimum distance

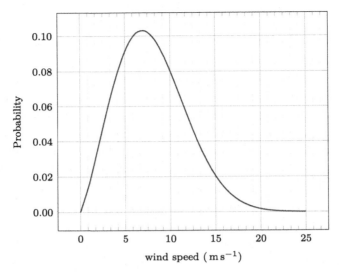

FIGURE 8.5 The Weibull PDF used for the studies.

of 200 m, or $5D$. The objective function is the same as shown in the pseudocode above.

8.4.1 Optimization for a constant-direction wind

We studied the farm from Section D.1 in Chapter 5 with different wind speeds and turbine distances. In those studies, the turbines were always located in a row. But for the present studies, the turbines have the possibility to be located in a two-dimensional field as shown in Fig. 8.1. Therefore, we seek the best placement of the turbines, for which the maximum yearly energy is obtained.

In the first scenario, we assume that the wind farm is located in a place where the wind always blows from one direction. For all the studies, we assume that the average wind velocity is $8.06 \, \mathrm{m \, s^{-1}}$, and the standard deviation of the wind is $s = 3.82$. For such a place, the shape factor of the Weibull PDF is

$$k = \left(\frac{s}{\bar{u}}\right)^{-1.086} = \left(\frac{3.82}{8.06}\right)^{-1.086} = 2.25, \qquad (8.3)$$

and its scale parameter is

$$c = 1.12\bar{u} = 1.12 \times 8.06 = 9.03. \qquad (8.4)$$

These values are chosen from a real field data and the PDF is plotted in Fig. 8.5.

From the explanations discussed before and as is illustrated by Fig. 8.4, we know that for any specific wind direction, there are many different optimum

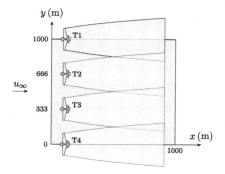

FIGURE 8.6 One of the possible layouts for gaining the maximum annual energy.

values. For example, for the west wind, two distinct possibilities are shown in the figure. Therefore, we expect the optimization software to give one of the solutions. Since PSO is a stochastic algorithm, it finds one of the solutions at each execution.

Example 8.6. Assume that the dimensions of the available field are $x_D = y_D = 1000$ m. What is the maximum annual energy of the farm when the wind blows from the west?

Answer. Since the minimum distance between the turbines is 200 m, the turbines can be arranged in the layout shown in Fig. 8.6. Such a situation yields the maximum annual energy since all the turbines are facing the undisturbed wind velocity, hence operating at their full power.

We are calculating the annual energy of the farm by the method of bins discussed in Section 5.7. In the present example, since the wind always blows from the west, we have only one bin. The calculations result in

$$E_{max} = 7.40657 \text{ GWh}.$$

This value is the maximum energy of the farm. □

Example 8.7. Assume the same field with the same dimensions as in Example 8.6. For the same conditions, use the PSO algorithm to find one of the best solutions.

Answer. The west wind corresponds to the standard wind direction of $WD = 270°$. For such a configuration, we have executed the PSO algorithm several times. The results are shown in Fig. 8.7. As it can be seen, all the layouts were able to gain $E_{max} = 7.40657$ GWh energy in a year. □

Example 8.8. Redo the problem for the situation when the wind blows from the north.

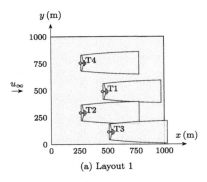 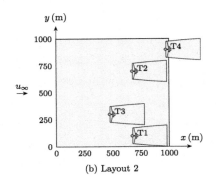

(a) Layout 1   (b) Layout 2

FIGURE 8.7 Optimization for the west wind.

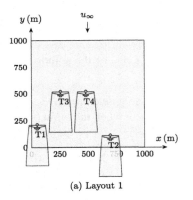 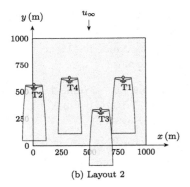

(a) Layout 1   (b) Layout 2

FIGURE 8.8 Optimization for the north wind.

Answer. The optimization process is the same as before; we just need to set $WD = 0°$. Again, we have executed the optimization code several times, and each time we got $E_{max} = 7.40657$ GWh energy in a year. The results for two different layouts are shown in Fig. 8.8.   ☐

Example 8.9. Redo the problem for the situation when the wind blows with angle of $WD = 45°$.

Answer. The results of calculations are shown in Fig. 8.9 for two possible solutions. Like in the other examples, there are infinitely many solutions for this problem.   ☐

In all the above examples, the dimensions of the farm were large enough such that the solutions of the code were able to deliver the maximum energy of the farm. But in some cases, we have a limited field; hence the results may not necessarily give the maximum annual energy. Here are some examples.

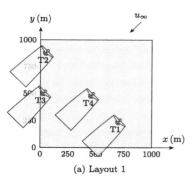
(a) Layout 1

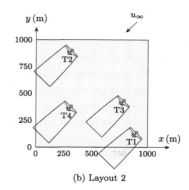
(b) Layout 2

FIGURE 8.9 Optimization for $WD = 45°$.

Example 8.10. If the dimensions of the farm are $x_D = y_D = 250\,\text{m}$, find the optimum solutions when the wind blows with angle of $WD = 45°$.

Answer. The results of calculations are shown in Fig. 8.10 for four possible solutions. Like in the other examples, there are infinitely many solutions to this problem. The only difference is that, since we do not have enough space to impose the 200-meter constraint, the turbines may be located in the wake of each other. Hence, the gained energy might not reach the maximum value.

As the figure illustrates, the solutions do not gain the maximum energy, and as is clear, Fig. 8.10c achieves the maximum output. Therefore, from the mathematical point of view, Fig. 8.10c shows the global maximum. In theory, the PSO algorithm must be able to find the maximum, but as we can see, by running the code, we find local maxima. The solution is to let the PSO algorithm run for more generations or start with more particles. For the present study, we have used 20 particles and 500 generations. Increasing these numbers always leads to the maximum of the problem. □

8.4.2 Optimization for a real case

In the real world, wind changes its direction. Hence we cannot find a layout where all the turbines are out of the wake of the others. Hence, the total annual energy of a real farm is not as large as in the cases studied in the previous subsection. This section uses actual field data for optimization and compares the results with the previous ones.

Example 8.11. Assume a 1000×1000-meter field is located in the city of Torbate Jaam, whose wind rose diagram is given in Fig. 2.7. Obtain the best layout and the maximum annual energy of the field.

Answer. The wind rose diagram of the city of Torbate Jaam shows that the wind blows from NNE most of the time. So we expect the optimized layout to set

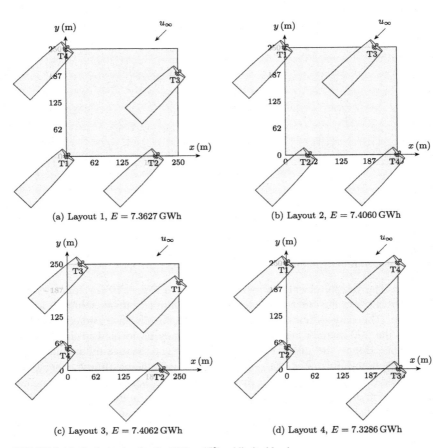

(a) Layout 1, $E = 7.3627\,\text{GWh}$

(b) Layout 2, $E = 7.4060\,\text{GWh}$

(c) Layout 3, $E = 7.4062\,\text{GWh}$

(d) Layout 4, $E = 7.3286\,\text{GWh}$

FIGURE 8.10 Optimization for the $WD = 45°$ and limited land.

the turbines from WNW to ESE. We have executed the PSO algorithm with 20 particles for 1000 generations. The results of optimization for four different runs is shown in Fig. 8.11. Some observations about the results are remarkable:

- First, the shown layouts are presented according to the dominant wind direction. In this place, the wind blows from all other directions, too.
- Second, in all the shown figures, the generated annual energy is substantially less than that in the previous examples. This is because, in the previous examples, the wind does not change direction, and also its average velocity is very high. Therefore, the present real case shows a much lesser amount of energy than in those theoretical examples.
- Third, the present obtained optimized layouts show that the turbines are located in the wake of each other in some cases. This is because their physical distance is enough for wake recovery, and also, the energy generation does not depend solely on the shown direction.

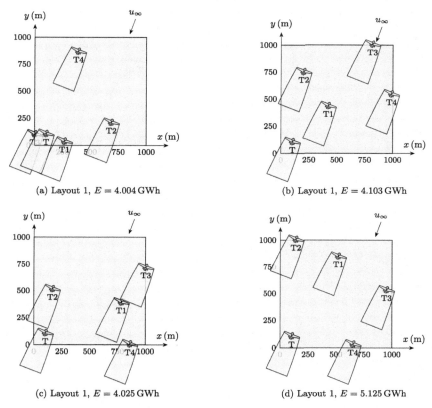

(a) Layout 1, $E = 4.004$ GWh

(b) Layout 1, $E = 4.103$ GWh

(c) Layout 1, $E = 4.025$ GWh

(d) Layout 1, $E = 5.125$ GWh

FIGURE 8.11 Optimization for Torbate Jaam city.

- Finally, comparing Figs. 8.11a to 8.11c, we see that the output energy in all the configurations is more or less the same. These data show that the optimization algorithm may find the local minima. But as we can see in Fig. 8.11d, the problem has some global maxima. Consequently, we need to let the optimization algorithm run for more generations.

This example shows that we need to rerun the optimization codes for real cases, again and again, to ensure we have found the optimized solution and to check the uniqueness of the solution. ☐

8.4.3 Optimization of a 4 × 4 wind farm

In this section, the wind farm introduced in Section D.2 is selected for verification. For this wind farm, we will obtain the best arrangement that produces the highest energy in a year. Practically, this problem is not different from the previous ones, except that the number of turbines is higher and it takes more

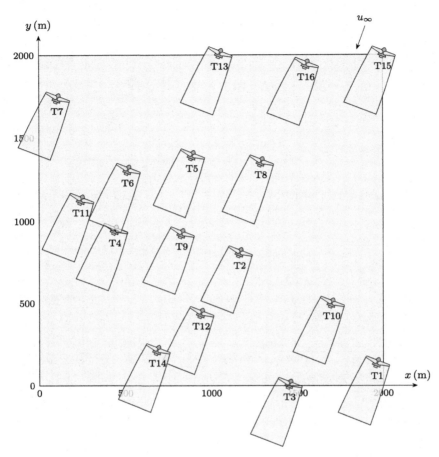

FIGURE 8.12 Optimization for a 4 × 4 wind farm.

computations. Since there are 16 wind turbines, we need to select a larger field to be able to fulfill the minimum-distance constraint.

Example 8.12. For the wind farm introduced in Section D.2, and the wind data of the city of Torbate Jaam, obtain the best farm layout.

Answer. As expected, the PSO algorithm is executed on the wind farm. This time, the number of design variables is 16. Other than that, there is no difference between the present study and the previous ones.

To reach the highest energy, we need to introduce the PSO particles and the number of generations. For the present calculations, we have selected 40 particles to run 2000 generations. The result is illustrated in Fig. 8.12 where the turbines are rotated toward the dominant wind in that location.

The result shows that the optimized layout is extending from WNW to ESE as expected. Moreover, some turbines are located in the wake of their upstream turbines. But as can be seen, their distance to their upstream turbines is large enough to avoid the wake. In the shown figure, the length of the wakes is 500 m, or about $12D$.

The calculations reveal that the total annual energy of the optimized farm is 15.3 GWh. Using this value, we can obtain the capacity factor of the whole farm. First, we need to calculate the theoretical value of the gained energy of the farm (see Section 2.6) as

$$E_{\text{theo}} = 8760 \times P_{eR} = 8760 \times 16 \times 500 = 70.08 \, \text{GWh}.$$

Then, using the net annual energy, we can find the capacity factor, which is

$$C_F = \frac{E_{\text{act}}}{E_{\text{theo}}} = \frac{15.3}{70.08} = 0.218.$$

This value is in the range of actual wind farms. □

Example 8.13. Compare the optimized layout with the original layout.

Answer. The original layout is shown in Section D.2 where the turbines are arranged in a 4×4 matrix form. Therefore, to have a proper comparison, we simulate the wind farm with the same dimensions used in Example 8.12.

The final result is shown in Fig. 8.13 where the turbines are rotated in the same direction as before. Although the arrangement seems to be very nice and the turbines have proper distance from each other, the final results show that the annual energy of this farm is 14.48 GWh. The energy loss compared to the optimized layout is

$$Loss = \frac{15.3 - 14.48}{15.3} = 0.0536 = 5.36\%.$$

Thus the optimized farm will produced 5% more energy in a year.

The capacity factor of such a configuration is

$$C_F = \frac{E_{\text{act}}}{E_{\text{theo}}} = \frac{14.48}{70.08} = 0.206,$$

which is obviously less than for the optimized layout. □

8.5 Applying additional constraints

In some cases, we need to apply some additional restrictions due to the different problems. For example, geographical features such as rivers, lakes, and hills may raise some problems in the layout design process. These restrictions can be added to the optimization algorithms in the same way we treated the minimum-distance constraint. In this section, we try to give such an example.

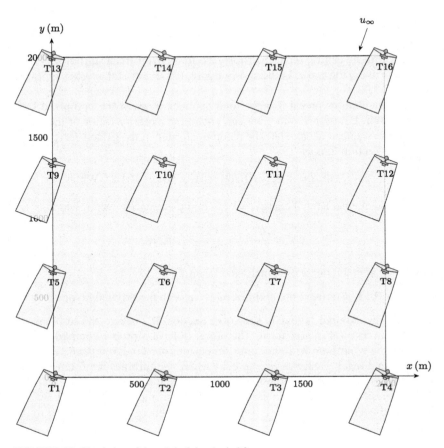

FIGURE 8.13 Simulation of the original 4 × 4 wind farm.

Example 8.14. Assume that there is lake in the field discussed in Examples 8.12 and 8.13. The location of the lake is shown in Fig. 8.14a. Discuss the constraints and perform optimization.

Answer. Since the lake is located inside the field, we have to apply additional constraints to the objective function. To implement the lake position, it would be more convenient to exclude the bounding polygon of the lake. Such a bounding polygon is shown in Fig. 8.14a, which is composed of two rectangles whose coordinates are shown in the figure.

The constraint is such that if a turbine is located inside the lake, we have to decrease the total energy by a significant amount. This idea adds a large penalty to the objective function, which, in return, makes the optimization algorithm remove that layout from the best candidates.

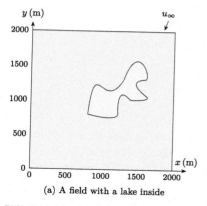

(a) A field with a lake inside

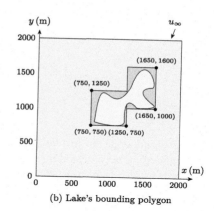

(b) Lake's bounding polygon

FIGURE 8.14 The limits of the lake.

The pseudocode of such an objective function is shown below:

```
1  double objFunc()
2  {
3  double objective = 0; // The final objective
4
5  double minDistance = 200; // The distance constraint
6  double actualDistance = 0; // A variable to store the minimum distance
7
8  energy = windFarmEnergy(); // Calculates the annual gained energy of the farm
9
10 // Calculates the distance between the turbines
11 for(int i=0; i<nTurbines-1; i++)
12 {
13 for(int j=i+1; j<nTurbines; j++)
14 {
15 actualDistance += max(0.0,minDistance-sqrt((pow((xTurbines[i]-xTurbines[j]),2)+(
       pow((yTurbines[i]-yTurbines[j]),2)))));
16 }
17 }
18
19 // If the turbine's position is located inside the lake, adds a large penalty to
       the objective function.
20 double Lake=0;
21 for(int i=0; i<nTurbines; i++)
22 {
23         if(xTurbines[i]>750 && xTurbines[i]<1250)
24         {
25                 if(yTurbines[i]>750 && yTurbines[i]<1250)
26                         Lake=1;
27         }
28         if(xTurbines[i]>1250 && xTurbines[i]<1650)
29         {
30                 if(yTurbines[i]>1000 && yTurbines[i]<1600)
31                         Lake=1;
```

```
32          }
33  }
34
35  // The objective is a combination of the annual energy and the constraint.
36  objective = -energy + 1000000*(actualDistance)+ 100000000*Lake;
37
38  return objective;
39  }
```

In this pseudocode, from line 20 to 33, we are examining all the turbines to see if any of them is located inside the lake bounding polygon or not. If a single turbine is located inside the lake, we add a huge penalty to the objective function. This penalty artificially reduces the total output energy. Hence, the optimization algorithm considers it as a suboptimal layout and, obviously, removes it from the *best answers*. Consequently, the optimization software finds a layout that has the optimum energy with no turbine located inside the lake.

The final results are shown in Fig. 8.15. As is clear, no turbine is located in the bounding polygon of the lake. Moreover, the optimized farm generates 15.1567 GWh energy in a year. □

8.6 Summary

In this chapter, the method described in Chapter 5 is coupled with the PSO algorithm to optimize the wind farm layout. It is emphasized that the coupling requires the design variables, objective function, and other constraints.

The optimization coupling is quite simple; hence, we can couple any simulation method to the code without any difficulty. Even the numerical methods discussed in Chapters 6 and 7 can be coupled with the optimization software. But it should be noted that the main calculation code is run over and over by the optimizing algorithm. Thus, using numerical methods as the main farm calculator will require lots of computational resources.

8.7 Problems

1. A 5-turbine wind farm is shown in Fig. 8.16. Use the data of Torbate Jaam as the main wind data. Calculate the energy of the farm using Jensen's model.
2. For the wind farm shown in Fig. 8.16, use the data of your own city as the primary wind data. Obtain the wind data from meteorological stations. Plot the statistical graphs of the wind. Then calculate the energy of the farm using Frandsen's model.
3. Assume that the wind farm shown in Fig. 8.16 is located in a windy city in your country. Name the city, obtain the wind data of that city from meteorological stations, and plot the statistical graphs of the wind. Then calculate the energy of the farm using Larsen's model.

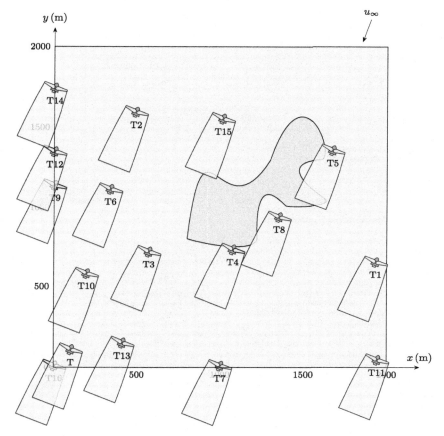

FIGURE 8.15 Optimization of the farm around a lake.

4. Using the model given in Chapter 5, repeat Problem 3, and calculate the energy of the farm.
5. Write the constraints and the objective function for a wind farm shown in Fig. 8.16.
6. For optimization of Problem 1, implement WOA discussed in Section E.2. Optimize the farm and compare the results.
7. Couple the Frandsen's model with CSA (Appendix E.1) and optimize the farm of Problem 2.
8. Couple Larsen's model with TBLO (Appendix E.3) and optimize the farm of Problem 3. Compare the net annual energy for the two cases.
9. Use the WOA algorithm to optimize the farm of Problem 4.
10. Apply different optimization algorithms to Problems 1 through 4 and compare the results.

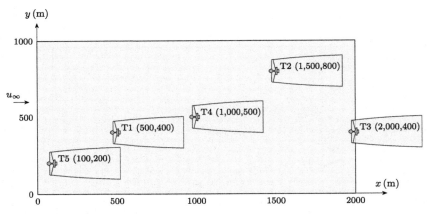

FIGURE 8.16 A sample 5-turbine farm.

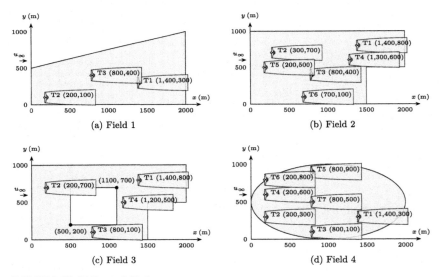

FIGURE 8.17 Different field shapes.

11. Four different fields are shown in Fig. 8.17. For each field,
    a. How many design variables do there exist?
    b. Define the limits for the design variables.
    c. Define proper constraints in the form of a pseudocode.
12. For the problems of Fig. 8.17,
    a. Calculate the farm energy using Jensen's model.
    b. Calculate the farm energy using Frandsen's model.
    c. Calculate the farm energy using Larsen's model.
    d. Calculate the farm energy using Ghadirian's model.

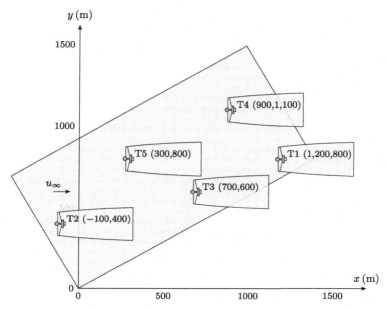

FIGURE 8.18 An inclined farm.

13. Assume the farms shown in Fig. 8.17 are located in a steady-state wind. The wind blows with a constant velocity from the west. Simulate each farm using the CFD code developed in Chapter 7 for different wind speeds from 5 to 25 m s^{-1}. Obtain the farm's power and plot the results in a single figure.

14. Do Problem 13 for the east wind. Are the results the same? Compare the results on a single figure and discuss.

15. For different wind angles, repeat Problem 13 and plot the farm's power in proper figures.

16. To compare the analytical models with the numerical ones, do Problem 13 using different wake models, namely Jensen's, Frandsen's, Larsen's, Ghadirian's, and the model developed in this book. Compare the results and discuss.

17. Obtain the annual energy of the farms shown in Fig. 8.17 using the analytical models.

18. To check the increase in annual energy for optimized farms, apply any desirable optimization method to Problem 17, and compare the final annual gained energy. Note that the constraints obtained in Problem 11 are required.

19. Fig. 8.18 shows an inclined rectangular field. Since the field has a rectangular shape, you can easily use the local coordinate system to convert it to a normal rectangular field. This process results in a simple normal problem, and we do not need any additional constraints.

For the shown field, and for a real wind data of a place of your choice,
   a. Define a proper local coordinate.
   b. Obtain the annual energy of the farm at its present layout.
   c. Perform optimization using different algorithms and obtain the best layout.
   d. Compare the results of the optimized layout with the original one.
20. Use the wind data for different places and perform optimization on the field discussed in Problem 19. Compare your results and discuss.

References

Frandsen, Sten, Barthelmie, Rebecca, Pryor, Sara, Rathmann, Ole, Larsen, S., Højstrup, J., Thøgersen, Morten, 2006. Analytical modelling of wind speed deficit in large offshore wind farms. Wind Energy 9 (1), 39–53.

Ghadirian, A., Dehghan, M., Torabi, F., 2014. Considering induction factor using BEM method in wind farm layout optimization. Journal of Wind Engineering and Industrial Aerodynamics 129, 31–39.

Hamedi, Razieh, Javaheri, Alireza, Dehghan, Omid, Torabi, Farshad, 2015. A semi-analytical model for velocity profile at wind turbine wake using blade element momentum. Energy Equipment and Systems 3 (1), 13–24.

Hosseini, Radmarz, Roohi, Reza, Ahmadi, Goodarz, 2019. Parametric study of a novel oscillatory wind turbine. Energy Equipment and Systems 7 (4), 377–387.

Javaheri, A., Canadillas, B., 2013. Wake modeling of an offshore wind farm using OpenFOAM. DEWI Magazin.

Mosetti, G., Poloni, C., Diviacco, B., 1994. Optimization of wind turbine positioning in large wind farms by means of a genetic algorithm. Journal of Wind Engineering and Industrial Aerodynamics 51 (1), 105–116. https://doi.org/10.1016/0167-6105(94)90080-9.

Naderi, Shayan, Torabi, Farschad, 2017. Numerical investigation of wake behind a HAWT using modified actuator disc method. Energy Conversion and Management 148, 1346–1357. https://doi.org/10.1016/j.enconman.2017.07.003.

Torabi, F., Hamedi, R., 2015. An analytical model for prediction of wind velocity profile and wind turbine wake. Renewable Energy 75, 945–955.

Appendix A

Ancient Persian wind turbines

The city of Nashtifan is located in the northeastern Iran near the border of Afghanistan (Atlas Obscura, 2021). It belongs to Khaf county from Razavi Khorasan province. The name of the city is composed of two words, namely Nash, which means sting, and Tifan, which means storm (tifan is very similar to typhoon in pronunciation and meaning). Some people believe that the name stands for the *the storm's sting*. This depicts the strength of the winds in this city. Other people believe that the name stands for the sting of scorpions found in hot and dry climates such as that of Nashtifan.

Although capturing wind power was known for thousands of years in sailing, converting its energy in the form of wind turbines was first attributed to ancient Iran. The oldest windmills have been installed in the city of Nashtifan (Howard, 2017). These windmills are known as *Asbad* in the local language. Asbad is a Persian word pronounced as "us"–"bud", and is composed of two distinct words, *As* (pronounced as "us") which is an acronym for the Persian word *Asiab* meaning "mill", and "bad" (pronounced as "bud") meaning "wind". Therefore, an Asbad means "windmill". The Persian empire invented the windmills thousands of years ago (Wikipedia, 2018). The invention of the windmills happened in the eastern provinces of Iran since these provinces have strong wind potential. Some documents reported that the first windmills were installed in the Persian empire about 3000 years ago.

The Sistan province in the southeast of Iran and the south Khorasan have strong potential for wind power. The Asbads of Nashtifan are designed upon the same experiences in that era. These windmills are still operational since their invention dating back from 500 to 900 AD. The turbines are categorized as vertical-axis windmills in which a vertical axis made of wood operates as the main shaft. Each shaft contains six wooden blades that capture the wind power using drag force. The rotating shaft is attached to a cylindrical grindstone for milling grain for flour. All the components are installed over a clay chamber. The blades are facing the wind on the rooftop, and on the chamber inside, the grindstone works. The chamber is shown in Fig. A.1.

As shown in Fig. A.2, the vertical blades or paddles are embraced with clay walls to ensure that the wind rotates the main shaft in a single direction. With this invention, the windmill can benefit from the wind blowing from any direction.

Hopefully, these windmills or Asbads are still operational in the city of Nashtifan. This fact shows that the windmills are pretty low-maintenance and can operate for a long time. Even a simple design as that of the ancient Asbads

FIGURE A.1 Nashtifan windmill's structure.

FIGURE A.2 The rotor of Nashtifan's windmills.

of Nashtifan can withstand the harsh and hot climates for thousands of years. In 2002 the windmills were recognized as a national heritage site by Iran.

Nashtifan is located in a dry and hot place. The rainfall or precipitation level is about 100 to 200 millimeters per year. Summers are very hot in this area, with the hottest month being June. This area experiences 120-day winds, or wind of 120 days. This is a strong, steady wind occurring in the eastern and southeastern Iran from late May to late September. The existence of the 120-day wind makes the area bearable. Otherwise, the weather becomes too hot for living. In winter, the weather becomes cold and rarely reaches icing conditions.

The city of Nashtifan extends from the North to the South, perpendicular to the 120-day winds. In this city, the northeastern winds are stronger than in the other places, just like the conditions of Torbate Jaam (see Fig. 2.7). These two cities are located in the same province hence benefit from the 120-day winds.

In Nashtifan, different major winds are active:

120-day winds also known as *black winds* blow from NNE to SSW. These winds start blowing from the beginning of June and end at the beginning of October. The black winds have a substantial effect on the nature and life of the eastern part of Iran. In Nashtifan, people have been using the power of these winds for building one of the earliest windmills in the world.

The main characteristics of these winds are their steadiness and strength. The speed of winds even reaches up to 120 km/h. Therefore, the strong and steady wind brings a good potential for the people of the place for constructing the Asbads. The invention of the Asbads shows the capability of the local people for coping with nature and converting the threats to benefits.

Farah wind blows during winters from the East, from the city Farah. This city is now located in Afghanistan, near the border with Iran. The wind has moderate speed and makes the cold weather become milder.

Neishaboor wind comes from the city of Neishaboor, which is a city in the eastern Iran. The wind has almost the same characteristics as those of Farah winds. It blows mostly in winter and makes the weather milder.

Southwestern winds are very important for this area since they bring humidity and rain.

Gavband wind is a cold wind blowing from the North.

Mountain winds are cold winds blowing from the East. The mountain winds cool down the weather temperature and sometimes cause freezing.

References

Atlas Obscura, 2021. Nashtifan windmills. https://www.atlasobscura.com/places/nashtifan-windmills. (Accessed 21 June 2021).

Howard, Brian Clark, 2017. See the 1,000-year-old windmills still in use today. https://www.nationalgeographic.com/science/article/nashtifan-iran-windmills. (Accessed 21 June 2021).

Wikipedia, 2018. Nashtifan. https://fa.wikipedia.org/wiki/2018. (Accessed 21 June 2021).

Appendix B

Wind turbine airfoils

The performance of a wind turbine is related to the performance of its blades. The blades are the primary energy capturing front; hence a proper knowledge of airfoils is essential. Consequently, improvements in the wind turbine industry largely depend on the progress of airfoils. Therefore, many companies and universities started to design proper airfoils optimized for wind turbines. Today, there are lots of different airfoil families that are used in this industry, which are well-studied and documented. The data of these airfoils (or family of airfoils) can be found in many different references.

An airfoil shows different behavior when the operating conditions change. This behavior makes it difficult to work with. For example, the lift and drag of an airfoil depend on the surface roughness, leading-edge contaminant, Reynolds number, and the variation of air pollution. Consequently, optimizing an airfoil is a very tedious task and requires lots of different investigations. Usually, the companies, universities, and other research centers try hard to improve their existing airfoils' performance. These improvements mean that the newly invented airfoils become less sensitive to surface roughness, leading-edge curvature, or other parameters. These factors are important since in a real operation the blade surfaces lose their smoothness due to the icing, dust, bird droppings, and similar items.

There are many different airfoil families and blade sections, for each of which many various studies are available (Chen et al., 2016; Chen, 2014; Grasso, 2014; Méndez et al., 2014). Therefore, different data are given in the shape of figures and tables. Usually, the drag and lift coefficients are the main important aerodynamic parameters used in blade design; but other parameters such as the pressure coefficient, momentum coefficient, and other dynamic coefficients are also important. There are lots of catalogs available in the open literature dedicated to these parameters. Here, some airfoil families are selected, and their lift and drag coefficients are given. Some notes are worth mentioning regarding the data of the present appendix:

- The selected airfoils are those that are used in the wind turbine industry. But it should be noted that they are not the only available ones. Also, it does not mean that they are the more important ones! These are just some typical selections used in different chapters of the book in various examples and problems.

- From different aerodynamic parameters, only lift and drag coefficients are given. These data are required for the simulation of wind turbines by means of the BEM method.
- The given data are valid for one Reynolds number (most of them are for $Re \sim \mathcal{O}(10^6)$) which means that the Reynolds number is of an order of one million. In many different references, the given data may be different due to the different Reynolds numbers.
- These data all are valid for a smooth surface. For an investigation of surface roughness, the reader should obtain proper data from valid references.

The lift and drag coefficients are strong functions of the angle of attack, or α. For most airfoil sections presented in this appendix, the data are available in the range of $-10 < \alpha < 25$, or even smaller. But in simulations, we need to have information in a wider range. In some examples, the angle of attack may reach up to 50 degrees. Hence, the present data are not suitable for these situations. There are two different ways to tackle this problem. We can either:

1. Obtain a wider range of data for the airfoils, or
2. Use an extrapolation method.

The former method is superior to the latter, but the problem is that a broader range of data may not be available. Therefore, many studies have been done to extend the range of the available data. The studies are based on either experiment or numerical methods. Many different applications are used to simulate the airfoils, and different approaches are applied to overcome turbulence modeling, which makes the simulation quite complicated.

The latter approach is used here by fitting proper polynomials to the aerodynamic data. For this purpose, the aerodynamic curves are split into different ranges. Then for each range, a simple polynomial is fitted so that the overall error becomes minimum. In the following figures, the actual data are plotted by symbols, and solid lines plot the fitted curves on the same figure. These extrapolations are not the best choices, but they give a reasonable value, which is important when the BEM algorithm is iterated. It is quite understandable that having correct data is vital in obtaining proper results.

B.1 NACA families

The advanced wind turbines were started to be designed using the airfoils that were available at the time. The conventionally used airfoils were NACA 23XXX, NACA 44XX, NACA 63-XX NACA 64-XX, and NACA 65-XX. The thicker airfoils were used for the root section where the structural loads are higher and the section bears more forces. Originally, these families were designed for airplanes and were not well-suited for wind turbines. There are lots of reasons why these sections show degradation in performance when being used in wind turbines. For example, they usually show unsuitable performance under the stall condition. This is not an issue for the airplane industry since airplane

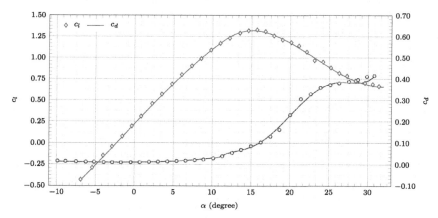

FIGURE B.1 NACA63415.

wings are usually designed to work quite far from stall point. But in wind turbines, stall-regulation is useful, and many turbines benefit from stall regions to maintain their rated power.

Although many different airfoils are designed that are specifically dedicated to the wind industry, NACA airfoils are still used in some applications (Bertagnolio et al., 2001; Fuglsang et al., 1998; Selig, 1995). Among the large NACA airfoil family, NACA 63-XX NACA 64-XX, and NACA 65-XX are more common. Figs. B.1–B.3 show the aerodynamic properties of NACA63415, NACA63421, and NACA63430 airfoils, respectively. Their corresponding curve-fitted functions are also given in Eqs. (B.1) to (B.6).

Curve-fitted data for NACA63415

$$
c_d = \begin{cases}
0.009 + 0.0001\alpha^2 + 6 \times 10^{-6}\alpha^3 & \alpha < 12 \\
-8.417 + 2.3316\alpha - 0.2503\alpha^2 + 0.0129941\alpha^3 & \\
-0.00032354\alpha^4 + 3.101 \times 10^{-6}\alpha^5 & \alpha > 12
\end{cases}
\tag{B.1}
$$

$$
c_l = \begin{cases}
0.22 + 0.097\alpha - 0.0002\alpha^2 - 7 \times 10^{-5}\alpha^3 & \alpha < 12 \\
-1.9 + 0.5288\alpha - 0.02827\alpha^2 + 0.000527\alpha^3 & \\
-2.56 \times 10^{-6}\alpha^4 & \alpha > 12
\end{cases}
\tag{B.2}
$$

Curve-fitted data for NACA63421

$$
c_d = \begin{cases}
0.0045 + 0.0017\alpha - 0.0004\alpha^2 + 3.5 \times 10^{-5}\alpha^3 & \alpha < 8.8 \\
-0.3686 + 0.09235\alpha - 0.00717\alpha^2 + 0.0001815\alpha^3 & \alpha > 8.8
\end{cases}
\tag{B.3}
$$

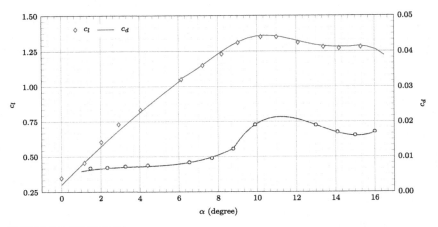

FIGURE B.2 NACA63421.

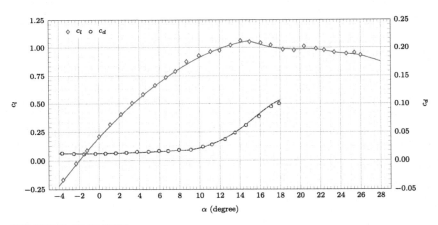

FIGURE B.3 NACA63430.

$$c_l = \begin{cases} 0.3 + 0.14\alpha - 0.003\alpha^2 & \alpha < 6 \\ -11.05 + 9.265\alpha - 2.9582\alpha^2 + 0.50108399\alpha^3 - 0.04806\alpha^4 \\ +0.00260762\alpha^5 - 7.422 \times 10^{-5}\alpha^6 + 8.55 \times 10^{-7}\alpha^7 & \alpha > 6 \end{cases}$$

(B.4)

Curve-fitted data for NACA63430

$$c_d = \begin{cases} 0.012 + 0.0002\alpha + 6 \times 10^{-5}\alpha^2 & \alpha < 8 \\ 0.53 - 0.217\alpha + 0.0365\alpha^2 - 0.003097\alpha^3 + 0.0001337\alpha^4 & \\ -2.285 \times 10^{-6}\alpha^5 & \alpha > 8 \end{cases}$$

(B.5)

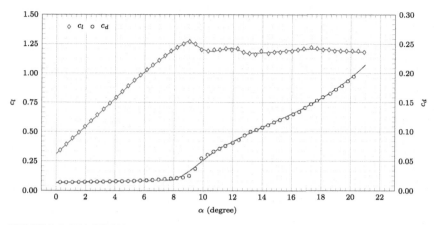

FIGURE B.4 FFA-W3-211.

$$
c_l = \begin{cases}
0.2 + 0.096\alpha - 0.00228\alpha^2 - 2 \times 10^{-5}x^3 & \alpha < 15 \\
-218.39 + 65.903\alpha - 8.1415\alpha^2 + 0.529938\alpha^3 & \\
-0.0191837\alpha^4 + 0.000366436\alpha^5 - 2.8874 \times 10^{-6}\alpha^6 & \alpha > 15
\end{cases}
\tag{B.6}
$$

B.2 FFA family

National Aeronautical Research Institute of Sweden is abbreviated FFA because, in Swedish, it is called Flygtekniska försöksanstalten. This institute was a former Swedish state governmental authority to conduct research in the aeronautical field. The institute has designed a series of airfoils for use in horizontal-axis wind turbines (Bertagnolio et al., 2001; Bjorck, 1990; Fuglsang et al., 1998). These airfoils are divided into three different series:

FFA-W1-xxx are airfoils with thickness-to-chord ratios from 12.8% to 27.1%. The lift coefficient of these airfoils ranges from 0.9 to 1.2.

FFA-W2-xxx airfoils have their lift coefficient approximately 0.15 lower than for the FFA-W1-xxx series.

FFA-W3-xxx are thick airfoils whose thickness-to-chord ratios are from 19.5% to 36%.

In all the above series, xxx represents the thickness of the airfoil. These blade sections were tested in the low speed wind tunnel L2000m located at KTH, Royal Institute of Technology, Stockholm.

Figs. B.4 and B.5 show the lift and drag coefficients of FFA-W3-211 and FFAW3-301, respectively, and the specifications of fitted curves are given in Eqs. (B.7) to (B.10).

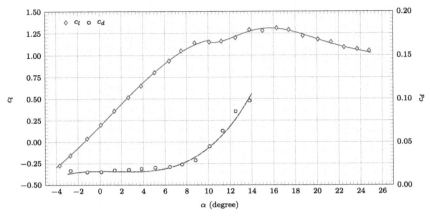

Curve-fitted data for FFA-W3-211

$$c_d = \begin{cases} 0.0135 + 0.00055\alpha & \alpha < 8 \\ 0.8722 - 0.2895\alpha + 0.03127\alpha^2 - 0.001053\alpha^3 & 8 < \alpha < 12 \\ -0.392 + 0.081\alpha - 0.00475\alpha^2 + 0.000108\alpha^3 & \alpha > 12 \end{cases} \quad (B.7)$$

$$c_l = \begin{cases} 0.31 + 0.113\alpha + 0.003\alpha^2 - 0.0004\alpha^3 & \alpha < 9 \\ -524.776 + 243.2796\alpha - 44.672706\alpha^2 + 4.07201\alpha^3 & \\ -0.1842959\alpha^4 + 0.003314\alpha^5 & 9 < \alpha < 13 \\ -7.44 + 3.449\alpha - 0.51957\alpha^2 + 0.037355\alpha^3 & \\ -0.00129265\alpha^4 + 1.733 \times 10^{-5}\alpha^5 & \alpha > 13 \end{cases} \quad (B.8)$$

Curve-fitted data for FFA-W3-301

$$c_d = 0.016 - 0.0002\alpha^2 + 4.7\alpha^3 \quad (B.9)$$

$$c_l = \begin{cases} 0.18 + 0.129\alpha + 0.001\alpha^2 - 0.0004\alpha^3 & \alpha < 10 \\ 23.2 - 7.445\alpha + 0.9935\alpha^2 - 0.0671\alpha^3 & \\ 0.00243236\alpha^4 - 4.5 \times 10^{-5}\alpha^5 + 3.3414 \times 10^{-7}\alpha^6 & \alpha > 10 \end{cases} \quad (B.10)$$

B.3   Risø family

Risø airfoils series A were developed by Risø National Laboratory for use in wind turbines (Bertagnolio et al., 2001; Fuglsang et al., 1998, 2004). These

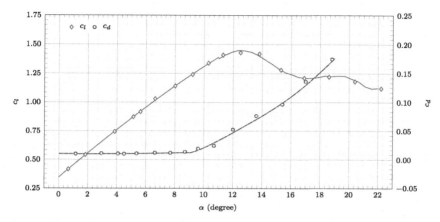

FIGURE B.6 RISO-A1-18.

airfoils were optimized for wind turbines, which means that they show good performance when employed in blades. All the airfoils were tested in the VELUX wind tunnel at Risø National Lab.

The aerodynamic coefficients of Risø-A1-18 and Risø-A1-21 are plotted in Figs. B.6 and B.7, respectively. These data are curve-fitted as before. The functions are given in Eqs. (B.11) to (B.14).

Curve-fitted data for RISØ-A1-18

$$
c_d = \begin{cases} 0.01 - 0.0002\alpha + 5 \times 10^{-5}\alpha^2 & \alpha < 9 \\ 1.532 - 0.584\alpha + 0.0858\alpha^2 - 0.00609\alpha^3 + 0.000213\alpha^4 \\ -2.916 \times 10^{-6}\alpha^5 & \alpha > 9 \end{cases} \quad (\text{B.11})
$$

$$
c_l = \begin{cases} 0.34500439 + 0.10722031\alpha - 0.00064668795\alpha^2 \\ -4.1066474 \times 10^{-5}\alpha^3 & \alpha < 11 \\ 241.28621 - 98.209095\alpha + 16.395834\alpha^2 - 1.4294394\alpha^3 \\ 0.068704644\alpha^4 - 0.0017284599\alpha^5 + 1.780636 \times 10^{-5}\alpha^6 & \alpha > 11 \end{cases}
$$
$$(\text{B.12})$$

Curve-fitted data for RISØ-A1-21

$$
c_d = \begin{cases} 0.01 + 0.0001\alpha + 8 \times 10^{-5}\alpha^2 & \alpha < 9 \\ 4.533 - 1.682\alpha + 0.23966\alpha^2 - 0.016397\alpha^3 + 0.000546\alpha^4 \\ -7.112 \times 10^{-6}\alpha^5 & \alpha > 9 \end{cases} \quad (\text{B.13})
$$

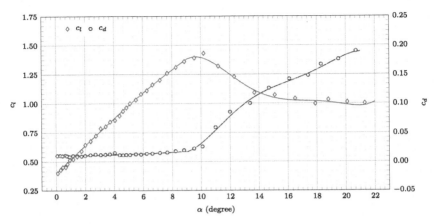

FIGURE B.7 RISO-A1-21.

$$c_l = \begin{cases} 0.4 + 0.12\alpha - 0.00014\alpha^3 & \alpha < 9 \\ -12.13 + 3.566\alpha - 0.2312\alpha^2 - 0.0131191\alpha^3 & \\ +0.002226\alpha^4 - 9.533 \times 10^{-6}\alpha^5 + 1.377 \times 10^{-6}\alpha^6 & \alpha > 9 \end{cases}$$ (B.14)

B.4 DU family

Delft University extensively worked on wind turbine airfoils and produced DU families (Bertagnolio et al., 2001; Timmer and Van Rooij, 2003). These airfoils were numerically designed and experimentally tested in low-speed wind tunnels of the university and at IAG Stuttgart. Finally, Delft University introduced DU sections with a generic name as DU-yy-W-xxx, in which:

DU stands for Delft University,

yy is the year when the airfoil was presented,

W means that the airfoil is specifically designed for wind turbines (and not for aviation or airplanes). After the letter W, there could be a number that indicates the version of the airfoil. For example, DU91-W2-25 is the second version (W2) of the airfoil that was designed in 1991.

xxx specifies 10 times the airfoil's maximum thickness in percent of the chord.

The aerodynamic coefficients of DU91-W2-250 are plotted in Fig. B.8, and the curve-fitted functions are given in Eqs. (B.15) and (B.16).

Curve-fitted data for DU91-W2-250

$$c_d = \begin{cases} +0.0091 + 0.0003\alpha & \alpha < 9 \\ +0.327 - 0.09\alpha + 0.00775\alpha^2 - 0.000183\alpha^3 & \alpha > 9 \end{cases}$$ (B.15)

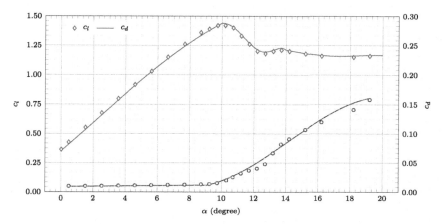

FIGURE B.8 DU91-W2-250.

$$c_l = \begin{cases} 0.35 + 0.12\alpha + 0.0016\alpha^2 - 0.00027\alpha^3 & \alpha < 10 \\ 455.987 - 210.2437\alpha - +38.46860\alpha^2 - 3.48055\alpha^3 & \\ +0.155744\alpha^4 - 0.0027586\alpha^5 - 7.9234 \times 10^{-5}\alpha^6 & 10 < \alpha < 14 \\ -13.77 + 3.832\alpha - 0.3605\alpha^2 + 0.0148\alpha^3 - 0.0002244\alpha^4 & \alpha > 14 \end{cases}$$

$$(B.16)$$

B.5 FX family

German aerodynamicist Franz Xaver Wortmann (September 24, 1921 – January 16, 1985) developed a series of laminar airfoils together with Richard Eppler and Dieter Althaus (Bertagnolio et al., 2001; Selig, 1995). The airfoil sections are named FX family, which was the abbreviation of Franz Xaver. Most of the FX airfoils are characterized by a wide low-drag range and a high maximum-lift coefficient.

The aerodynamic coefficients of FX66-S196-V1 are plotted in Fig. B.9, and the curve-fitted functions are given in Eqs. (B.17) and (B.18).

Curve-fitted data for FX66-S196-V1

$$c_d = \begin{cases} 0.008 + 0.001\alpha - 0.0003\alpha^2 + 3.1e - 05\alpha^3 & \alpha < 6 \\ -1.795 + 1.1762\alpha - 0.304439\alpha^2 + 0.039124\alpha^3 & \\ -0.002496\alpha^4 + 6.323 \times 10^{-5}\alpha^5 & \alpha > 6 \end{cases}$$

$$(B.17)$$

$$c_l = \begin{cases} 0.51 + 0.115\alpha - 0.0008\alpha^2 & \alpha < 7.6 \\ -10.47 + 3.849\alpha - 0.448\alpha^2 + 0.02227\alpha^3 & \\ -0.000403\alpha^4 & \alpha > 7.6 \end{cases}$$

$$(B.18)$$

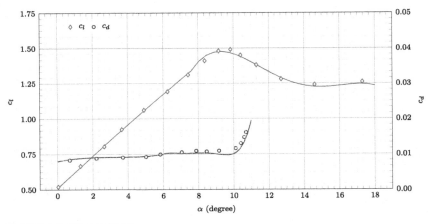

FIGURE B.9 FX66-S196-V1.

B.6 NREL family

National Renewable Energy Laboratory (NREL), formerly the Solar Energy Research Institute (SERI), started developing a series of airfoils that are insensitive to surface roughness (Bertagnolio et al., 2001; Gregorek et al., 1995; Jonkman et al., 2009; Ramsay et al., 1996; Somers, 2005; Tangler and Somers, 1995). These airfoils are specially designed for horizontal-axis wind turbines. The designed airfoils are optimized for use in stall-regulated, pitch-controlled, and variable-rpm wind turbines. NREL has used Eppler Airfoil Design software and has tested the final designs in various wind tunnels. Some sections have been tested in Delft University's low-turbulence wind tunnel.

The aerodynamic coefficients of S809-D and S814 are plotted in Figs. B.10 and B.11, respectively. These data are curve-fitted as before, and their respected functions are given in Eqs. (B.19) to (B.22).

Curve-fitted data for S809

$$c_d = 0.008 + 0.0023\alpha - 0.00074\alpha^2 + 6.6 \times 10^{-5}\alpha^3 \tag{B.19}$$

$$c_l = \begin{cases} 0.14 + 0.1\alpha + 0.009\alpha^2 - 0.0011\alpha^3 & \alpha < 8.1 \\ -55.6 + 26.112\alpha - 4.8852\alpha^2 + 0.47367\alpha^3 & \\ -0.0250989\alpha^4 + 0.0006898\alpha^5 - 7.695 \times 10^{-6}\alpha^6 & \alpha > 8.1 \end{cases} \tag{B.20}$$

Curve-fitted data for S814

$$c_d = 0.008 - 0.00011\alpha^2 + 2.5 \times 10^{-5}\alpha^3 \tag{B.21}$$

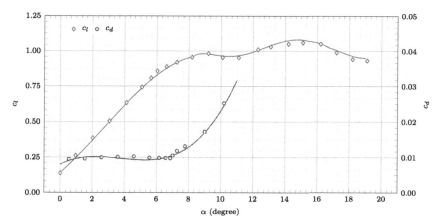

FIGURE B.10 S809-D.

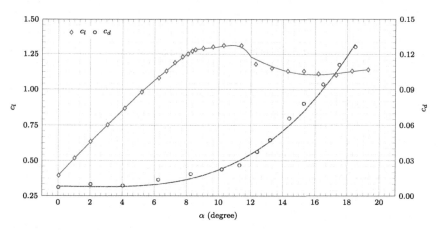

FIGURE B.11 S814.

$$
c_l = \begin{cases}
0.3932 + 0.1199\alpha - 0.00064567\alpha^2 - 0.0001244\alpha^3 & \alpha < 8.4 \\
-38.31 + 16.41\alpha - 2.5484\alpha^2 + 0.17565\alpha^3 - 0.00453\alpha^4 & 8.5 < \alpha < 12 \\
-2.7501483 + 1.3589159\alpha - 0.15705809\alpha^2 \\
\quad +0.0074744161\alpha^3 - 0.00012662303\alpha^4 & \alpha > 12
\end{cases}
$$
$$(B.22)$$

B.7   Summary

In this appendix, for some selected airfoils, c_l and c_d are plotted, and their corresponding curve-fitted functions are given. The fitted functions are useful for use in numerical codes such as BEM or others. These functions provide accurate data within the range at which the functions are obtained. Also, they give reasonable values at outer ranges where there is no available data.

It should be kept in mind that these airfoils are selected for the educational purposes of the present book. For real calculations, one should not rely on the present data, and for any specific calculation, a correct table of data must be selected. The reason is that aerodynamic coefficients strongly depend on Reynold's number, which varies from case to case.

All the airfoil data used in this appendix are taken from Bertagnolio et al. (2001). More data is available in the literature in many forms, such as tables, figures, functions, etc. Some of the references are given in the bibliography, but there are many others in the open literature.

References

Bertagnolio, Franck, Sørensen, Niels, Johansen, Jeppe, Fuglsang, Peter, 2001. Wind Turbine Airfoil Catalogue.

Bjorck, Anders, 1990. Coordinates and calculations for the FFA-WL-xxx, FFA-W2-xxx, FFA-W2-xxx and FFA-W3-xxx series of airfoils for horizontal axis wind turbines. Technical Report FFA Tn, 15.

Chen, Jin, Wang, Quan, Zhang, Shiqiang, Eecen, Peter, Grasso, Francesco, 2016. A new direct design method of wind turbine airfoils and wind tunnel experiment. Applied Mathematical Modelling 40 (3), 2002–2014.

Chen, Xiaomin, 2014. Optimization of wind turbine airfoils/blades and wind farm layouts. Applied Mathematical Modelling.

Fuglsang, Peter, Antoniou, Ioannis, Dahl, Kristian S., Madsen, Helge Aa., 1998. Wind tunnel tests of the FFA-W3-241, FFA-W3-301 and NACA 63-430 airfoils. RISO-REPORTS-RISO R.

Fuglsang, Peter, Bak, Christian, Gaunaa, Mac, Antoniou, Ioannis, 2004. Design and verification of the Risø-B1 airfoil family for wind turbines. J. Sol. Energy Eng. 126, 1002–1010.

Grasso, Francesco, 2014. Design of a family of new advanced airfoils for low wind class turbines. Journal of Physics: Conference Series 555, 012044.

Gregorek, G.M., Hoffmann, M.J., Ramsay, R.R., Janiszewska, J.M., 1995. Study of pitch oscillation and roughness on airfoils used for horizontal axis wind turbines. Technical report. National Renewable Energy Lab. (NREL), Golden, CO, United States.

Jonkman, Jason, Butterfield, Sandy, Musial, Walter, Scott, George, 2009. Definition of a 5-MW reference wind turbine for offshore system development. Technical report. National Renewable Energy Lab. (NREL), Golden, CO, United States.

Méndez, B., Munduate, X., San Miguel, U., 2014. Airfoil family design for large offshore wind turbine blades. Journal of Physics: Conference Series 524, 012022.

Ramsay, R., Janiszewska, J., Gregorek, G., 1996. Wind tunnel testing of three S809 aileron configurations for use on horizontal axis wind turbines: airfoil performance report. National Renewable Energy Laboratory NREL.

Selig, Michael S., 1995. Summary of Low Speed Airfoil Data. SOARTECH Publications.

Somers, Dan M., 2005. S833, S834, and S835 airfoils: November 2001 – November 2002. Technical report. National Renewable Energy Lab. (NREL), Golden, CO, United States.

Tangler, James L., Somers, Dan M., 1995. NREL airfoil families for HAWTs. Technical report. National Renewable Energy Lab., Golden, CO, United States.

Timmer, W.A., Van Rooij, R.P.J.O.M., 2003. Summary of the Delft University wind turbine dedicated airfoils. J. Sol. Energy Eng. 125 (4), 488–496.

Appendix C

Some wind turbine specifications

There are many active companies related to the wind industry. Many of them manufacture wind turbine parts, and some of them produce the whole turbine. These companies and wind turbine manufacturers have produced lots of different wind turbines with different technologies. The specifications of the turbines can be found in their catalogs and also on the manufacturers' websites. In the present appendix, a selection of different wind turbines is summarized. These turbines are selected just to be used for different purposes in different chapters of the book. The selected turbines cover a wide range from a few kilowatts to large-scale ones producing megawatts of electrical power.

C.1 Enercon E-16

Enercon GmbH is a German manufacturer founded in 1984 and still alive. The headquarters is located in Aurich, Germany. The products of this company include a broad range of power from 10-kilowatt to megawatt turbines.

Enercon E-16 (Enercon, 2021a) is a 55-kilowatt, 3-bladed wind turbine that is suitable for onshore purposes. The power curve of this turbine is shown in Fig. C.1. As the power curve indicates, it is a pitch-regulated turbine since its power remains unchanged after the rated velocity of $u_R = 12 \, \mathrm{m \, s^{-1}}$. Other specifications are tabulated in the following tables.

Power		Rotor	
Rated power	55.0 kW	Diameter	16.2 m
Cut-in wind speed	$3.0 \, \mathrm{m \, s^{-1}}$	Swept area	206.1 m^2
Rated wind speed	$12.0 \, \mathrm{m \, s^{-1}}$	Number of blades	3
Cut-out wind speed	$25.0 \, \mathrm{m \, s^{-1}}$	Rotor speed, max	50.0 rpm
Survival wind speed	$59.5 \, \mathrm{m \, s^{-1}}$	Tipspeed	$42 \, \mathrm{m \, s^{-1}}$
Offshore	no	Type	16WPX-TAR
Onshore	yes	Material	Glass fiber
		Manufacturer	Aerpac
		Power density 1	266.9 W m^{-2}
		Power density 2	3.7 m^2 kW^{-1}

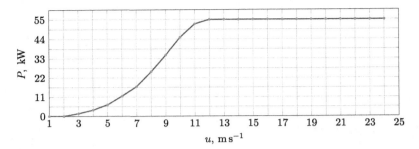

FIGURE C.1 Power curve for Enercon E-16 (Enercon, 2021a).

Gearbox	
Type	spur
Ratio	1:20

Generator	
Type	synchronous
Number	1
Speed, max	1,000.0 rpm
Voltage	690.0 V
Grid connection	IGBT
Grid frequency	50 Hz
Manufacturer	Weier

Tower	
Hub height	22.5/28.5 m
Type	lattice
Corrosion protection	galvanized

C.2 Enercon E-18

Enercon E-18 (Enercon, 2021b) is an 80-kilowatt, 3-bladed wind turbine that is suitable for onshore purposes. Its specifications and the power curve of the turbine are given in Fig. C.2 and the following tables. The power curve of the turbine reveals that it is a pitch-regulated turbine. Also, the c_p-curve of the turbine is plotted in the same figure.

Power	
Rated power	80.0 kW
Cut-in wind speed	2.5 m s^{-1}
Rated wind speed	12.0 m s^{-1}
Cut-out wind speed	25.0 m s^{-1}
Survival wind speed	67.0 m s^{-1}
Offshore	no
Onshore	yes

Rotor	
Diameter	18.0 m
Swept area	254.5 m^2
Number of blades	3
Rotor speed, max	52.0 rpm
Tipspeed	49 m s^{-1}
Type	FX-84W
Material	GFK
Manufacturer	LM Glasfiber
Power density 1	314.3 W m^{-2}
Power density 2	3.2 m^2 kW^{-1}

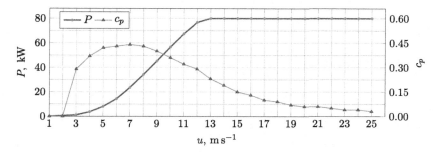

FIGURE C.2   Power curve for Enercon E-18 (Enercon, 2021b).

Gearbox			Generator	
Type	helical with parallel shafts		Type	synchronous
			Number	1
Stages	3		Speed, max	1,284.0 rpm
Ratio	1:25		Voltage	380.0 V
			Grid connection	inverters
			Grid frequency	50 Hz
			Manufacturer	Weier

Tower			Weight	
Hub height	28.5/30.1/ 32.5/34.1/ 36.7 m		Single blade	0.3 t
			Nacelle	5.0 t
			Tower, max	38.0 t
Type	concrete or steel tubes		Total weight	48.0 t
Shape	cylindrical or tripod			
Corrosion protection	painted			
Manufacturer	Pfleiderer			

C.3   Nordtank NTK 150

Nordtank Energy Group, or shortly Nordtank, was a Danish company with many different products. Nordtank NTK 150 (Nordtank, 2021a) is a 150-kilowatt turbine for onshore purposes. The details of the turbine are tabulated in the following tables. And its power curve is shown in Fig. C.3.

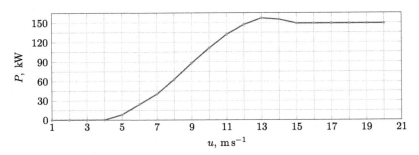

FIGURE C.3   Power curve for Nordtank NTK 150 (Nordtank, 2021a).

Power		Rotor	
Rated power	150.0 kW	Diameter	24.6 m
Cut-in wind speed	4.0 m s^{-1}	Swept area	475.0 m^2
Rated wind speed	13.5 m s^{-1}	Number of blades	3
Cut-out wind speed	25.0 m s^{-1}	Rotor speed, max	37.0 rpm
Survival wind speed	60.0 m s^{-1}	Tipspeed	48 m s^{-1}
Offshore	no	Type	LM 12 HHT
Onshore	yes	Material	GFK
		Manufacturer	LM Glasfiber
		Power density 1	315.8 W m^{-2}
		Power density 2	3.2 m^2 kW^{-1}
Gearbox		**Generator**	
Type	spur	Type	asynchronous
Stages	2	Number	1.0
Ratio	1:40	Speed, max	1,520.0 rpm
Manufacturer	Flender	Voltage	400.0 V
		Grid connection	thyristor
		Grid frequency	50.0 Hz
		Manufacturer	ABB
Tower		**Weight**	
Hub height	32.5 m	Single blade	0.8 t
Type	steel tube	Nacelle	6.0 t
Shape	conical	Tower, max	11.0 t
Corrosion protection	epoxy painted or galvanized	Total weight	23.0 t
Manufacturer	Nordtank		

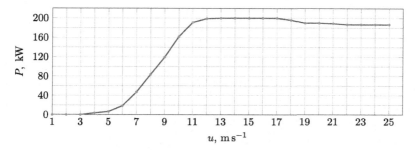

FIGURE C.4 Power curve for Nordtank NTK 200 (Nordtank, 2021b).

C.4 Nordtank NTK 200

Nordtank NTK 200 (Nordtank, 2021b) is a 200-kilowatt turbine for onshore purposes. The details of the turbine are tabulated in the following tables. And its power curve is shown in Fig. C.4.

Power		Rotor	
Rated power	200.0 kW	Diameter	28.0 m
Cut-in wind speed	4.0 m s^{-1}	Swept area	615.8 m^2
Rated wind speed	12.5 m s^{-1}	Number of blades	3
Cut-out wind speed	25.0 m s^{-1}	Type	LM
		Material	GFRP
		Manufacturer	LM Glasfiber A/S
		Power density 1	324.8 W m^{-2}
		Power density 2	3.1 m^2 kW^{-1}

Gearbox		Generator	
Type	parallel shafts	Type	induction
Stages	2	Number	1
		Voltage	690.0 V
		Grid frequency	50 Hz
		Manufacturer	ABB

Tower		Weight	
Hub height	30 m	Nacelle	12.5 t
Type	steel tube	Tower, max	14.0 t
Shape	cylindrical		

C.5 Vestas V27

Vestas Wind Systems A/S, or shortly Vestas, is a Danish manufacturer established in 1979 and still alive. The company makes onshore turbines while the offshore turbines are produced at MHI Vestas Offshore Wind A/S founded in 2014.

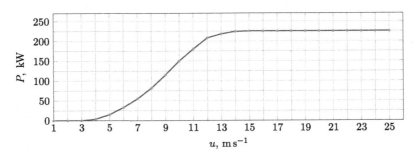

FIGURE C.5 Power curve for Vestas V27 (Vestas, 2021a).

Vestas V27 (Vestas, 2021a) is a 225-kilowatt turbine for onshore purposes. The datasheet of the turbine is given in the following tables and Fig. C.5.

Power		Rotor	
Rated power	225.0 kW	Diameter	27.0 m
Cut-in wind speed	3.0 m s^{-1}	Swept area	573.0 m^2
Cut-out wind speed	25.0 m s^{-1}	Number of blades	3
Offshore	no	Power density 1	392.7 W m^{-2}
Onshore	yes	Power density 2	$2.5 \text{ m}^2 \text{ kW}^{-1}$
Generator		**Tower**	
Grid connection	double wound asynchronous 480	Hub height	110 ft m

C.6 Vestas V29

Vestas V29 (Vestas, 2021b) is another 225-kilowatt turbine for onshore purposes. The datasheet of the turbine is given in the following tables and Fig. C.6.

Power		Rotor	
Rated power	225.0 kW	Diameter	29.0 m
Cut-in wind speed	3.5 m s^{-1}	Swept area	661.0 m^2
Rated wind speed	14.0 m s^{-1}	Number of blades	3
Cut-out wind speed	25.0 m s^{-1}	Rotor speed, max	40.5 rpm
Survival wind speed	56.0 m s^{-1}	Tipspeed	61 m s^{-1}
Offshore	yes	Type	NACA 63214
Onshore	yes	Material	GFK
		Manufacturer	Vestas
		Power density 1	340.4 W m^{-2}
		Power density 2	$2.9 \text{ m}^2 \text{ kW}^{-1}$

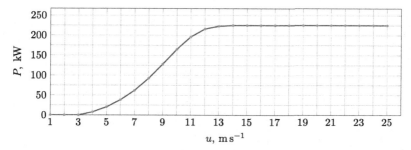

FIGURE C.6 Power curve for Vestas V29 (Vestas, 2021b).

Gearbox		Generator	
Type	spur	Type	asynchronous
Stages	2	Number	1.0
Ratio	1:25	Speed, max	1,008.0 rpm
Manufacturer	Valmet	Voltage	400.0 V
		Grid connection	thyristor
		Grid frequency	50.0 Hz
		Manufacturer	Siemens

Tower		Weight	
Hub height	31.5 / 51.5 m	Single blade	0.6 t
Type	steel	Nacelle	7.9 t
	tube/lattice	Tower, max	14.0 t
		Total weight	30.0 t

C.7 Micon M 530

Micon A/S (shortly Micon) was a Danish company established in 1983 and active until 1997.

Micon M 530 (Micon, 2021) is a 250 kW, 3-bladed turbine whose properties are summarized in the following tables and figure. Fig. C.7 indicates that this turbine is a stall-regulated one. The power curve is not flat even if the wind exceeds the rated velocity.

Power		Rotor	
Rated power	250.0 kW	Diameter	26.0 m
Cut-in wind speed	4.0 m s^{-1}	Swept area	530.0 m^2
Rated wind speed	14.5 m s^{-1}	Number of blades	3
Cut-out wind speed	25.0 m s^{-1}	Rotor speed, max	41.5 rpm
Survival wind speed	65.0 m s^{-1}	Tipspeed	56 m s^{-1}
Offshore	no	Material	fiberglas
Onshore	yes		reinforced
		Manufacturer	AeroStar
		Power density 1	471.7 W m^{-2}
		Power density 2	2.1 m^2 kW^{-1}

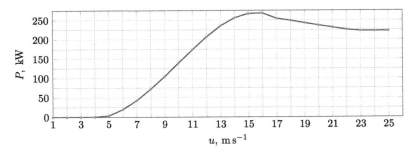

FIGURE C.7 Power curve for Micon M 530 (Micon, 2021).

Gearbox		Generator	
Type	coaxial	Type	induction
Stages	2	Speed, max	1,500.0 rpm
Ratio	1:41	Voltage	400.0 V
Manufacturer	Hansen/Flender	Grid connection	250.0
		Grid frequency	50.0 Hz
Tower		**Weight**	
Hub height	30.0 m	Rotor	4.8 t
Type	24 edged steel plate	Nacelle	8.0 t
		Tower, max	13.0 t
Shape	conical	Total weight	26.0 t
Corrosion protection	hot dip galvanized		

C.8 Enercon E-30

Enercon E-30 (Enercon, 2020) is a 300-kilowatt, 3-bladed wind turbine that is suitable for onshore purposes. The specifications and power curve of the turbine are given in the following tables and Fig. C.8.

Power		Rotor	
Rated power	300.0 kW	Diameter	29.6 m
Cut-in wind speed	2.5 m s^{-1}	Swept area	707.0 m^2
Rated wind speed	13.5 m s^{-1}	Number of blades	3
Cut-out wind speed	25.0 m s^{-1}	Rotor speed, max	48.0 rpm
Survival wind speed	70.0 m s^{-1}	Tipspeed	74 m s^{-1}
Offshore	No	Type	AERO E-30
Onshore	Yes	Material	GFK/Epoxy
		Manufacturer	Enercon
		Power density 1	424.3 W m^{-2}
		Power density 2	2.4 m^2 kW^{-1}

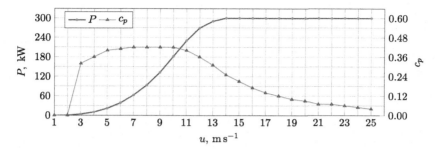

FIGURE C.8 Power curve for Enercon E-30 (Enercon, 2020).

Gearbox			Generator	
Type	with out. direct drive		Type	synchronous
			Number	1.0
			Speed, max	48.0 rpm
			Voltage	440.0 V
			Grid connection	WR
			Grid frequency	50.0 Hz
			Manufacturer	Enercon
Tower			**Weight**	
Hub height	50.0 m		Single blade	0.3 t
Type	Steel tube		Rotor	3.8 t
Shape	conical		Nacelle	12.6 t
Corrosion protection	painted		Tower, max	27.3 t
Manufacturer	CAL/SAM		Total weight	44.0 t

C.9   Nordtank NTK 400

Nordtank NTK 400 (Nordtank, 2020) is a 400-kilowatt turbine for onshore purposes. The details of the turbine are tabulated in the following tables. Also its power curve is shown in Fig. C.9.

Power		Rotor	
Rated power	400.0 kW	Diameter	35.0 m
Cut-in wind speed	4.5 m s^{-1}	Swept area	961.0 m^2
Rated wind speed	18.0 m s^{-1}	Number of blades	3
Cut-out wind speed	25.0 m s^{-1}	Rotor speed, max	35.0 rpm
Survival wind speed	53.0 m s^{-1}	Tipspeed	64 m s^{-1}
Offshore	no	Type	LM 17 HHT
Onshore	yes	Material	glassfiber reinforced polyester
		Manufacturer	LM Glasfiber A/S
		Power density 1	416.2 W m^{-2}
		Power density 2	$2.4 \text{ m}^2 \text{ kW}^{-1}$

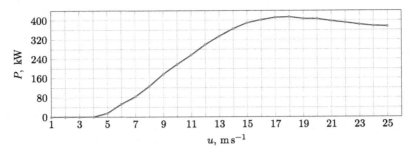

FIGURE C.9 Power curve for Nordtank NTK 400 (Nordtank, 2020).

Gearbox	
Type	spur
Stages	3
Manufacturer	Flender

Generator	
Type	induction
Number	1
Speed, max	1,800.0 rpm
Voltage	400.0 V
Grid connection	thyristor
Grid frequency	60 Hz
Manufacturer	Siemens

Tower	
Hub height	35 m
Type	steel tube
Shape	conical
Corrosion protection	painted

C.10   Vestas V39

Vestas V39 (Vestas, 2021c) is a 500-kilowatt turbine for both onshore and off-shore purposes. The datasheet of the turbine is given in the following tables and Fig. C.10.

Power	
Rated power	500.0 kW
Cut-in wind speed	4.0 m s^{-1}
Rated wind speed	15.0 m s^{-1}
Cut-out wind speed	25.0 m s^{-1}
Survival wind speed	52.0 m s^{-1}
Offshore	yes
Onshore	yes

Rotor	
Diameter	39.0 m
Swept area	1,195.0 m^2
Number of blades	3
Rotor speed, max	30.0 rpm
Tipspeed	61 m s^{-1}
Type	NACA 63-600
Material	GFK/Epoxy
Manufacturer	Vestas
Power density 1	418.4 W m^{-2}
Power density 2	2.4 m^2 kW^{-1}

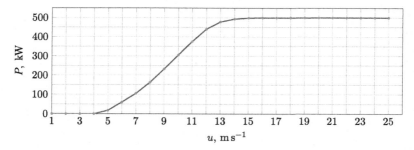

FIGURE C.10 Power curve for Vestas V39 (Vestas, 2021c).

Gearbox		Generator	
Type	spur/planetary	Type	asynchronous
Stages	3	Number	1.0
Ratio	1:50	Speed, max	1,522.0 rpm
Manufacturer	Hansen	Voltage	690.0 V
		Grid connection	thyristor
		Grid frequency	50.0 Hz
		Manufacturer	Siemens

Tower		Weight	
Hub height	40.5 / 53 m	Single blade	1.1 t
Type	steel tube	Nacelle	18.0 t
Shape	conical	Tower, max	58.0 t
Corrosion protection	painted	Total weight	85.0 t
Manufacturer	CAL/Roug		

C.11 Nordtank NTK 500/41

Nordtank NTK 500/41 (Nordtank, 2021c) is a 500-kilowatt turbine for both onshore and offshore purposes. The datasheet of the turbine is given in the following tables and Fig. C.11.

Power		Rotor	
Rated power	500.0 kW	Diameter	41.0 m
Cut-in wind speed	4.0 m s^{-1}	Swept area	1,325.0 m^2
Rated wind speed	14.0 m s^{-1}	Number of blades	3
Cut-out wind speed	25.0 m s^{-1}	Rotor speed, max	27.0 U min^{-1}
Survival wind speed	60.0 m s^{-1}	Tipspeed	58 m s^{-1}
Offshore	no	Type	LM 19.1
Onshore	yes	Material	GFK/CFK
		Manufacturer	LM Glasfiber
		Power density 1	377.4 W m^{-2}
		Power density 2	2.7 m^2 kW^{-1}

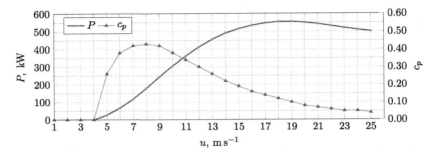

FIGURE C.11 Power curve for Nordtank NTK 500/41 (Nordtank, 2021c).

TABLE C.1 Blade characteristics of NTK 500/41.

Radius, m	Twist, degrees	Chord, m	Radius, m	Twist, degrees	Chord, m
4.5	20.0	1.63	5.5	16.3	1.597
6.5	13.0	1.540	7.5	10.05	1.481
8.5	7.45	1.420	9.5	5.85	1.356
10.5	4.85	1.294	11.5	4.00	1.229
12.5	3.15	1.163	13.5	2.60	1.095
14.5	2.02	1.026	15.5	1.36	0.955
16.5	0.77	0.881	17.5	0.33	0.806
18.5	0.14	0.705	19.5	0.05	70.545
20.3	0.02	0.265			

Gearbox	
Type	spur/planetary
Stages	3
Ratio	1:56
Manufacturer	Flender

Generator	
Type	asynchronous
Number	1.0
Speed, max	1,500.0 $U\,min^{-1}$
Voltage	690.0 V
Grid connection	thyristor
Grid frequency	50.0 Hz
Manufacturer	Siemens

Tower	
Hub height	42.1 / 50 m
Type	steel tube
Shape	conical
Corrosion protection	painted
Manufacturer	Nordtank

Weight	
Single blade	1.9 t
Nacelle	16.8 t
Tower, max	42.0 t
Total weight	65.0 t

As the power curve indicates, NTK 500/41 is a stall-regulated turbine with no pitch controller. The blades' characteristics are tabulated in Table C.1, but there is no additional data available for the used airfoils.

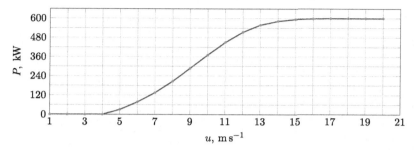

FIGURE C.12 Power curve for Vestas V44 (Vestas, 2021d).

C.12 Vestas V44

Vestas V44 (Vestas, 2021d) is a 600-kilowatt turbine for onshore purposes. The datasheet of the turbine is given in the following tables and Fig. C.12.

Power		Rotor	
Rated power	600.0 kW	Diameter	44.0 m
Cut-in wind speed	4.0 m s^{-1}	Swept area	1,521.0 m^2
Rated wind speed	16.0 m s^{-1}	Number of blades	3
Cut-out wind speed	20.0 m s^{-1}	Rotor speed, max	28.0 rpm
Survival wind speed	52.0 m s^{-1}	Tipspeed	65 m s^{-1}
Offshore	no	Type	FFA
Onshore	yes		W3/NACA
		Material	GFK/Epoxy
		Manufacturer	Vestas
		Power density 1	394.5 W m^{-2}
		Power density 2	2.5 m^2 kW^{-1}

Gearbox		Generator	
Type	planetary spur gear	Type	asynchronous
		Number	1.0
Stages	3	Speed, max	1,650.0 rpm
Ratio	1:51	Voltage	690.0 V
Manufacturer	Hansen	Grid connection	thyristor
		Grid frequency	50.0 Hz
		Manufacturer	Weier

Tower		Weight	
Hub height	40.5/53/63 m	Single blade	1.5 t
Type	steel tube	Rotor	8.4 t
Shape	conical	Nacelle	19.8 t
Corrosion protection	painted	Tower, max	58.0 t
Manufacturer	CAL/Roug	Total weight	87.0 t

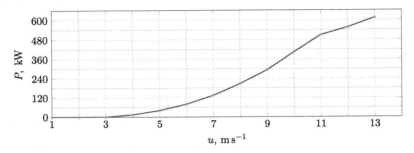

FIGURE C.13 Power curve for Enercon E-40/6.44 (Enercon, 2021c).

C.13 Enercon E-40/6.44

Enercon E-40/6.44 (Enercon, 2021c) is a 600-kilowatt, 3-bladed wind turbine that is suitable for onshore purposes. The specifications and power curve of the turbine are given in the following tables and Fig. C.13.

Power

Rated power	600.0 kW
Cut-in wind speed	2.5 m s^{-1}
Rated wind speed	12.0 m s^{-1}
Cut-out wind speed	28.0 m s^{-1}
Offshore	no
Onshore	yes

Rotor

Diameter	43.7 m
Swept area	1,521.0 m^2
Number of blades	3
Rotor speed, max	34.0 rpm
Tipspeed	78 m s^{-1}
Type	AERO E-40
Material	GFK
Manufacturer	Enercon
Power density 1	394.5 W m^{-2}
Power density 2	2.5 m^2 kW^{-1}

Gearbox

Type	with out. direct drive

Generator

Type	synchronous
Number	1
Speed, max	34.0 rpm
Voltage	440.0 V
Grid connection	WR
Grid frequency	50 Hz
Manufacturer	Enercon

Tower

Hub height	50/58/65/78 m
Type	steel tube
Shape	conical
Corrosion protection	painted
Manufacturer	SAM

Weight

Rotor	8.7 t
Nacelle	20.5 t
Tower, max	99.0 t

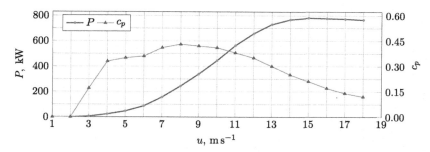

FIGURE C.14 Power curve for Wincon W755/48 (Wincon, 2021).

C.14 Wincon W755/48

Wincon West Wind A/S, shortly Wincon, was a Danish turbine manufacturer founded in 1986 and active until 2000.

Wincon W755/48 (Wincon, 2021) is a 755-kilowatt turbine for onshore installation. The datasheet of the turbine is tabulated in the following tables, and its power curve is shown in Fig. C.14.

Power		Rotor	
Rated power	755.0 kW	Diameter	48.0 m
Cut-in wind speed	4.0 m s^{-1}	Swept area	$1,810.0 \text{ m}^2$
Rated wind speed	14.0 m s^{-1}	Number of blades	3
Cut-out wind speed	25.0 m s^{-1}	Rotor speed, max	25.0 rpm
Offshore	no	Tipspeed	63 m s^{-1}
Onshore	yes	Type	LM 23.0
		Material	GFK
		Manufacturer	LM
		Power density 1	417.1 W m^{-2}
		Power density 2	$2.4 \text{ m}^2 \text{ kW}^{-1}$

Gearbox		Generator	
Type	planetary/helical	Type	asynchronous
		Number	1
Stages	3	Speed, max	1,000.0 rpm
Ratio	1:67	Voltage	690.0 V
Manufacturer	Flender	Grid connection	thyristor
		Grid frequency	50 Hz
		Manufacturer	Siemens

Tower		Weight	
Hub height	45–73 m	Rotor	21.0 t
Type	steel tube	Nacelle	21.6 t
Shape	conical		
Corrosion protection	painted		

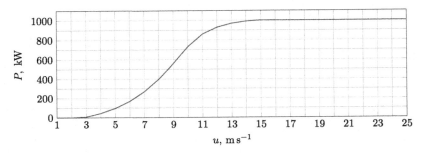

FIGURE C.15 Power curve for Vergnet GEV HP 1000/62 (Vergnet, 2021).

C.15 Vergnet GEV HP 1000/62

Vergnet Eolien SA is an active French manufacturer, with the short name Vergnet.

Vergnet GEV HP 1000/62 (Vergnet, 2021) is a 2-bladed wind turbine specifically designed for onshore installation. The turbine rated power is 1-MW, and its power curve is plotted in Fig. C.15.

Power		Rotor	
Rated power	1,000.0 kW	Diameter	62.0 m
Flexible power ratings	–	Swept area	3,019.0 m^2
		Number of blades	2
Cut-in wind speed	3.0 m s^{-1}	Rotor speed, max	23 rpm
Rated wind speed	15.0 m s^{-1}	Tipspeed	7541 m s^{-1}
Cut-out wind speed	25.0 m s^{-1}	Type	ACO 30
Survival wind speed	52.5 m s^{-1}	Material	GFK
Wind zone (DIBt)	–	Manufacturer	Vergnet
Wind class (IEC)	IIIa	Power density 1	331.2 W m^{-2}
Offshore	no	Power density 2	3.0 m^2 kW^{-1}
Onshore	yes		

Gear box		Generator	
Type	epicyclic & parallel	Type	asynchronous
		Number	1
Stages	3	Speed, max	–
Ratio	–	Voltage	690.0 V
Manufacturer	Winergy	Grid connection	IGBT
		Grid frequency	50/60 Hz
		Manufacturer	ABB

Tower		Weight	
Hub height	70 m	Single blade	4.5 t
Type	guyed wire;	Hub	1.3 t
	tubular or	Rotor	14.8 t
	lattice	Nacelle	65.0 t
Shape	conical/guyed	Tower, max	78.0 t
	wire	Total weight	158.0 t
Corrosion protection	marine		
	painting C5		
Manufacturer	Vergnet		

C.16   Siemens SWT-1.3-62

Siemens Wind Power A/S is a Danish manufacturer established in 2004 and an active turbine producer in 2021. The company produces both onshore and offshore turbines with a broad range of power.

Siemens SWT-1.3-62 (Siemens, 2021a) is a 1.3 megawatt turbine whose technical data is given in the following tables.

Power		Rotor	
Rated power	1,300.0 kW	Diameter	62.0 m
Cut-in wind speed	3.0 m s^{-1}	Number of blades	3
Rated wind speed	14.0 m s^{-1}	Rotor speed, max	19.0 rpm
Cut-out wind speed	25.0 m s^{-1}	Tipspeed	62 m s^{-1}
Offshore	no	Material	GFK
Onshore	yes	Manufacturer	LM
Gear box		**Generator**	
Stages	3	Type	asynchronous
Ratio	1:78	Number	1
Manufacturer	Winergy	Speed, max	1,500.0 rpm
		Voltage	690.0 V
		Grid connection	–
		Grid frequency	50 Hz
		Manufacturer	ABB
Tower		**Weight**	
Hub height	68–90 m	Rotor	30.0 t
Type	steel	Nacelle	50.0 t
		Tower, max	188.0 t
		Total weight	248.0 t

C.17   Vestas V80 2 MW

Vestas V80 has different variants, including 1.8- and 2-megawatt versions. Here, the specifications of the 2-megawatt turbine are tabulated. Vestas V80 (Vestas,

2021e) is designed for onshore purposes. The datasheet of the turbine is given in the following tables.

Power	
Rated power	2000.0 kW
Cut-in wind speed	4.0 m s^{-1}
Rated wind speed	16.0 m s^{-1}
Cut-out wind speed	25.0 m s^{-1}
Offshore	yes
Onshore	no

Rotor	
Diameter	80.0 m
Swept area	5,027.0 m^2
Rotor speed, max	19.1 rpm
Tipspeed	80 m s^{-1}
Power density 1	379.9 W m^{-2}
Power density 2	2.5 m^2 kW^{-1}

Gearbox	
Type	spur/planetary
Stages	3

Generator	
Type	asynchronous
Voltage	690 V

Tower	
Hub height	80 m
Type	steel tower
Shape	conical

C.18   Vestas V90 2 MW

Vestas V90 has different variants, including 1.8- and 2-megawatt versions. Here, the specifications of the 2-megawatt turbine are tabulated. Vestas V90 (Vestas, 2021e) is designed for onshore purposes. The datasheet of the turbine is given in the following tables and Fig. C.16.

Power	
Rated power	2000.0 kW
Cut-in wind speed	4.0 m s^{-1}
Rated wind speed	13.5 m s^{-1}
Cut-out wind speed	25.0 m s^{-1}
Offshore	no
Onshore	yes

Rotor	
Diameter	90.0 m
Swept area	6,362.0 m^2
Rotor speed, max	16.9 rpm
Tipspeed	80 m s^{-1}
Power density 1	314.4 W m^{-2}
Power density 2	3.2 m^2 kW^{-1}

Gearbox	
Type	one planetary stage
Stages	two helical stages

Generator	
Grid connection	permanent magnet generator

Tower	
Hub height	80–125 m
Type	tubular steel tower

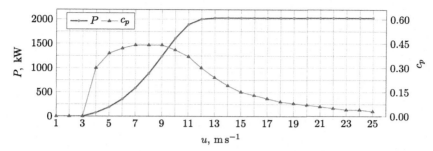

FIGURE C.16 Power curve for Vestas V90 Gridstreamer (Vestas, 2021e).

C.19   Eno Energy Eno 100

Eno Energy Eno 100 (Eno Energy, 2021) is a 2.2 MW turbine whose data is tabulated in the following tables. The power curve and c_p variation is plotted in Fig. C.17.

Power	
Rated power	2200.0 kW
Cut-in wind speed	3.0 m s^{-1}
Rated wind speed	12.0 m s^{-1}
Cut-out wind speed	25.0 m s^{-1}
Wind zone (DIBt)	II
Wind class (IEC)	IIIa
Offshore	no
Onshore	yes

Rotor	
Diameter	100.5 m
Swept area	7,926.0 m^2
Number of blades	3
Rotor speed, max	14.2 rpm
Tipspeed	75 m s^{-1}
Type	LM 49.1
Material	GRP
Manufacturer	LM Glasfiber
Power density 1	277.6 W m^{-2}
Power density 2	3.6 m^2 kW^{-1}

Gearbox	
Type	spur/planetary
Stages	3
Ratio	1:111

Generator	
Type	synchronous
Number	1.0
Speed, max	1,576.0 rpm
Voltage	690.0 V
Grid connection	IGBT
Grid frequency	50.0 Hz

Tower	
Hub height	99.0 m
Type	steel tube
Shape	conical
Corrosion protection	painted

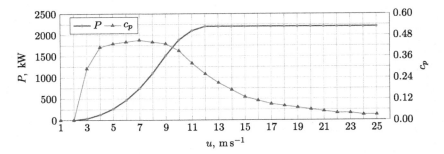

FIGURE C.17   Power curve for Eno Energy Eno 100 (Eno Energy, 2021).

C.20   Siemens SWT-2.3-93 Offshore

Siemens SWT-2.3-93 (Siemens, 2021b) is a 2.3-megawatt turbine whose technical data is given in the following tables. This turbine is suitable for offshore installation.

Power		Rotor	
Rated power	2,300.0 kW	Diameter	93.0 m
Cut-in wind speed	4.0 m s^{-1}	Number of blades	3
Rated wind speed	13.0 m s^{-1}	Rotor speed, max	16.0 rpm
Cut-out wind speed	25.0 m s^{-1}	Tipspeed	78 m s^{-1}
Survival wind speed	59.5 m s^{-1}	Material	GRE
Offshore	yes	Manufacturer	Siemens
Onshore	no		

Gear box		Generator	
Type	spur/planetary	Type	asynchronous
Stages	3	Number	1
Ratio	1:91	Voltage	690.0 V
Manufacturer	Win-ergy/Hansen	Grid frequency	50 Hz
		Manufacturer	Siemens

Tower		Weight	
Hub height	80 m	Rotor	62.0 t
Type	steel tube	Nacelle	82.0 t
Shape	cylindric/tapered tubular		
Corrosion protection	painted		

C.21   Mapna MWT2.5-103-I

Mapna Group is the largest wind turbine company in Iran. Not only do they produce 2.5 MW wind turbines, but they have also installed the largest wind farm in Iran. Therefore, they are also considered the largest wind energy operators.

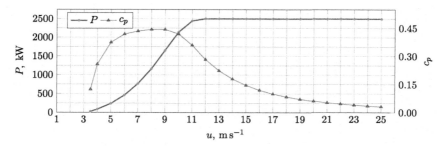

FIGURE C.18   Power curve for Mapna MWT2.5-103-I.

Power

Rated power	2500.0 kW
Cut-in wind speed	3.5 m s^{-1}
Rated wind speed	12.0 m s^{-1}
Cut-out wind speed	25.0 m s^{-1}
Wind zone (DIBt)	–
Wind class (IEC)	IIa & IIIa
Offshore	No
Onshore	Yes

Gearbox

Type	spur/planetary
Stages	3
Ratio	1:79.2
Nominal power	≈ 2.6 MW
Oil type	VG 320

Tower

Hub height	85.0 m
Type	steel tube
Shape	conical
Corrosion protection	epoxy resin surface coating

Rotor

Diameter	103.2 m
Swept area	8,332.0 m^2
Number of blades	3
Rotor speed, max	14.5 rpm
Tipspeed	89 m s^{-1}
Material	fiberglass-reinforced plastic (FRP)
Power density	334.0 W m^{-2}

Generator

Type	asynchronous
Number	1.0
Speed, max	1,310.0 rpm
Voltage	690.0 V
Grid connection	IGBT
Grid frequency	50.0 Hz

MWT2.5-103-I IEC IIA constitutes a 3-bladed, upwind, and variable-speed wind turbine. The turbine operates with a horizontal axis, 103-m rotor diameter and a nominal power output of 2,500 kW. It is designed to meet the requirements of IEC 61400-1, wind class 2A. The power curve of MWT2.5-103-I is shown in Fig. C.18.

The specifications of blades are tabulated in Table C.2. The rotor blades are made of high-grade fiberglass-reinforced plastic (FRP). No more information is available about the blades; therefore, the airfoil class ore shapes used for making the blades are unknown.

TABLE C.2 Blade geometry.

Radius r (m)	Twist angle γ (deg)	Chord c (m)	Radius r (m)	Twist angle γ (deg)	Chord c (m)
2.5	11.55	2.758	26.250	2.93	2.109
3.750	13.03	2.831	28.125	2.45	1.997
5.625	14.82	3.262	30.000	2.03	1.861
7.500	15.13	3.693	31.875	1.60	1.756
9.375	14.42	3.982	33.750	1.22	1.660
11.250	11.73	3.901	35.625	0.86	1.569
13.125	9.69	3.662	37.500	0.54	1.480
15.000	8.08	3.399	39.375	0.24	1.388
16.875	6.78	3.126	41.250	−0.04	1.295
18.750	5.72	2.866	43.125	−0.30	1.201
20.625	4.85	2.635	45.000	−0.54	1.106
22.500	4.11	2.435	46.875	−0.74	1.011
24.375	3.48	2.260	48.750	−0.86	0.917

C.22   Siemens SWT-4.0-130

Siemens SWT-4.0-130 (Siemens, 2021c) is a 4-megawatt turbine whose technical data is given in the following tables. This turbine is suitable for onshore installation.

Power

Rated power	4,000.0 kW
Flexible power ratings	–
Cut-in wind speed	5.0 m s^{-1}
Rated wind speed	12.0 m s^{-1}
Cut-out wind speed	25.0 m s^{-1}
Offshore	no
Onshore	yes

Rotor

Diameter	130.0 m
Swept area	$13,273.0 \text{ m}^2$
Number of blades	3
Rotor speed, max	13.0 rpm
Tipspeed	88 m s^{-1}
Type	B63
Material	GFK/Epoxy
Manufacturer	Siemens
Power density 1	301.4 W m^{-2}
Power density 2	$3.3 \text{ m}^2 \text{ kW}^{-1}$

Gear box

Type	planetary/helical
Stages	3
Ratio	1:119

Generator

Type	squirrel cage Induction generator
Number	1
Voltage	690.0 V
Grid connection	IGBT
Grid frequency	50 Hz
Manufacturer	Siemens

Tower

Hub height	89.5 m
Type	steel tube
Shape	conical
Corrosion protection	painted

C.23   Siemens SWT-6.0-154

Siemens SWT-6.0-154 (Siemens, 2021d) is a 6-megawatt turbine whose technical data is given in the following tables. This turbine is suitable for both onshore and offshore installation.

Power

Rated power	6,000.0 kW
Cut-in wind speed	4.0 m s^{-1}
Rated wind speed	13.0 m s^{-1}
Cut-out wind speed	25.0 m s^{-1}
Survival wind speed	70.0 m s^{-1}
Offshore	yes
Onshore	yes

Rotor

Diameter	154.0 m
Swept area	18,600.0 m^2
Number of blades	3
Rotor speed, max	11.0 rpm
Tipspeed	89 m s^{-1}
Type	B75
Material	GRE
Manufacturer	Siemens
Power density 1	7322.6 W m^{-2}
Power density 2	3.1 m^2 kW^{-1}

Gear box

Type	with out. direct drive

Generator

Type	synchronous/ PMG
Number	1
Speed, max	11.0 rpm
Voltage	690.0 V
Grid connection	IGBT
Grid frequency	50 Hz
Manufacturer	Siemens

Tower

Hub height	site-specific m
Type	cylindrical and/or tapered tubular
Shape	conical
Corrosion protection	painted
Manufacturer	Welcon

Weight

Nacelle	360.0 t

C.24   Aerodyn-8.0MW

Aerodyn Energiesysteme GmbH, or in short Aerodyn, is a German turbine manufacturer established in 1983 and still an active company in 2021.

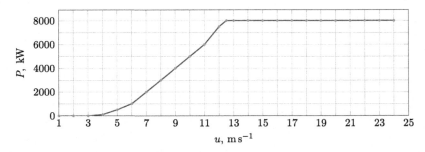

FIGURE C.19 Power curve for Aerodyn SCD 8.0/168 (Aerodyn, 2021).

Aerodyn-8.0MW (Aerodyn, 2021) is an 8-megawatt wind turbine specifically designed for both onshore and offshore installation. This turbine is 2-bladed and starts at $3.5\,\mathrm{m\,s^{-1}}$ wind. The technical data of Aerodyn-8.0MW are summarized in the following tables. The power curve of this pitch-regulated turbine is also plotted in Fig. C.19.

Power		Rotor	
Rated power	8,000.0 kW	Diameter	168.0 m
Cut-in wind speed	$3.5\,\mathrm{m\,s^{-1}}$	Swept area	22,167.0 m^2
Cut-out wind speed	$25.0\,\mathrm{m\,s^{-1}}$	Number of blades	2
Wind class (IEC)	Ib	Rotor speed, max	11.4 rpm
Offshore	Yes	Tipspeed	$100\,\mathrm{m\,s^{-1}}$
Onshore	Yes	Material	GFRP
		Power density 1	$360.9\,\mathrm{W\,m^{-2}}$
		Power density 2	$2.8\,\mathrm{m^2\,kW^{-1}}$

Gear box		Generator	
Type	planetary	Type	synchronous
Stages	2	Number	1
Ratio	1:27	Speed, max	308.0 rpm
		Grid connection	IGBT
		Grid frequency	50/60 Hz

Tower		Weight	
Hub height	100 m (or site specific) m	Nacelle	395.0 t
Type	steel tube		
Shape	conical		
Corrosion protection	painted		

C.25 AMSC wt10000dd SeaTitan

AMSC American Superconductor, or AMSC, is an American company which produces wind turbines. The company was founded in 1987 and still an active manufacturer in 2020.

AMSC wt10000dd SeaTitan (AMSC, 2021) is a 10-megawatt 3-bladed wind turbine which can be installed both onshore and offshore. This was the largest wind turbine in 2020 whose specifications are tabulated in the following tables.

Power		Rotor	
Rated power	10,000.0 kW	Diameter	190.0 m
Cut-in wind speed	4.0 m s^{-1}	Swept area	28,353.0 m^2
Rated wind speed	11.5 m s^{-1}	Number of blades	3
Cut-out wind speed	30.0 m s^{-1}	Rotor speed, max	10.0 rpm
Onshore	yes	Tipspeed	99 m s^{-1}
Offshore	yes	Power density 1	352.7 W m^{-2}
		Power density 2	2.8 m^2 kW^{-1}

Gear box		Generator	
Type	with out. direct drive	Type	HTS synchronous
		Number	1
		Speed, max	10.0 rpm
		Voltage	12,000.0 V
		Grid connection	IGBT
		Grid frequency	50/60 Hz
		Manufacturer	AMSC

Tower	
Hub height	125 m
Type	steel tube
Shape	conical
Corrosion	protection painted

C.26   Summary

The specifications of some selected turbines are given in this appendix. The data is used in different parts of the present book. It is understandable that many different important aspects of a wind turbine are not provided here. For example, the noise level, power curve for nonstandard air density, sensitivity of the turbine to different parameters such as icing, surface roughness, etc., and many other parameters are not considered. The data of the present appendix are just for some examples of a vast range of products.

References

Aerodyn, 2021. Aerodyn SCD 8.0/168. https://en.wind-turbine-models.com/turbines/1116-aerodyn-scd-8.0-168. (Accessed 12 May 2021).

AMSC, 2021. AMSC wt10000dd SeaTitan. https://en.wind-turbine-models.com/turbines/425-amsc-wt10000dd-seatitan. (Accessed 12 May 2021).

Enercon, 2020. Enercon E-30. https://en.wind-turbine-models.com/turbines/376-enercon-e-30#powercurve. (Accessed 12 May 2021).

Enercon, 2021a. Enercon E-16. https://en.wind-turbine-models.com/turbines/605-enercon-e-16# powercurve. (Accessed 12 May 2021).

Enercon, 2021b. Enercon E-18. https://en.wind-turbine-models.com/turbines/353-enercon-e-18# powercurve. (Accessed 12 May 2021).

Enercon, 2021c. Enercon E-40/6.44. https://en.wind-turbine-models.com/turbines/68-enercon-e-40-6.44#powercurve. (Accessed 12 May 2021).

Eno Energy, 2021. Eno Energy Eno 100. https://en.wind-turbine-models.com/turbines/651-eno-energy-eno-100#powercurve. (Accessed 12 May 2021).

Micon, 2021. Micon M 530. https://en.wind-turbine-models.com/turbines/257-micon-m-530# powercurve. (Accessed 12 May 2021).

Nordtank, 2020. Nordtank NTK 400. https://en.wind-turbine-models.com/turbines/1353-nordtank-ntk-400. (Accessed 12 May 2021).

Nordtank, 2021a. Nordtank NTK 150. https://en.wind-turbine-models.com/turbines/166-nordtank-ntk-150. (Accessed 12 May 2021).

Nordtank, 2021b. Nordtank NTK 200. https://en.wind-turbine-models.com/turbines/1355-nordtank-ntk-200. (Accessed 12 May 2021).

Nordtank, 2021c. Nordtank NTK 500/41. https://en.wind-turbine-models.com/turbines/384-nordtank-ntk-500-41. (Accessed 12 May 2021).

Siemens, 2021a. Siemens SWT-1.3-62. https://en.wind-turbine-models.com/turbines/1457-siemens-swt-1.3-62. (Accessed 12 May 2021).

Siemens, 2021b. Siemens SWT-2.3-93 Offshore. https://en.wind-turbine-models.com/turbines/50-siemens-swt-2.3-93-offshore. (Accessed 12 May 2021).

Siemens, 2021c. Siemens SWT-4.0-130. https://en.wind-turbine-models.com/turbines/601-siemens-swt-4.0-130. (Accessed 12 May 2021).

Siemens, 2021d. Siemens SWT-6.0-154. https://en.wind-turbine-models.com/turbines/657-siemens-swt-6.0-154. (Accessed 12 May 2021).

Vergnet, 2021. Vergnet GEV HP 1000/62. https://en.wind-turbine-models.com/turbines/435-vergnet-gev-hp-1000-62#powercurve. (Accessed 12 May 2021).

Vestas, 2021a. Vestas V27. https://en.wind-turbine-models.com/turbines/9-vestas-v27. (Accessed 12 May 2021).

Vestas, 2021b. Vestas V29. https://en.wind-turbine-models.com/turbines/273-vestas-v29. (Accessed 12 May 2021).

Vestas, 2021c. Vestas V39. https://en.wind-turbine-models.com/turbines/383-vestas-v39# powercurve. (Accessed 12 May 2021).

Vestas, 2021d. Vestas V44. https://en.wind-turbine-models.com/turbines/272-vestas-v44# powercurve. (Accessed 12 May 2021).

Vestas, 2021e. Vestas V90 Gridstreamer. https://en.wind-turbine-models.com/turbines/248-vestas-v90-gridstreamer. (Accessed 12 May 2021).

Wincon, 2021. Wincon W755/48. https://en.wind-turbine-models.com/turbines/441-wincon-w755-48. (Accessed 12 May 2021).

Appendix D

Sample wind farms

There are many different wind farms all around the world. Also, many researchers have simulated theoretical wind farms to investigate different physical concepts or compare the results of their simulations. In the present appendix, some selected wind farms are introduced. These examples are suitable for any desirable investigations.

D.1   A 4-in-a-row wind farm

The first theoretical wind farm is shown in Fig. D.1. The wind farm was introduced by Ghadirian et al. (2014) in a simple form. Here, the farm is extended to become more applicable.

As can be seen, there are four identical wind turbines located in a row, extending from the west to the east. The diameter of each turbine is considered to be D_0 and is used for normalizing the distances. One can adjust the distance between each turbine simply by varying the x factor.

As illustrated, the wind blows from the west, or according to the standard notations, the wind angle is 270°. However, this is not a fixed parameter. It means that, for simulation of the wind farm, any wind direction can be imposed.

The wind farm is a theoretical example; hence it does not belong to any place. Therefore, it can be simulated using any available wind data.

D.2   A 4 × 4 wind farm

Ghadirian et al. (2014) also introduced a 4 × 4 wind farm, which was an extension of the 4-in-a-row farm. Here, their proposed model is modified to become more practical.

The matrix form wind farm is illustrated in Fig. D.2. The farm contains 16 identical turbines with a diameter of D_0. Their horizontal distance can be adjusted by the x_D factor and their vertical distance by y_D. For example, if we

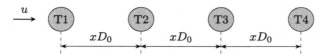

FIGURE D.1   A wind farm with 4 turbines in a row.

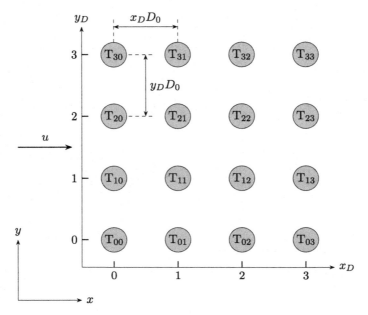

FIGURE D.2   Wind blowing with $\delta = 0°$.

set $x_D = 5$ and $y_D = 7$, the horizontal distances would be $5D_0$, and their corresponding vertical distances would be $7D_0$. Thus, the horizontal and vertical distances can be differently set up.

The turbines are named by two indices, one for x_D and the other for y_D directions. These indices are enumerated such that they are consistent with their x- and y-positions. For example, T_{01} is located at $x = 0$, and $y = 1y_D D_0$. As another example, T_{23} is located at $x = 2x_D D_0$, and $y = 3y_D D_0$. Such a notation facilitates the positioning of the turbines. The axis labels shown in the figure indicate the indices.

The wind can blow from any direction. What is shown in the figure is a West wind that blows from the west to the east. This direction, of course, is not the only possible one. In a typical wind farm, the wind direction can be any desired value.

Again, this is a theoretical wind farm which does not belong to any specific place. Hence, it can be modeled with any available or desired wind data.

D.3   Horns Rev wind farm

Horns Rev wind farm was the first large-scale wind farm (Gaumond et al., 2014; Naderi and Torabi, 2017). This farm is located in the North Sea in Danish waters. The farm was built in three different phases.

For the construction of the first phase of the farm, Vestas V80 turbines were used. Each turbine produces 2 MW of power. The first phase, Hornes Rev 1, ben-

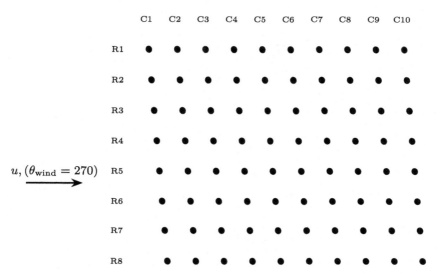

FIGURE D.3 Horns Rev 1 offshore wind farm.

efited from 80 turbines that were installed by monopile technology. The second phase used 91 Siemens SWT 2.3-93 turbines with a total capacity of 209 MW. Finally, the third phase used 49 MHI Vestas V164-8.3 MW with a total capacity of 406.7 MW.

Fig. D.3 shows the Horns Rev 1 wind farm, which consists of 10 columns in 8 rows. As mentioned, the turbines are Vestas V80, whose diameters are 80 m, and the hub height is 70 m. The horizontal and vertical distance between each turbine is 7D. As shown in the figure, the rows are horizontal, but the columns have an inclination angle of about 7°.

The wind farm is located in the North Sea, around 15 km off the westernmost point of Denmark. Hence the whether data of the place must be used for any simulation and comparison.

D.4 Aghkand wind farm in Iran

Aghkand is the name of a city in the northeastern Iran. It belongs to the Mianeh county of Eastern Azerbaijan province. The location of the town is shown in Fig. D.4a. On November 22, 2019, Mapna group opened a 50-megawatt wind farm near the city of Aghkand. The location of the wind farm and the turbines are shown in Fig. D.4b. Mapna currently owns the largest wind turbine manufacturing facilities in Iran and also is known as the largest Iranian owner and operator of wind energy.

The Aghkand wind farm contains 20 wind turbines made by Mapna. The specifications of the 2.5-megawatt turbines are summarized in Appendix C.21.

(a) The location of the farm

(b) Satellite image indicating the location of the turbines

FIGURE D.4 The map of Aghkand wind farm.

(a) Mapna 2.5 MW turbines used in Aghkand farm

(b) A view of the farm and the installation terrain

FIGURE D.5 Some photos of Aghkand wind farm.

These turbines (shown in Fig. D.5a) are installed in the field as shown in Fig. D.5b.

References

Gaumond, M., Réthoré, P.-E., Ott, Søren, Pena, Alfredo, Bechmann, Andreas, Hansen, Kurt Schaldemose, 2014. Evaluation of the wind direction uncertainty and its impact on wake modeling at the horns rev offshore wind farm. Wind Energy 17 (8), 1169–1178.

Ghadirian, A., Dehghan, M., Torabi, F., 2014. Considering induction factor using BEM method in wind farm layout optimization. Journal of Wind Engineering and Industrial Aerodynamics 129, 31–39.

Naderi, Shayan, Torabi, Farschad, 2017. Numerical investigation of wake behind a HAWT using modified actuator disc method. Energy Conversion and Management 148, 1346–1357. https://doi.org/10.1016/j.enconman.2017.07.003.

Appendix E

Optimization methods

Mehrzad Alizadeh
Department of Mechanical Engineering, Graduate School of Engineering, Osaka University, Osaka, Japan

This appendix provides a theoretical background to optimization techniques by reviewing the optimization concept in reference to the theory and methods of several well-known algorithms. In general, optimization refers to the procedure of arriving at the "best" or "optimum" result (the latter of which is a more technical word) amid all probable outcomes under specific conditions (if any). It is a branch of mathematics that focuses on the quantitative study of optima and the methods of discovering them. An optimization problem could be expressed as a maximization or minimization process of a mathematical function. In practice, the function has a given set of adjustable parameters. In optimization literature, the optimized function and the adjustable parameters are known as *objective function* and *decision variables*, respectively. The objective function of a maximization problem is usually called *fitness*; while in the case of a minimization problem, it is commonly known as the *cost function*. All the decision variables together form the dimension (d) of the problem space. The ultimate goal of any optimization procedure is to find a set of decision variables by which either the minimum or maximum feasible value of the objective function could be obtained such that a set of predetermined conditions are met. System cost minimization, system efficiency maximization, and production capacity maximization could be named as some common optimization problems in diverse engineering applications. Nevertheless, optimization is not limited to engineering practices. A variety of fields, like economics, management, planning, physics, and chemistry, use this mathematical tool to reach the optimum solutions. The optimization process begins with the formulation of the objective by converting the problem statements to a mathematical function and defining whether the function should be maximized or minimized. From a mathematical standpoint, any maximization problem can be defined as a corresponding minimization problem using the negative of the objective function, and vice versa. Next, the constraints of the decision variables' changes, which specify the search space, have to be expressed as inequality or equality formulas.

Example E.1. A factory uses two different raw materials, $R1$ and $R2$, to produce three different paints, $P1$, $P2$, and $P3$. The information about the profit margin of each product, the required amount of raw materials for producing each product, and the maximum yearly availability of raw materials are stated in the following table.

Raw material	Tons of raw material per ton of paint			Maximum yearly availability (ton)
	$P1$	$P2$	$P3$	
$R1$	5	3	1	6500
$R2$	3	4	5	9200
Profit margin ($ per ton)	7000	8000	5000	

The results of a market survey showed that the maximum yearly demand for $P2$ is 1500 tons. Moreover, due to some limitations in the production line, the total yearly produced amount of $P1$ and $P3$ cannot exceed 1000 tons. The factory's managers are trying to adjust the production configuration in a way to achieve the maximum possible under the conditions obligated by the market demand and limitations of the production line.

Answer. Proper adjustment of the production capacity of each paint, in this case, is an optimization problem in which the ultimate goal is the maximization of the profit. Supposing x_1, x_2, and x_3 as the yearly production amount in tons of $P1$, $P2$, and $P3$, respectively, the total yearly profit of the factory can be expressed as

$$f = 7000x_1 + 8000x_2 + 5000x_3.$$

The optimization objective is to maximize the value of f. Besides, based on the results of the market survey, it is known that x_2 is confined to the maximum value of 1500 tons per year; otherwise, the extra produced amount of $P2$ cannot be sold in the market. This condition, which is dictated by the market demand, can be given by an inequality constraint as

$$x_2 \leq 1500.$$

In addition, the constraint obligated by the practical conditions of the production line is an inequality type that is stated as

$$x_1 + x_3 \leq 1000.$$

Knowing the amount of production capacity of each paint (x_1, x_2, and x_3), the consumed amounts of both raw materials (C_{R_1} and C_{R_2}) are calculated by the following equations:

$$C_{R_1} = 5x_1 + 3x_2 + x_3,$$
$$C_{R_2} = 3x_1 + 4x_2 + 5x_3.$$

Due to the limitations in the supply of the raw materials, the consumed amount of these materials cannot surpass their maximum yearly availability.

Hence, two other restrictions are imposed as follows:

$$5x_1 + 3x_2 + x_3 \leq 6500,$$
$$3x_1 + 4x_2 + 5x_3 \leq 9200.$$

Evidently, none of the production capacities take negative values. Therefore, the nonnegativity constraints are stated as: $x_1 \geq 0$, $x_2 \geq 0$, and $x_3 \geq 0$. The complete formulation of this optimization problem is given by:

$$\underset{x_1, x_2, x_3}{\text{maximize}} \quad f = 7000x_1 + 8000x_2 + 5000x_3$$

$$\text{subject to} \qquad\qquad x_2 \leq 1500,$$
$$x_1 + x_3 \leq 1000,$$
$$5x_1 + 3x_2 + x_3 \leq 6500,$$
$$3x_1 + 4x_2 + 5x_3 \leq 9200,$$
$$x_1, x_2, x_3 \geq 0.$$

□

Over the past few centuries, from the time the first optimization methods emerged to the present time, a wide range of algorithms have been developed, each of which has its pros and cons. However, the major breakthroughs within this field began in the 20th century, when the first digital electronic computing machines were introduced. The optimization methods are classified into two main categories, namely classical (deterministic) and heuristic methods. The former category is beneficial for optimization problems whose search spaces are linear and is suitable for differentiable objective functions. The classical methods' requirement for calculation of first- or second-order derivatives of the function, which increases the complexity of optimization procedure, on the one hand, and their proneness of local optima entrapment in problems with non-linear search spaces, on the other hand, make them inefficient techniques for solving real-world optimization problems. Linear and nonlinear programming, the simplex method, and Newton's steepest descent method are some of the well-known methods of the classical category. On the contrary, unconventional heuristic methods are designed to look for an acceptable approximation of the global optimum for an immediate aim. That is why they are sometimes called *approximate methods* as well. A group of heuristic methods, named metaheuristic algorithms, can be applied to a broad range of problems with only minor modifications due to their problem-independent nature. Most of the metaheuristic algorithms are inspired by the physical phenomena, the behaviors of animals, or the basic concepts of the evolution process. The algorithms of this group are split into five general classes based on inspiration source:

1. Evolutionary algorithms,
2. Swarm intelligence algorithms,

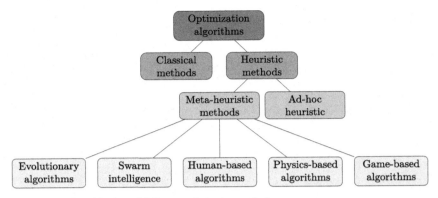

FIGURE E.1 Classification of different optimization methods.

3. Human-based algorithms,
4. Physics-based algorithms, and
5. Game-based algorithms.

A complete schematic of optimization methods classification is illustrated in Fig. E.1. Primarily, metaheuristic methods try to reach the optimal solution by generating some random solutions and leading them in a heuristic way to the best solution. Using a stochastic approach makes these algorithms simple, flexible, and gradient-free. Genetic algorithm (GA), differential evolution (DE) algorithm, particle swarm optimization (PSO) algorithm, crow search algorithm (CSA), simulated annealing (SA), teaching–learning-based optimization (TLBO), dice game optimizer (DGO), and tabu search (TS) algorithm are some of the popular metaheuristic methods, to mention a few.

The evolutionary algorithms, like GA, are a group of metaheuristic optimization methods inspired by the biological evolution described in Darwinian evolutionary theory. According to this theory, the natural selection of species is controlled by the *survival of the fittest* principle. Different operators, such as cross-over and mutation, are used in these algorithms to generate enhanced solutions in the subsequent generations. Swarm intelligence algorithms are originated from the collective behavior of animals in their communities. These types of algorithms are established based on the information sharing between the populations of a swarm. The next category of metaheuristic algorithms, human-based methods, is derived from humans' social behavior and their mutual interplay. For instance, the TLBO algorithm is developed in accordance with learner–teacher and learner–learner interactions in the learning process. Physics-based algorithms have thrived in line with the governing rules of physical processes, such as thermodynamics and Newton's gravitational laws. Finally, game-based methods are a bunch of algorithms that emerged based on the rules of different games. Considering the relative simplicity of metaheuristic algorithms and their wide applicability to various real-world problems, some of the recent methods of this kind are introduced in the following sections. Although the fundamental concepts of algorithms are described briefly, the main focus is the implementation procedure of each algorithm.

E.1   Crow search algorithm

Crow search algorithm, shortly known as CSA, is a nature-inspired, metaheuristic optimization method proposed by Askarzadeh (2016). This algorithm is designed in accordance with the natural characteristics of crows. Crows are medium-size, black, aggressive birds whose retentive memory, capability of communicating with their group members, and tool-making ability made them well-known as some of the smartest animals. Their brain-to-body size ratio is large, and they are commonly found in different countries around the world. Using their memorizing capability, crows are able to hide extra food in the foraging process and retrieve it after a long time. In addition, to protect their food's hiding place against other crows, they try to fool any greedy chaser crow, trying to steal their food by finding and attacking its hiding place. By exploiting their experience in the role of a thief crow, they can assess and determine the behavior of any potential pilferer when being threatened. In such a case, when the chased crow gets aware of being followed by another crow, the former goes to a place of the environment other than its hiding place to prevent losing the concealed food. The explained concept of crow's behavior in the foraging process is the basis of CSA development, which made it easy to understand and implement the technique. In short, the main characteristics of crows from which CSA is originated are:

1. Crows live together,
2. Crows remember the position of their food hiding place,
3. Crows try to steal each other's food by following them and finding its hiding place, and
4. Crows try to protect their food's hiding place by deceiving other crows who are chasing them.

The crow, food source, environment, food source quality, and the best food source are corresponding to a search agent, potential solution, problem search space, objective function value (fitness), and best global solution, respectively, in the case of the CSA algorithm. Based on the aforementioned equivalency, the primary principles of the CSA are taken from the natural behaviors of crows. Considering the fact that the crows live together, and each of them remembers the place where it hides food, several agents are involved in the CSA, whose hiding place is stored in their memory. From an optimization point of view, the hiding place of each crow at each iteration is the best solution (position) found by this search agent so far. As explained before, any crow tries to steal others' food by following them. In case the followed crow is aware of being chased by another crow, the first crow fools the second one to prevent being robbed. In the same way, each agent's position in the search space is updated according to the hiding place of another agent only if the chased agent is not aware of being trailed.

First, to implement the CSA, all the problem decision variables, the search space of each decision variable, the maximum number of algorithm iterations, and the number of optimization agents (crows) should be specified for all the agents. Additionally, two other parameters, including flight length and awareness probability, must be identified for all agents. The flight length parameter specifies how far the chaser crow (say, i) moves forward in the direction of the hiding position of the chased crow (say, j) and whether it can reach the target position or would go beyond it ($i \neq j$). This concept is graphically shown in Fig. E.2. The awareness probability, chosen randomly, determines if crow i can successfully find the hiding position of crow j, which is the best individual solution found by agent j so far. If crow j becomes aware of being chased, it will change its position to a random place to deceive the chaser.

After defining the problem, the initial position of each crow should be specified randomly, and the fitness of each initialized solution has to be calculated according to the problem objective function. Considering that the crows have no searching history at the initialization step yet, the hiding position of all agents is considered the same as their initial position. In the following iterations (say, $iter$), the position of any agent i, who is chasing a random agent j to find its hiding position, is updated as:

$$X_i^{iter+1} = X_i^{iter} + r_i \times l_i^{iter} \times (H_j^{iter} - X_i^{iter}) \qquad \text{if } r_k \geq AP_j^{iter},$$
$$= \text{random position} \qquad \text{if } r_k < AP_j^{iter}, \qquad (E.1)$$

where r_i and r_k are uniformly distributed random numbers in the interval $[0, 1]$, AP_j^{iter} is the awareness probability of crow j, and superscript $iter$ indicates the current iteration number. In this optimization algorithm, crow j is only aware of being chased by crow i if a randomly generated number r_k is less than its awareness probability. In such a case, the position of the crow i should be updated randomly. The fitness value of the agent i should be calculated based on the given objective function, and in case the new position gives a better fitness compared with the existing hiding position of this agent, its hiding position is also updated. The position updating process has to be performed for all search agents in each iteration. After performing the position updating procedure for all the crows, the global best solution is redetermined based on the newly obtained solutions. This process is repeated until the algorithm termination criterion (for instance, reaching the maximum iteration number) is met.

Example E.2. The sphere function is a widely used benchmark problem in the optimization field. The function is defined according to the following equation, where $d = 4$ is the problem dimensions size and $X = [x_1, x_2, \ldots, x_d]$ is a vector of decision variables. Assuming that each decision variable ranges from -5 to

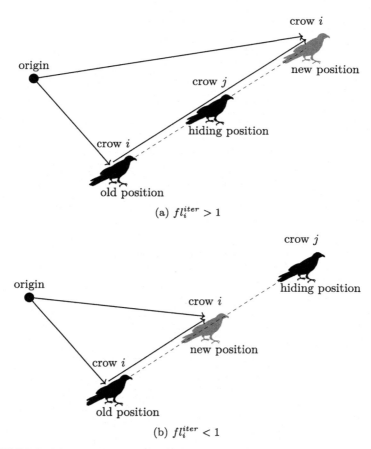

(a) $fl_i^{iter} > 1$

(b) $fl_i^{iter} < 1$

FIGURE E.2 Position updating procedure of CSA.

5, the objective is to find the minimum sphere function value of

$$f_{\text{sphere}} = \sum_{n=1}^{d} x_n^2.$$

Answer. Intuitively, we know that the minimum value of this function is zero. To solve this optimization problem through CSA, first, the number of crows (search agents) and the maximum iterations' number should be specified. Let us assume the number of crows to be 10 and the maximum iterations' number to be 100 (only the results of one iteration are demonstrated here). Moreover, the flight length and awareness probability of all crows are considered to be $fl = 0.8$ and $AP = 0.4$, respectively. Next, a population of the 10 crows has to be positioned randomly. The obtained population is enlisted in the following table.

Crows	Initial position				Fitness
	x_1	x_2	x_3	x_4	
1	4.40123	−4.96063	2.55058	0.152745	50.5075
2	−3.29859	4.7058	−4.41252	−0.28062	52.5744
3	−2.38945	−3.78414	−0.85528	2.38121	26.4308
4	−3.16706	−4.69512	−3.52809	2.02506	48.6227
5	−3.20551	−2.56584	−2.03207	2.10837	25.4334
6	−1.83859	2.48527	3.34864	0.402997	20.9328
7	1.863	−0.05875	2.3513	−1.86941	12.4975
8	−2.07053	3.62087	−3.21253	−4.94232	52.1447
9	−3.53786	0.023042	1.75436	−2.89209	23.9589
10	1.04144	−0.6502	−1.1803	−1.94082	**6.66728**

As it can be seen from this table, the minimum fitness, which is 6.66728, is achieved by crow $n = 10$. In the next step, the position of all crows should be updated one by one. Staring with crow $n = 1$ and choosing a random crow as the chased agent (let us assume $n = 4$), the position of the first crow has to be updated according to Eq. (E.1). Two other parameters, including r_i and r_k, have to be selected randomly. We assume these factors to be $r_i = 0.1$ and $r_k = 0.3$ (chosen randomly). Since r_k is smaller than the awareness probability of the chased crow, the chased agent can fool the first crow and take it to a random position. Hence, the position of the crow $n = 1$ is updated randomly. The same process should be repeated for other crows as well. The updated position of all crows, the selected chased crow, and the value of r_i and r_k are showed in the table below.

Crow	Chased crow	r_i	r_k	Updated position				Fitness
				x_1	x_2	x_3	x_4	
1	4	0.1	0.3	1.88223	−4.89715	−1.54347	−1.71072	32.8337
2	4	0.9	0.8	−3.20389	−2.06286	−3.77573	1.37947	30.6794
3	2	0.1	0.9	−2.46218	−3.10495	−1.13986	2.16826	21.7037
4	9	0.1	0	2.3571	2.12516	−0.22263	3.99106	26.0503
5	9	0.5	0.1	−3.93124	1.68691	2.96838	1.88009	30.6464
6	5	0.8	0.3	−3.80062	−2.21366	−1.59597	−0.25254	21.9559
7	4	0.6	0	0.558031	3.48537	4.0701	2.20054	33.8673
8	1	0.2	0.6	−1.03505	2.24783	−2.29043	−4.12711	28.4032
9	4	1	0.9	−3.24122	−3.75149	−2.4716	1.04163	31.773
10	5	0.6	0.7	−0.9971	−1.56971	−1.58915	0.002789	**5.9836**

According to the data of this table, the second crow is again chasing the crow $n = 4$. However, since the value of r_k is greater than the awareness probability ($AP = 0.4$), the position of this crow is updated according to the hiding position of crow $n = 4$ by means of Eq. (E.1). For instance, the new value of x_1 is computed as

$$x_1 = −3.29859 + [0.9 × 0.8 × (−3.16706 + 3.29859)] = −3.20389.$$

The same calculations shall be done for all crows to get the updated position of all search agents. Then, the hiding position of each crow has been updated just in case the new fitness is better than the existing hiding fitness. The obtained global best fitness is 5.9836. This paradigm is repeated until reaching the maximum number of iterations. The global best fitness achieved at the last iteration is the optimal fitness value. □

It is noteworthy that despite the good performance of CSA in solving optimization problems, two significant shortcomings restrict its efficiency. The first problem refers to its methodology in dealing with cases where the chased crow is aware of being followed by another crow. In these cases, the position of the chaser is updated randomly, which might result in producing weak solutions. Moreover, according to the principles of this algorithm, the search agents are not obligated to follow the found optimum solution so far (global best solution) for updating their position at each iteration. These two factors are mainly known as defects of this optimization technique.

E.2 Whale optimization algorithm

In 2016, Mirjalili and Lewis (2016) proposed a nature-inspired optimization method called whale optimization algorithm (WOA) based on the foraging behavior of a species of baleen whale. These whales are the so-called *humpback whales*, and their social behavior during the hunting process, known as a bubble-net feeding strategy, is the basis of this optimization method.

Due to the slowness of humpback whales, they are not able to catch and consume fish in a chasing process. Instead, they employ a unique trapping technique by making bubbles. By moving in a spiral-shaped path and making baubles simultaneously, a humpback whale (or a group of them) is able to guide a school of fish to the surface of the ocean, where it attacks the school (see Fig. E.3).

WOA is a metaheuristic, swarm-based algorithm, and like any other population-based method, its search process includes exploration and exploitation stages. The whole optimization process implements three mechanisms:

1. Encircling the prey,
2. Bubble-net attacking (exploitation phase), and
3. Searching for the pray (exploration phase).

The encircling mechanism updates the position of each search agent based on the location of the best position found by search agents so far. Actually, this method considers that the current best position is the optimum position, or at least, it is close to the target solution. Therefore, the mathematical representation of the encircling mechanism is as follows:

$$\vec{X}(t+1) = \vec{X}^*(t) - \vec{A}.\vec{D}, \tag{E.2}$$

in which

$$\vec{D} = |\vec{C}.\vec{X}^*(t) - \vec{X}(t)|. \tag{E.3}$$

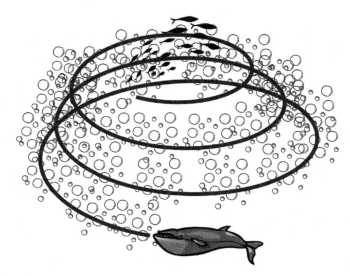

FIGURE E.3 Bubble-net foraging technique of a humpback whale.

In Eqs. (E.2) and (E.3), t is the iteration counter, while $\vec{X}$ and $\vec{X}^*$ are the position vectors of the searching agent and the best solution found so far. In addition, $\vec{A}$ and $\vec{C}$ are two coefficient vectors defined as:

$$\vec{A} = 2\vec{a}.\vec{r} - \vec{a} \tag{E.4}$$

and

$$\vec{C} = 2\vec{r}, \tag{E.5}$$

where $\vec{a}$ is a vector in which the values of its arrays decrease linearly from 2 to 0 over the course of iterations. Also, $\vec{r}$ is a random vector in [0, 1]. In Eqs. (E.2) to (E.5), $|\cdot|$ is absolute value operator and . is element-by-element multiplication operator (or inner vector product). Furthermore, a method inspired by the bubble-net attacking strategy used by humpback whales is used for exploitation purposes. Searching for the optimum solution is performed in the exploitation stage using two mechanisms: (a) shrinking encircling and (b) spiral updating position. The former mechanism works by decreasing the value of $\vec{a}$, which results in decline of the alteration range of $\vec{A}$. The spiral updating position mechanism tries to move the search agents in a helix-shaped trajectory toward the prey (current best solution). From the mathematical point of view, this movement is given by

$$\vec{X}(t+1) = \vec{D}'.\exp(bl).\cos(2\pi l) + \vec{X}^*(t), \tag{E.6}$$

where b is a shape coefficient, l is a random number in $[-1, 1]$, and

$$\vec{D}' = |\vec{X}^*(t) - \vec{X}(t)|. \tag{E.7}$$

Considering the fact that humpback whales use both mechanisms simultaneously, in the WOA algorithm, 50% probability is considered for selecting either position updating procedure. The aforementioned process is mathematically expressed as follows:

$$\vec{X}(t+1) = \begin{cases} \vec{X}^*(t) - \vec{A}.\vec{D} & \text{if } p < 0.5, \\ \vec{D}'.\exp{(bl)}.\cos{(2\pi l)} + \vec{X}^*(t) & \text{if } p \geq 0.5, \end{cases} \tag{E.8}$$

in which p denotes the probability and is chosen randomly in [0, 1]. In addition to the exploitation phase, according to WOA, any search agent tries to look for the prey in a more stochastic way based on the position of other whales. This process is called the exploration phase, which takes place when the absolute magnitude of coefficient vector $\vec{A}$ is greater than unity ($|\vec{A}| > 1$) and is an attempt for global search in the total search space. In this case, the position of a search agent is updated according to the position of another randomly chosen agent. The new position of an agent-based on this updating scheme is given by Eqs. (E.9) and (E.10):

$$\vec{D} = |\vec{C}.\vec{X}_{\text{rand}}(t) - \vec{X}(t)|, \tag{E.9}$$

$$\vec{X}(t+1) = \vec{X}_{\text{rand}}(t) - \vec{A}.\vec{D}. \tag{E.10}$$

To put it together, the respective steps of implementation of the WOA algorithm are as follows. First, the number of search agents, humpback whales, should be specified. Next, the position of each agent should be randomly initialized, and its fitness should be calculated. Having the fitness values of all search agents, the current best position and fitness could be determined. Afterwards, the iterations start. The values of different parameters, including a, A, C, l, and p, should be specified for each search agent. In the next step, the position of each search agent should be updated using Eqs. (E.2), (E.6), or (E.10). If the value of probability parameter, p, and the absolute magnitude of $\vec{A}$ are less than 0.5 and 1, respectively, Eq. (E.2) should be used for updating the position of a search agent. However, if $p < 0.5$ and $|\vec{A}| \geq 1$, the position of a search agent should be updated according to Eq. (E.10). Finally, in a case $p > 0.5$, Eq. (E.6) should be employed to get a new position. After updating the position of all agents, the fitness values should be calculated. In case there is a better solution compared to the current best solution, the best solution should be replaced. This process is repeated till meeting the termination criteria.

Example E.3. Ackley function is a well-known test function widely used to assess the performance and characteristics of an optimization algorithm. The two-dimensional form of this nonconvex function contains two independent variables and is given by

$$f(x, y) = -20\exp[-0.2\sqrt{0.5 \times (x^2 + y^2)}]$$

$$- \exp[0.5 \times (\cos(2\pi x) + \cos(2\pi y)] + e + 20,$$

in which e is Euler's number. Also, x and y are the two independent variables. The objective is to find the global minimum value of this function.

Answer. We consider that the search space for both x and y is $[-5, 5]$. To begin the optimization process, we assume the number of search agents (whales) and iterations to be 10 and 300, respectively. By performing the initialization, the position of each agent and their fitness values would be specified. These data are given in the following table.

Whales	Initial position		Fitness
	x	y	
1	4	−2	10.4435
2	−4	5	12.9838
3	0	5	11.209
4	−4	2	10.4442
5	5	−2	11.7307
6	4	−2	10.4435
7	−1	4	9.90681
8	2	−5	11.7302
9	−2	5	11.7317
10	1	−2	**6.49139**

It can be seen that the agent number 10 has the minimum objective function. Hence, this agent has the best solution found so far (at the initialization step). Next, the position of each agent should be updated according to the aforementioned procedure. The updated positions and the values of some other parameters are shown in the table below.

Whales	p	l	A	C	$\|A\|$	Updated position		Fitness
						x	y	
1	0.3	0.2	[−1.6, 2]	[0.2, 2]	2.56125	−2.08	−5	11.8688
2	0.1	0.4	[2, −0.8]	[2, 0.6]	2.15407	−2.4	1.4	8.32137
3	0.3	−0.2	[1.6, 0.8]	[1.8, 1.4]	1.78885	−5	−5	13.7112
4	0.3	1	[0.4, −0.8]	[1.2, 0.6]	0.894427	−1.08	0.56	4.59007
5	0.8	−0.6	[−2, −1.6]	[0, 0.2]	2.56125	1	−2	6.49139
6	0.7	−1	[−1.2, 0.8]	[0.4, 1.4]	1.44222	1	−2	6.49139
7	0.1	0	[0.8, 2]	[1.4, 2]	2.15407	−0.92	−5	11.4228
8	0.6	−0.6	[2, −0.8]	[2, 0.6]	2.15407	1	−2	6.49139
9	0.7	0	[1.2, 1.2]	[1.6, 1.6]	1.69706	4	5	12.9838
10	0.8	−0.6	[−0.8, 1.2]	[0.6, 1.6]	1.44222	1	−2	6.49139

For instance, the value of p and magnitude of A for agent 4 are 0.3 and 0.894427, respectively, the new position of this agent should be calculated through Eq. (E.2). Knowing from initialization step that $\vec{X}^* = [1, -2]$, parameter $\vec{D}$ could be computed as

$$\vec{D} = |([1.2, 0.6].[1, -2]) - [-4, 2]| = [5.2, 3.2].$$

Then, according to Eq. (E.2), the updated position of this agent is given by

$$\vec{X}(t+1) = [1, -2] - ([0.4, -0.8].[5.2, 3.2]) = [-1.08, 0.56].$$

Putting the new values of x and y in the objective function, the fitness could be obtained as 4.59007. By repeating this process till reaching the final iteration, the optimum value of the objective function could be found to be 0.624045 (for $x = 1.29 \times 10^{-06}$ and $y = 0.202407$). □

E.3 Teaching–learning-based optimization algorithm

The teaching–learning-based optimization (TLBO) is an efficient human-based optimization algorithm proposed by Rao et al. (2011) who observed the learning process of a group of students in a class. Generally, a student's learning process is occurred in two main ways:

1. Learning from the teacher, and
2. Learning from interactions with other students.

Hence, the optimization procedure is divided into two phases: *teacher phase* and *learner phase*. The former is related to gaining knowledge from the teacher, and the latter stimulates learning through interaction with other learners. The better a teacher is, the better the students perform in terms of the average obtained mark. In this algorithm, students are search agents, and the student with the best fitness is considered the teacher. The teacher will try to improve the class mean mark by teaching them what he/she knows. However, due to the effects of other factors, students' knowledge depends on the quality of the teaching, but is also affected by the students' quality. This randomness of the learning process is essentially implemented mathematically in both teacher and learner phases.

As explained before, in the teacher phase, the teacher tries to improve the mean of the class, M_i, to its level by the following updating scheme:

$$X_{\text{new},i} = X_{\text{old},i} + d_i, \tag{E.11}$$

in which

$$d_i = r_i (M_{\text{new}} - T_F M_i), \tag{E.12}$$

X is the position of a search agent, M_i is the mean values of different decision variables (among a group of search agents), M_{new} is the new mean values, T_F is a teaching factor, and r_i is a random number in the range [0, 1]. Actually, the new mean (M_{new}) is the teacher's knowledge level (or X_{teacher}). According to TLBO algorithm, the teaching factor (T_F) is either 1 or 2. This factor is chosen randomly with equal probability. The new solution achieved in the teacher phase is only accepted if it gives better fitness. Otherwise, the previous solution will be kept unchanged.

In addition to the teacher, interactions between the students help them learn more. Different means, such as group work and presentations, help students increase their knowledge from random interactions with other students. This behavior is mathematically modeled as follows. For any student X_i, another student X_j, $i \neq j$, is randomly chosen. The position of search agent X_i in this stage is updated by

$$X_{\text{new},i} = \begin{cases} X_{\text{old},i} + r_i(X_i - X_j) & \text{if } fit_i < fit_j, \\ X_{\text{old},i} + r_i(X_j - X_i) & \text{if } fit_i \geq fit_j, \end{cases} \tag{E.13}$$

where fit_i and fit_j are the objective function values of search agents X_i and X_j, respectively. Similar to the teacher phase, the updated solution is only accepted if it provides better fitness compared to the old one. This process is continued until the satisfaction of termination criteria. Since TLBO does not require any algorithm parameters, it could be considered simpler compared to other optimization algorithms from a tuning point of view.

Example E.4. Himmelblau's function, named after David Mautner Himmelblau, is a multimodal standard function used for assessing the performance of an optimization algorithm. The function is expressed as

$$f(x, y) = (x^2 + y - 11)^2 + (x + y^2 - 7)^2.$$

This function has four local minima at $(3, 2)$, $(-2.805118, 3.283186)$, $(-3.779310, -3.283186)$, and $(3.584458, -1.848126)$, for all of which the function value equals zero. The objective is to find the minimum value of this function by the TLBO algorithm on $x, y \in [-6, 6]$.

Answer. Like any other population-based method, to begin with, we need to specify the number of iterations and the search agents. In this example, we consider the number of iterations to be 200 and the number of search agents (students) to be 10. Next, the solutions should be initialized randomly, and the fitness values should be determined. By conducting the initialization step, search agents are given as indicated in the table below.

Students	Initial solution		Fitness
	x	y	
1	−6	−3.6	457.962
2	−3.6	2.4	42.4352
3	2.4	−3.6	148.035
4	−4.8	4.8	409.923
5	4.8	2.4	221.187
6	4.8	1.2	175.875
7	0	−6	1130
8	6	6	2186
9	6	−3.6	601.002
10	−3.6	3.6	**36.4832**

According to the data of the initialization step, agent 10 has the minimum objective function value. Therefore, this agent is considered to be the teacher in the teacher phase. Afterwards, all solutions (students) should be updated according to the previously explained procedure. To conduct the teacher phase, the mean of both x and y should be computed. The mean values of x and y are -0.6 and -0.36, respectively. For instance, the updated values of (x, y) for agent number 1 are computed based on Eqs. (E.11) and (E.12) as follows (assuming r_i to be 0.23 and T_F to be 2):

$$X_{\text{new},1} = (-6, -3.6) + 0.23 \times [(-3.6, 3.6) - 2 \times (-0.6, -0.36)]$$
$$= (-6.552, -2.6064).$$

The fitness value for $(-6.552, -2.6064)$ is 1238.35, which is greater than the initial fitness value 457.962). Since updating the solution in this phase gives a higher fitness, the values of x and y should be kept unchanged. To proceed with the leaner phase, a random agent (for example, student 6) should be chosen. According to Eq. (E.13), the updated values of x and y for the first student are computed as

$$X_{\text{new},1} = (-6, -3.6) + 0.23 \times [(4.8, 1.2) - (-6, -3.6)] = (-3.516, -2.496).$$

By calculating the objective function for this new solution, the obtained fitness will be 453.77. Since the new solution gives a lower fitness value (compared to the old solution), the values of x and y for the first student are updated. This process should be repeated for all students until reaching the last iteration. The obtained optimum solution for this case could be $(-2.80512, 3.13131)$. □

E.4 Particle swarm optimization algorithm

The particle swarm optimization (PSO) algorithm, which applies to a broad range of optimization problems, was first introduced in 1995 by Eberhart and Kennedy (1995). This algorithm is an optimization method for continuous non-linear functions inspired by a simplified social model. It consists of a simple concept that makes it a fast algorithm and only uses primitive mathematical operators. Despite its simplicity, it could be implemented to a vast array of optimization problems with only a few lines of computer code (Eberhart and Kennedy, 1995). The advantages of using the PSO algorithm are high convergence speed, an average level of complexity, average dependency on an initial guess, and a high ability to locate global optimum.

A particle is a group of decision variables within the search space of a problem with particular characteristics, including position and velocity. The value of the objective function could be determined using each particle's position data, which is, in fact, representative of the value of all decision variables. After assigning the number of particles used for an optimization problem in the first

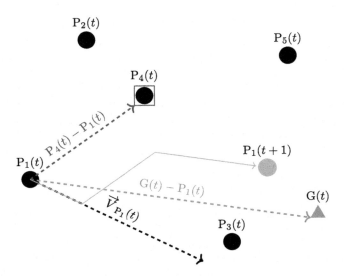

FIGURE E.4 Schematic of a particle swarm optimization problem.

iteration, the PSO algorithm solves the problem by moving the swarm particles with simple mathematical formulas over their position and velocity in an iterative process. Randomly generated values acquire the initial values of position and velocity of each particle within the search space in an early step called initialization. In other subsequent iterations, the movement of each particle is controlled by its local best and is guided toward the global best positions, which are updated as other particles find better positions. The term "local best" has been used to refer to the position of the particle with the best result in every single iteration, and the term "global best" could be described as the position of the particle with the best result over all of the previous iterations. In this sense, the best result is defined as a maximum or minimum objective function value based on the problem type (maximization or minimization problem). The schematic of a swarm with five particles is illustrated in Fig. E.4 for the tth iteration. As aforementioned, the new position of each particle (such as $P_1(t)$ in this figure) in the next iteration is a combination of its current position, its present velocity, its distance from the local best ($P_4(t)$ in this figure), and its distance from the global best ($G(t)$).

Prior to the PSO implementation, an objective function that will be optimized in further stages shall be defined. A general minimization optimization problem is shown as:

$$
\begin{aligned}
\text{minimize} \quad & F = F(x_1, x_2, \ldots, x_n), \\
\text{such that} \quad & x_i \in [L_i, U_i],
\end{aligned}
\tag{E.14}
$$

where F is the objective function and x_is are decision variable. First, the lower and upper bound (L_i and U_i) of each decision variable should be specified to confine the search space. Next, the number of total iterations and particles are selected. Note that the increment of the number of iterations and quantity of particles in each swarm will increase the computation cost and time. At the initialization step, a random position is allocated to each particle. Velocities are considered to be zero in this step for all particles. The position and velocity of the particle P_j at the first iteration ($t = 0$) are expressed as follows:

$$\overrightarrow{P}_j(t = 0) = r_1 \times (\overrightarrow{U}_P - \overrightarrow{L}_P) + \overrightarrow{L}_P \qquad \text{(E.15)}$$

and

$$\overrightarrow{V}_j(t = 0) = 0, \qquad \text{(E.16)}$$

where $\overrightarrow{P}_j(t = 0)$ and $\overrightarrow{V}_j(t = 0)$ are the position and velocity of particle P_j at the initialization step, $\overrightarrow{L}_P$ and $\overrightarrow{U}_P$ are lower and upper bound vectors of the decision variables (position bounds), and r_1 is a random number between 0 and 1. Based on $\overrightarrow{L}_P$ and $\overrightarrow{U}_P$, the lower and upper ranges of the velocity are defined as:

$$\overrightarrow{L}_V = V_{\text{coeff}} \times (\overrightarrow{U}_P - \overrightarrow{L}_P) \qquad \text{(E.17)}$$

and

$$\overrightarrow{U}_V = -\overrightarrow{L}_V, \qquad \text{(E.18)}$$

where $\overrightarrow{L}_V$ and $\overrightarrow{U}_V$ are lower and upper bound vectors of the velocity, and V_{coeff} is the velocity coefficient. The value of the objective function for each particle is evaluated by means of Eq. (E.14), and consequently, the best local and global solutions could be determined by comparing the objective function values for the whole swarm. For the particular case of the initialization step, the local and global best are the same. By completing this step, the process of updating the velocity and the position of each particle of the swarm is then carried out. The velocity and the position of each particle are updated based on their previous position and velocity using Eqs. (E.17) and (E.18), respectively:

$$\overrightarrow{V}_j(t + 1) = w \times \overrightarrow{V}_j(t) + C_1 \times r_2 \times \left[\overrightarrow{P}_{\text{local}} - \overrightarrow{P}_j(t)\right]$$
$$+ C_2 \times r_3 \times \left[\overrightarrow{P}_{\text{global}} - \overrightarrow{P}_j(t)\right], \qquad \text{(E.19)}$$

$$\overrightarrow{P}_j(t + 1) = \overrightarrow{P}_j(t) + \overrightarrow{V}_j(t + 1). \qquad \text{(E.20)}$$

In these equations, $\overrightarrow{V}_j(t + 1)$ is the updated velocity, $\overrightarrow{V}_j(t)$ is the velocity of the old particle, $\overrightarrow{P}_j(t)$ is its old position, $\overrightarrow{P}_j(t + 1)$ is the position of the updated particle, $\overrightarrow{P}_{\text{local}}$ is the best local position, $\overrightarrow{P}_{\text{global}}$ is the best global position, w is inertia coefficient, C_1 is personal acceleration coefficient, C_2 is social acceleration coefficient, while r_2 and r_3 are random numbers between 0 and 1. It is notable that if the updated value of position or velocity exceeds its lower or upper bounds, the updated values should be adjusted by substituting the appropriate lower or upper limit. It is known that the performance of this algorithm strongly depends on the values of control coefficients. Moreover, a number of the control coefficient values that work for a problem may not be as useful in another problem. Over the past years since the introduction of this algorithm, researchers have conducted a significant number of studies with the purpose of finding the proper configuration of the PSO control coefficient and investigating the dependency of these parameters on the nature of the problem. In order to achieve a rational estimation of w, C_1, and C_2, Clerc and Kennedy (2002) proposed a configuration for PSO parameters, known as constriction coefficients, which is defined by the following equation:

$$\chi = \frac{2 \times \kappa}{\left| 2 - \phi - \sqrt{\phi^2 - 4 \times \phi} \right|}, \tag{E.21}$$

in which κ is a value between 0 and 1, ϕ is a value greater than 4, calculated by

$$\phi = \phi_1 + \phi_2. \tag{E.22}$$

Commonly, ϕ and κ values are set to 4.1 (assuming $\phi_1 = \phi_2 = 2.05$) and 1, respectively (Poli et al., 2007). According to Clerc and Kennedy (2002), PSO parameters can be calculated using Eqs. (E.23) to (E.25):

$$w = \chi, \tag{E.23}$$
$$C_1 = \chi \phi_1, \tag{E.24}$$
$$C_2 = \chi \phi_2. \tag{E.25}$$

After calculating the updated positions and velocities of all the swarm particles, similar to the initialization step, the objective function value for all of the new particles should be assessed. Again, the local and global best could be found by examination of the measured objective functions. The global best position is only updated if the recognized local best position in the tth iteration leads to a better solution (lesser objective function value for the case of minimization problem) compared to the local best values of all previous iterations. The position and velocity updating paradigm and the evaluation of the objective function value will be continued until the final iteration. Evidently, the optimum solution is the global best position of the last iteration. The explained procedure is demonstrated graphically in Fig. E.5.

Example E.5. Any curve fitting process could be interpreted as a minimization problem, where the goal is to minimize the distance between the given data and

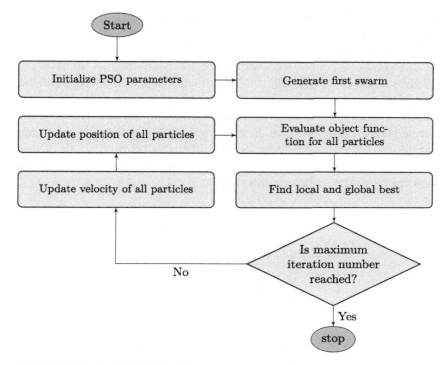

FIGURE E.5 Flowchart of PSO algorithm.

modeled curve. Assuming that y is a function of x only and some data points are given as indicated in the table below, the objective is to find a third-degree polynomial that best fits the given data.

x	0	2	4	6	8
y	10	16	29	55	123

Answer. Knowing that y is a function of x only, we have $y = f(x)$. In general, a third-degree polynomial could be expressed as

$$f = a_3 x^3 + a_2 x^2 + a_1 x + a_0.$$

The goal is finding the unknown coefficients, a_0 to a_3, in a way that the function f fits the given data points with minimum error. To do so, an error function (objective function) should be defined as

$$err = \sum_{i=1}^{N} (f(x_i) - y_i)^2,$$

in which N is the number of data points (in this example, $N = 5$). The objective of this optimization problem is to minimize the error function. The lower the value of the object function (error), the better the obtained function (f) fits the data points. To confine the search space, we assume that all unknown coefficients would range in $[-10, 10]$. To start implementing the PSO algorithm for solving this problem, we assume the number of particles to be 10 and the maximum iterations number to be 200. However, only the calculations of the first iteration are indicated here. In the initialization step, the position of each particle is specified randomly, as shown in the table below. Also, the fitness value of each particle is determined according to the specified positions and are mentioned in the last column of this table.

Particle	Initial position				Fitness
	a_0	a_1	a_2	a_3	
1	0	2	4	8	2.17E+07
2	10	6	6	8	2.36E+07
3	10	−10	−8	−2	3.77E+06
4	−10	8	−8	0	**4.51E+05**
5	8	−8	−6	−2	3.15E+06
6	8	6	−2	−10	3.37E+07
7	−2	−2	−8	−6	1.69E+07
8	10	−4	8	−2	4.78E+05
9	2	−6	−2	−6	1.37E+07
10	6	−6	8	−10	2.69E+07

At the initialization step, the local best of each particle is equal to the fitness of that particle in this step, and the global best is the lowest fitness value obtained by all particles. According to the data mentioned in the table for the initialization step, in this example, the best fitness value (lowest error value) obtained at the initialization step belongs to particle 4, which is 4.51×10^5. As aforementioned, in this step, the velocities of all particles are considered to be zero. Considering $\phi_1 = \phi_2 = 2.05$, the values of w, C_1, and C_2 are 0.729844, 1.49618, and 1.49618, respectively. Using Eq. (E.20), the updated position of each particle is calculated, as shown in the table below. Afterwards, the new fitness value of each particle is computed. For each particle, if the new fitness value is lower than that of the previous step, the local best of that particle should be updated. The global best fitness is also updated in case any better solution is found among the local best compared to the global best of the previous step. As it can be seen in the table, the best local best is still 4.51×10^5, and no better solution is found in this step.

Particle	Updated position				Fitness	Local best
	a_0	a_1	a_2	a_3		
1	−4	6	0	4	4.65E+06	4.65E+06
2	10	7.19694	2	4	5.46E+06	5.46E+06
3	7.00764	−6	−8	−2	3.63E+06	3.63E+06
4	−10	8	−8	0	4.51E+05	**4.51E+05**
5	4	−4	−6.59847	−2	3.19E+06	3.15E+06
6	4	7.19694	−5.59083	−6	1.47E+07	1.47E+07
7	−6	−0.50382	−8	−2	3.49E+06	3.49E+06
8	6	0	4	−1.70076	6.14E+05	4.78E+05
9	−1.59083	−2	−5.59083	−6	1.55E+07	1.37E+07
10	3.60611	−6	4	−6	1.06E+07	1.06E+07

By counting the calculations till the last iteration, we reach the best solution to be $a_3 = 0.346918$, $a_2 = -1.69968$, $a_1 = 5.4743$, and $a_0 = 10$. The value of the objective function (error) for this solution is 7.91. Hence, the obtained equation for the fitted curve is given by

$$f(x) = 0.346918x^3 - 1.69968x^2 + 5.4743x + 10.$$

The fitted curve obtained by the optimization process and the data points are depicted in Fig. E.6. Evidently, using the PSO algorithm for finding the unknown coefficients of the polynomial resulted in an appropriate agreement between the fitted curve and the given data points.   □

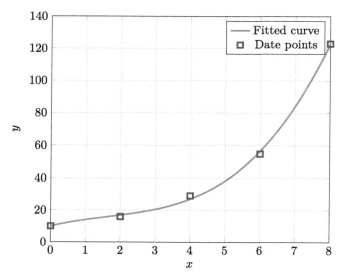

FIGURE E.6 Comparison between the input data and the fitted curve.

E.5 Genetic algorithm

Genetic algorithm (GA) is a search method, the concept of which was inspired by natural biological evolution. This algorithm was first published by Holland and his colleagues (Holland, 1975). According to the natural selection mechanism, weak species face extinction. However, the strong ones gain more chances to pass their genes to the next generations in the reproduction process. In the long term, species that carry the correct combination in their genes become dominant. On the other hand, some random changes may also happen in genes during the evolution process. In case these changes lead to benefits in the challenges of species for survival, new species evolve from the old ones. Negative changes are terminated by nature (Konak et al., 2006). The explained idea forms the basis of the GA algorithm. The advantages of the GA algorithms can be discussed under different headings, such as handling noisy functions, resistance against trapping in the local optima, and acceptable performance in large-scale optimization problems. Furthermore, not being useful for smooth unimodal functions, the requirement for a vast number of objective function calculations, and the requirement for choosing a lot of various control coefficients could be counted as some of the drawbacks of this algorithm.

GA solves a problem by generating a group of possible solutions, known as chromosomes, and evolving them with some mathematical formulas. These initial chromosomes, which are also known as parents, start producing new solutions (children). Some control coefficients control the percentage of reproduction. After the evaluation of parents' fitness, those that are closer to the optimal solution are more likely to reproduce. The broad use of the term "fitness function" in the literature is equated with the objective function you are willing to minimize or maximize, and the term "fitness" is thus considered as the value of the objective function for a particular solution. GA uses two main operators for the reproduction process, including crossover and mutation. The crossover technique combines features (genes) of two selected parent chromosomes in order to reproduce the chromosomes of two new children (new solutions). In this method, parents are chosen through three different procedures, including the random selection, the tournament selection, and the roulette wheel selection.

The first method picks out the parents in a completely random manner. However, in the tournament method, some parents are selected randomly on the basis of the tournament size, which has to be specified at the beginning of the implementation process, and then two best fits will be selected for crossover. In the third method, the so-called roulette wheel method, the chance of parents' selection is dependent on their fitness. The more a parent's chromosome fits, the higher the chance of selection. The roulette wheel selection method is the most commonly deployed. However, a combination of all three methods can be used in the optimization procedure. Here, only the roulette wheel procedure will be explained. After finishing the parents' selection procedure, crossover takes place. The crossover method is divided into two types, arithmetical and linear crossover. In the arithmetical crossover, two parents reproduce two children, while in the linear crossover, two parents reproduce three children. Two of these three which have better fit are substituted with their parents in the population (Michalewicz et al., 1994). On the other hand, random changes during the evolution occur by the mutation operator. This operator changes several genes of a randomly selected chromosome. These genes are selected randomly and substituted with new feasible ones. The percentage of mutation and the number of genes that change are controlled separately by two different coefficients. A detailed explanation of the GA step-by-step procedure and its mathematical formulation is provided below.

To accomplish optimization using the GA algorithm, similar to the PSO algorithm, at the beginning of the optimization process, the minimum and maximum values of each decision variable (the search space), the number of algorithm iterations, and the size of chromosome population have to be specified. In addition, the control parameters, including crossover percentage, mutation percentage, mutation rate, and selection pressure, might be designated. A recommended set of control parameters are indicated in Table E.1. Based on these parameters, the number of parents for the crossover process and the number of mutants could be derived by

$$nc = 2 \times \left\lceil \frac{pc \times npop}{2} \right\rceil \tag{E.26}$$

and

$$nm = \lfloor pm \times npop \rceil, \tag{E.27}$$

in which nc is the number of parents, pc is the crossover percentage, $npop$ is the size of chromosome population, nm is the number of mutants, and pm is the mutation percentage. Since the numbers of parents and mutants should be integers, the rounding function has been used in their formula. The values of the genes of each chromosome need to be initialized randomly under the minimum and maximum limit constraints. Afterwards, the fitness of each chromosome has to be explored using Eq. (E.14). As the next step, by sorting the obtained fitness values, the best and worst fit chromosomes could be distinguished. Henceforth, using the roulette wheel selection method, parent chromosomes are elected for the crossover process. In this method, the probability of selecting each chromosome is determined based on its fitness. The probability of each chromosome can be determined based on Boltzmann probability distribution

$$P_i = \exp\left[-\beta \times \frac{F_i}{F_W}\right] \tag{E.28}$$

In Eq. (E.28), P_i is the probability of a chromosome, β is the selection pressure, F_i is the fitness of a chromosome, and F_W is the worst fitness of the whole

TABLE E.1 Values of genetic algorithm parameters.

Parameter	Value
Crossover percentage (pc)	80%
Mutation percentage (pm)	5%
Mutation rate (mu)	0.1
Selection pressure (β)	10

population in this iteration. Using Eq. (E.29), probabilities can be normalized,

$$q_i = \frac{P_i}{\sum\limits_{n=1}^{npop} P_i}, \tag{E.29}$$

in which q_i is the normalized probability of a chromosome. After that, two chromosomes will be selected randomly as parents considering the normalized probability. Now, the two selected parents shall be used for crossover to reproduce two new children. The genes of the new children are calculated based on

$$x_{ch1,i} = \alpha \times x_{pr1,i} + (1 - \alpha) \times x_{pr2,i}, \tag{E.30a}$$

$$x_{ch2,i} = \alpha \times x_{pr2,i} + (1 - \alpha) \times x_{pr1,i}, \tag{E.30b}$$

where x_i is the value of the ith gene, α is the crossover coefficient which has been selected randomly, and subscripts ch1, ch2, pr1, and pr2 indicate the first and second child, as well as the first and second parent, respectively. The crossover process is repeated until reaching the number of parents to be crossed over. To start the mutation process, one of the initial chromosomes is chosen randomly. Further, to apply the mutation process to the selected chromosome, the number of genes to be mutated is computed through mutation percentage and number of genes using Eq. (E.31) as follows:

$$nMu = \lceil mu \times n_g \rceil, \tag{E.31}$$

where nMu is the number of mutant genes, mu is the mutation rate, and n_g is the number of genes of each chromosome. Next, the genes are selected randomly based on the computed number of mutant genes. Gene mutation could be performed by

$$x_{m,i} = x_{pr,i} + r \times \eta \times (x_{Max,i} - x_{Min,i}), \tag{E.32}$$

in which $x_{m,i}$ is the value of the ith gene of mutant chromosome, $x_{pr,i}$ is the value of the ith gene of the parent chromosome, r is a random number, η is a controllable coefficient, and $x_{Max,i}$ and $x_{Min,i}$ are the maximum and minimum

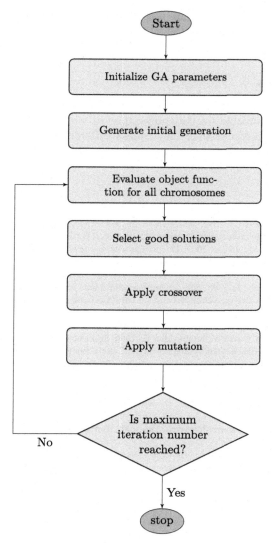

FIGURE E.7 Flowchart of GA algorithm.

possible values of the ith gene. The mutation operation is executed repeatedly for the number of parents that should be mutated. At this point, the fitness of all reproduced children by crossover and mutation processes are evaluated using the objective function of the problem. Following that, the chromosomes with the best and worst fitness are recognized. Among all the initial, crossovered and mutated chromosomes, those with the number of chromosomes equal to the size of initial population with the best fit are selected as the survived chromosomes for the next generation. The entire procedure explained above will be repeated for the predetermined number of iterations. The flow chart of the GA algorithm implementation process is displayed in Fig. E.7.

E.6 Summary

In this appendix, different optimization algorithms are discussed in short. All the methods are suitable for the optimization of wind turbines and farms and may give different results since they employ different algorithms to search for the solution. Therefore, for good practice, it would be better to use different methods and compare the results.

References

Askarzadeh, Alireza, 2016. A novel metaheuristic method for solving constrained engineering optimization problems: crow search algorithm. Computers & Structures 169, 1–12.

Clerc, Maurice, Kennedy, James, 2002. The particle swarm – explosion, stability, and convergence in a multidimensional complex space. IEEE transactions on Evolutionary Computation 6 (1), 58–73.

Eberhart, R.C., Kennedy, J., 1995. Particle swarm optimization. In: Proceedings of ICNN'95 – International Conference on Neural Networks. Perth, Australia, pp. 1942–1948.

Holland, John, 1975. Adaptation in natural and artificial systems: an introductory analysis with application to biology. In: Control and Artificial Intelligence.

Konak, Abdullah, Coit, David W., Smith, Alice E., 2006. Multi-objective optimization using genetic algorithms: a tutorial. Reliability Engineering & System Safety 91 (9), 992–1007.

Michalewicz, Zbigniew, Logan, Thomas, Swaminathan, Swarnalatha, 1994. Evolutionary operators for continuous convex parameter spaces. In: Proceedings of the 3rd Annual conference on Evolutionary Programming. World Scientific, pp. 84–97.

Mirjalili, Seyedali, Lewis, Andrew, 2016. The whale optimization algorithm. Advances in Engineering Software 95, 51–67.

Poli, Riccardo, Kennedy, James, Blackwell, Tim, 2007. Particle swarm optimization. Swarm Intelligence 1 (1), 33–57.

Rao, R. Venkata, Savsani, Vimal J., Vakharia, D.P., 2011. Teaching–learning-based optimization: a novel method for constrained mechanical design optimization problems. Computer-Aided Design 43 (3), 303–315.

Appendix F

Implementing optimization methods in C++

Optimization algorithms are widely used in many industrial and academic sectors. Among the many different methods, genetic algorithm (GA) and particle swarm optimization (PSO) are quite famous and used in many different applications, including the optimization of wind farms. These two algorithms are explained in detail in Appendix E. In the present appendix, the implementation of these two methods in the C++ language is given. We tried to write the code as simple as possible. For this purpose, the function and variables' names are selected such that they resemble their functions. Moreover, proper comments are applied to the codes to emphasize the task of each function.

We have also used the object-oriented programming capability of the C++ language to make the codes look clearer. However, to make them more educational, individual functions are also implemented.

Finally, the two codes are implemented with a uniform algorithm. We tried to make them as close to each other as possible. Therefore, by studying the codes and concentrating on the differences, both codes would become clearer. It is worth mentioning that other optimization algorithms mentioned in Appendix E, such as crow search and whale optimization algorithms, can be implemented with the same concepts.

F.1 Genetic algorithm (GA)

The GA algorithm is implemented in the following code. The `main` program is located at line 241. As can be seen, it starts defining the initial parameters at lines 246–248 and defines the decision parameters at lines 250–257. The main GA object is defined at line 259 by calling the constructor of the GA class. Then it solves the problem simply by following the following four steps:

1. Initializing the particles on the domain by calling `GA_.initialization()` method,
2. Doing optimization by calling `GA_optimiztion()` method,
3. Obtaining the final optimized decision variables by calling `GA_.ouptu()`, and
4. Announcing the final results to the user by calling `GA_.result()` function.

The implementation of different methods is straightforward in the following code. The code also is commented to become understandable.

```cpp
1  #include <iostream>
2  #include <fstream>
3  #include <stdio.h>
4  #include <stdlib.h>
5  #include <math.h>
6  #include <time.h>
7
8  // This class holds the main information about the particles
9  class particle
10 {
11 public:
12
13     double *position, *velocity, cost, *bestPosition, bestCost;
14     bool sortBool = true;
15
16     particle()
17     {
18     }
19
20     // Allocates the vectors to the number of decision variables
21     void allocation(int numbDecisionVariable)
22     {
23         position = new double[numbDecisionVariable];
24         velocity = new double[numbDecisionVariable];
25         bestPosition = new double[numbDecisionVariable];
26     }
27 };
28
29 // This function performs crossover
30 double **crossOver(double *position1, double *position2, int nVar, double *minVar,
        double *maxVar, double gamma)
31 {
32     double alpha, **crossPosition;
33
34     crossPosition = new double * [2];
35
36     crossPosition[0] = new double[nVar];
37     crossPosition[1] = new double[nVar];
38
39     srand(time(NULL));
40
41     for(int ii=0; ii<nVar; ii++)
42     {
43         alpha = ((rand()%11)/10.0)*((1+gamma)-(-gamma))+(-gamma);
44         crossPosition[0][ii] = alpha*position1[ii]+(1-alpha)*position2[ii];
45         crossPosition[1][ii] = alpha*position2[ii]+(1-alpha)*position1[ii];
46     }
47
48     for(int ii=0; ii<2; ii++)
49     {
50         for(int jj=0; jj<nVar; jj++)
51         {
52             if(crossPosition[ii][jj]>maxVar[ii])
```

```
53                {
54                        crossPosition[ii][jj] = maxVar[ii];
55                }
56
57                if(crossPosition[ii][jj]<minVar[ii])
58                {
59                        crossPosition[ii][jj] = minVar[ii];
60                }
61            }
62        }
63
64      return crossPosition;
65  }
66
67  // This function performs mutation
68  double *mutation(double *position, double mu, int nVar, double *minVar, double *
        maxVar)
69  {
70      int nmu, tmp1;
71      int *j;
72      bool *selection;
73      double sigma, tmp2;
74      double *mutPosition;
75
76      mutPosition = new double[nVar];
77
78      tmp2 = ceil(mu*nVar);
79      nmu = (int) tmp2;
80
81      selection = new bool[nVar];
82      j = new int[nmu];
83
84      for(int ii=0; ii<nVar; ii++)
85      {
86          selection[ii] = false;
87      }
88
89      srand(time(NULL));
90
91      for(int ii=0; ii<nmu; ii++)
92      {
93          tmp1 = (int) rand()%nVar;
94
95          while(selection[tmp1])
96          {
97              tmp1 = (int) rand()%nVar;
98          }
99
100         j[ii] = tmp1;
101         selection[tmp1] = true;
102     }
103
104     for(int ii=0; ii<nVar; ii++)
```

```
105          {
106              mutPosition[ii] = position[ii];
107          }
108
109      srand(time(NULL));
110
111      for(int ii=0; ii<nmu; ii++)
112      {
113          tmp1 = j[ii];
114          sigma = 0.1*(maxVar[tmp1] - minVar[tmp1]);
115          mutPosition[tmp1] = position[tmp1]+sigma*((rand()%11)/10.0);
116      }
117
118      for(int ii=0; ii<nVar; ii++)
119      {
120          if(mutPosition[ii]>maxVar[ii])
121          {
122              mutPosition[ii] = maxVar[ii];
123          }
124
125          if(mutPosition[ii]<minVar[ii])
126          {
127              mutPosition[ii] = minVar[ii];
128          }
129      }
130
131      return mutPosition;
132  }
133
134  // Generates a random number produced by roulette wheel
135  int rouletteWheelSelection(double *prob, int nPop)
136  {
137      int indice;
138      double r, tmp;
139
140      srand(time(NULL));
141
142      r = ((rand()%11)/10.0);
143      tmp = 0;
144
145      for(int qq=0; qq<nPop; qq++)
146      {
147          tmp = tmp+prob[qq];
148
149          if((r-tmp)<0.01)
150          {
151              indice = qq;
152              break;
153          }
154      }
155
156      return indice;
157  }
```

```
158
159
160    // Defines the objective function
161    double objFunc(double *position, int numbDecisionVariable)
162    {
163        double error = 0;
164
165        for (int i=0; i<numbDecisionVariable; i++)
166        {
167            error = error+(position[i]*position[i]);
168        }
169
170        return error;
171    }
172
173    // This is the main genetic algorithm class
174    class GA
175    {
176        int nc, nm, nPop, nVar, ite;
177        double pc, pm, mu, gamma, beta;
178        double *prob, probSum, *minVar, *maxVar, *globalBestPosition, **crossPosition,
               *mutPosition;
179        double globalBest, globalWorst;
180        particle *par, *crossPar, *mutPar, *mergePar, *mergeSortPar;
181
182    public:
183        // The main constructor that builds the GA object
184        GA(int numbIteration, int numbChromosomes, int numbDecisionVariable, double *
           minDecisionVariable , double *maxDecisionVariable)
185        {
186            pc = 0.8; pm = 0.05; mu = 0.1; gamma = 0.05; beta = 10;
187            ite = numbIteration; nPop = numbChromosomes; nVar = numbDecisionVariable;
188            minVar = minDecisionVariable; maxVar = maxDecisionVariable;
189
190            nc = (int) 2*ceil(pc*nPop/2);
191            nm = (int) ceil(pm*nPop);
192
193            par = new particle[nPop];
194            crossPar = new particle[nc];
195            mutPar = new particle[nm];
196            mergePar = new particle[nPop+nc+nm];
197            mergeSortPar = new particle[nPop+nc+nm];
198            globalBestPosition = new double[nVar];
199            prob = new double[nPop];
200
201            globalBest = 1e20;
202            globalWorst = -1e20;
203
204            for(int ii=0; ii<nPop; ii++)
205            {
206                par[ii].allocation(nVar);
207            }
208
```

```
209          for(int ii=0; ii<nc; ii++)
210          {
211              crossPar[ii].allocation(nVar);
212          }
213
214          for(int ii=0; ii<nm; ii++)
215          {
216              mutPar[ii].allocation(nVar);
217          }
218
219          for(int ii=0; ii<(nPop+nc+nm); ii++)
220          {
221              mergePar[ii].allocation(nVar);
222          }
223
224          for(int ii=0; ii<(nPop+nc+nm); ii++)
225          {
226              mergeSortPar[ii].allocation(nVar);
227          }
228      }
229  // The initialization method sets up the initial position of the particles.
230  // The method is defined at line 269
231      void initialization();
232  // The main optimization function. Defined at line 302
233      void optimization();
234  // The result function prints the results to the console. Defined at line 449
235      void result();
236  // Returns the optimized decision variables. Defined at line 461
237      double *output();
238  };
239
240  // Main program
241  int main()
242  {
243      double *minDecisionVariable, *maxDecisionVariable, *optimizedSolution;
244      int numbDecisionVariable, numbIterations, numbChromosomes;
245
246      numbDecisionVariable = 10;
247      numbIterations = 5000;
248      numbChromosomes = 200;
249
250      minDecisionVariable = new double[numbDecisionVariable];
251      maxDecisionVariable = new double[numbDecisionVariable];
252
253      for(int i=0; i<numbDecisionVariable; i++)
254      {
255          minDecisionVariable[i] = -1;
256          maxDecisionVariable[i] = 1;
257      }
258
259      GA GA_(numbIterations, numbChromosomes, numbDecisionVariable,
           minDecisionVariable, maxDecisionVariable);
260      GA_.initialization(); // Initializes the code
```

```
261    GA_.optimization();    // Performs optimization
262    optimizedSolution = GA_.output(); // Returns the final best positions
263    GA_.result();          // Prints the necessary information to the console
264
265    return 0;
266  }
267
268  // Initialization method (see line 231).
269  void GA :: initialization()
270  {
271    srand(time(NULL));
272
273    for(int ii=0; ii<nPop; ii++)
274    {
275
276        for (int jj=0; jj<nVar; jj++)
277        {
278            par[ii].position[jj] = ((rand()%11)/10.0)*(maxVar[jj]-minVar[jj]) +
       minVar[jj];
279            par[ii].bestPosition[jj] = par[ii].position[jj];
280        }
281
282        par[ii].cost = objFunc(par[ii].position, nVar);
283        par[ii].bestCost = par[ii].cost;
284
285        if(par[ii].cost<globalBest)
286        {
287            globalBest = par[ii].cost;
288
289            for(int kk=0; kk<nVar; kk++)
290            {
291                globalBestPosition[kk] = par[ii].position[kk];
292            }
293        }
294        if(par[ii].cost>globalWorst)
295        {
296            globalWorst = par[ii].cost;
297        }
298    }
299  }
300
301  // Optimization code
302  void GA :: optimization()
303  {
304    srand(time(NULL));
305
306    for(int ii=0; ii<ite; ii++)
307    {
308        for(int jj=0; jj<nPop; jj++)
309        {
310            prob[jj] = exp(-beta*(par[jj].cost/globalWorst));
311        }
312
```

```
313         probSum = 0;
314
315         for(int jj=0; jj<nPop; jj++)
316         {
317             probSum = probSum + prob[jj];
318         }
319
320         for(int jj=0; jj<nPop; jj++)
321         {
322             prob[jj] = prob[jj]/probSum;
323         }
324
325         int nc2 = (int) nc/2;
326
327         for(int jj=0; jj<nc2; jj++)
328         {
329             int i1, i2;
330
331             srand(time(NULL));
332             i1 = rouletteWheelSelection(prob, nPop);
333             srand(time(NULL));
334             i2 = rouletteWheelSelection(prob, nPop);
335
336             crossPosition = crossOver(par[i1].position, par[i2].position, nVar,
        minVar, maxVar, gamma);
337
338             for(int kk=0; kk<nVar; kk++)
339             {
340                 crossPar[2*jj].position[kk] = crossPosition[0][kk];
341                 crossPar[2*jj+1].position[kk] = crossPosition[1][kk];
342             }
343
344             crossPar[2*jj].cost = objFunc(crossPar[2*jj].position, nVar);
345             crossPar[2*jj+1].cost = objFunc(crossPar[2*jj+1].position, nVar);
346         }
347
348         srand(time(NULL));
349
350         for(int jj=0; jj<nm; jj++)
351         {
352             int i;
353             double tmp;
354
355             tmp = (rand()%nPop);
356             i = (int) tmp;
357             mutPosition = mutation(par[i].position, mu, nVar, minVar, maxVar);
358
359             for(int kk=0; kk<nVar; kk++)
360             {
361                 mutPar[jj].position[kk] = mutPosition[kk];
362             }
363
364             mutPar[jj].cost = objFunc(mutPar[jj].position, nVar);
```

```
365            }
366
367            for(int jj=0; jj<nPop; jj++)
368            {
369                for(int kk=0; kk<nVar; kk++)
370                {
371                    mergePar[jj].position[kk] = par[jj].position[kk];
372                }
373
374                mergePar[jj].cost = par[jj].cost;
375                mergePar[jj].sortBool = true;
376            }
377
378            for(int jj=nPop; jj<nPop+nc; jj++)
379            {
380                for(int kk=0; kk<nVar; kk++)
381                {
382                    mergePar[jj].position[kk] = crossPar[jj-nPop].position[kk];
383                }
384
385                mergePar[jj].cost = crossPar[jj-nPop].cost;
386                mergePar[jj].sortBool = true;
387            }
388
389            for(int jj=nPop+nc; jj<nPop+nc+nm; jj++)
390            {
391                for(int kk=0; kk<nVar; kk++)
392                {
393                    mergePar[jj].position[kk] = mutPar[jj-(nPop+nc)].position[kk];
394                }
395
396                mergePar[jj].cost = mutPar[jj-(nPop+nc)].cost;
397                mergePar[jj].sortBool = true;
398            }
399
400            for(int jj=0; jj<nPop+nc+nm; jj++)
401            {
402                mergeSortPar[jj].cost = 1e20;
403            }
404
405            int tempIndice;
406
407            for(int jj=0; jj<nPop+nc+nm; jj++)
408            {
409                for(int kk=0; kk<nPop+nc+nm; kk++)
410                {
411                    if(mergePar[kk].sortBool)
412                    {
413                        if(mergePar[kk].cost<mergeSortPar[jj].cost)
414                        {
415                            mergeSortPar[jj].cost = mergePar[kk].cost;
416                            mergeSortPar[jj].position = mergePar[kk].position;
417                            tempIndice = kk;
```

```
418                         }
419                     }
420                 }
421
422             mergePar[tempIndice].sortBool = false;
423         }
424
425         for(int jj=0; jj<nPop; jj++)
426         {
427             par[jj].position = mergeSortPar[jj].position;
428             par[jj].cost = mergeSortPar[jj].cost;
429         }
430
431         if(globalWorst<mergeSortPar[nPop+nc+nm-1].cost)
432         {
433             globalWorst = mergeSortPar[nPop+nc+nm-1].cost;
434         }
435
436         if(globalBest>mergeSortPar[0].cost)
437         {
438             globalBest = mergeSortPar[0].cost;
439
440             for(int jj=0; jj<nVar; jj++)
441             {
442                 globalBestPosition[jj] = mergeSortPar[0].position[jj];
443             }
444         }
445     }
446 }
447
448 // Prints the results
449 void GA :: result()
450 {
451     cout<<"********** Optimization Result **********"<<endl;
452     cout<<"Optimum objective function value is: "<<globalBest<<endl;
453
454     for(int ii=0; ii<nVar; ii++)
455     {
456         cout<<"Decision Variable No. ["<<ii+1<<"] is: "<<globalBestPosition[ii]<<
        endl;
457     }
458 }
459
460 // Returns the final best position parameters
461 double * GA :: output()
462 {
463     return globalBestPosition;
464 }
```

F.2 Particle swarm optimization (PSO)

The PSO algorithm is implemented in the following code. We tried to make the present code become as close to the GA as possible. The `main` program is located at line 101. As can be seen, it follows the same steps as GA by defining proper variables. Then it defies the main `PSO` object at line 119 by calling the constructor of the `PSO` class. Then it follows the same four steps just like GA:

1. Initializing the particles on the domain by calling `PSO_.initialization()` method,
2. Doing optimization by calling `PSO_optimiztion()` method,
3. Obtaining the final optimized decision variables by calling `PSO_.ouptu()`, and
4. Announcing the final results to the user by calling `PSO_.result()` function.

The rest of the code is quite clear and is commented where necessary.

```cpp
1  #include <iostream>
2  #include <math>
3  #include <stdlib.h>
4
5  using namespace std;
6
7  // Class particle to define particles
8  class particle
9  {
10 public:
11
12     double *position, *velocity, cost, *bestPosition, bestCost;
13     bool sortBool = true;
14
15     particle()
16     {
17     }
18     // Allocates the positions, velocity, and the best position for each decision
       variable
19     void allocation(int numbDecisionVariable)
20     {
21         position = new double[numbDecisionVariable];
22         velocity = new double[numbDecisionVariable];
23         bestPosition = new double[numbDecisionVariable];
24     }
25 };
26
27 // Defines the objective function
28 double objFunc(double *position , int numbDecisionVariable)
29 {
30     double error = 0;
31
32     for (int i=0; i<numbDecisionVariable; i++)
33     {
34         error = error+(position[i]*position[i]);
35     }
```

```
36
37        return error ;
38   }
39
40   // The main PSO class
41   class PSO
42   {
43        double kappa, phi1, phi2, phi, chi;
44        double tmp ;
45        double w, wdamp, c1, c2, velCoeff;
46        double * globalBestPosition ;
47        double globalBest = 1e20 ;
48        int nPop, nVar, ite;
49        double *minVar, *maxVar, *minVelocity, *maxVelocity;
50        particle * par ;
51
52   public:
53        // Defines default constructor
54        PSO(int numbIteration , int numbPopulation , int numbDecisionVariable , double
             *minDecisionVariable , double *maxDecisionVariable)
55        {
56            kappa = 1; phi1 = 2.05; phi2 = 2.05;
57
58            phi = phi1+phi2 ;
59            chi = (2*kappa)/fabs(2-phi-sqrt(pow(phi,2)-4*phi));
60            w = chi;
61            wdamp = 1;
62            c1 = chi*phi1;
63            c2 = chi*phi2;
64            velCoeff = 0.2;
65
66            ite = numbIteration; nPop = numbPopulation; nVar = numbDecisionVariable;
67
68            minVar = minDecisionVariable; maxVar = maxDecisionVariable;
69
70            globalBestPosition = new double[nVar];
71
72            minVelocity = new double[nVar];
73
74            maxVelocity = new double[nVar];
75
76            for(int ii=0 ; ii<nVar ; ii++)
77            {
78                maxVelocity[ii] = velCoeff * (maxVar[ii] - minVar[ii]);
79                minVelocity[ii] = -maxVelocity[ii];
80            }
81
82            par = new particle[nPop];
83
84            for(int ii=0 ; ii<nPop ; ii++)
85            {
86                par[ii].allocation(nVar);
87            }
```

```
88        }
89   // The initialization method sets up the initial position of the particles.
90   // The method is defined at line 129
91        void initialization();
92   //   The main optimization function. Defined at line 160
93        void optimization();
94   //   The result function prints the results to the console. Defined at line 230
95        void result();
96   //   Returns the optimized decision variables. Defined at line 242
97        double * output();
98   };
99
100  // Main program
101  int main()
102  {
103       double *minDecisionVariable, *maxDecisionVariable, *optimizedSolution;
104       int numbDecisionVariable, numbIterations, numbParticles;
105
106       numbDecisionVariable = 10;
107       numbIterations = 50;
108       numbParticles = 10;
109
110       minDecisionVariable = new double[numbDecisionVariable];
111       maxDecisionVariable = new double[numbDecisionVariable];
112
113       for(int i=0; i<numbDecisionVariable; i++)
114       {
115           minDecisionVariable[i] = -1;
116           maxDecisionVariable[i] = 1;
117       }
118
119       PSO PSO_(numbIterations, numbParticles, numbDecisionVariable,
               minDecisionVariable, maxDecisionVariable);
120       PSO_.initialization();
121       PSO_.optimization();
122       optimizedSolution = PSO_.output();
123       PSO_.result();
124
125       return 0;
126  }
127
128  // Initialization method (see line 91)
129  void PSO :: initialization()
130  {
131       srand(time(NULL));
132
133       for(int ii=0 ; ii<nPop ; ii++)
134       {
135           for (int jj=0 ; jj<nVar ; jj++)
136           {
137               par[ii].position[jj] = ((rand()%11)/10.0)*(maxVar[jj]-minVar[jj])+
               minVar[jj];
138
```

```
139          par[ii].velocity[jj] = 0;
140
141          par[ii].bestPosition[jj] = par[ii].position[jj];
142      }
143
144      par[ii].cost = objFunc(par[ii].position , nVar);
145      par[ii].bestCost = par[ii].cost;
146
147      if(par[ii].bestCost < globalBest)
148      {
149          for(int kk=0 ; kk<nVar ; kk++)
150          {
151              globalBestPosition[kk] = par[ii].bestPosition[kk];
152          }
153
154          globalBest = par[ii].bestCost;
155      }
156    }
157 }
158
159 // Optimization code (see line 93)
160 void PSO :: optimization()
161 {
162     srand(time(NULL));
163
164     for(int ii=0 ; ii<ite ; ii++)
165     {
166         for(int jj=0 ; jj<nPop ; jj++)
167         {
168             for(int kk=0 ; kk<nVar ; kk++)
169             {
170                 tmp = (w*par[jj].velocity[kk]) + (c1*((rand()%11)/10.0)*(par[jj].
         bestPosition[kk]-par[jj].position[kk])) + (c2*((rand()%11)/10.0)*(
         globalBestPosition[kk]-par[jj].position[kk])) ;
171
172                 if(tmp < minVelocity[kk])
173                 {
174                     par[jj].velocity[kk] = minVelocity[kk];
175                 }
176                 else if(tmp > maxVelocity[kk])
177                 {
178                     par[jj].velocity[kk] = maxVelocity[kk];
179                 }
180                 else
181                 {
182                     par[jj].velocity[kk] = tmp;
183                 }
184
185                 tmp = par[jj].position[kk] + par[jj].velocity[kk];
186
187                 if(tmp < minVar[kk])
188                 {
189                     par[jj].position[kk] = minVar[kk];
```

```
190                    }
191                    else if(tmp > maxVar[kk])
192                    {
193                            par[jj].position[kk] = maxVar[kk];
194                    }
195                    else
196                    {
197                            par[jj].position[kk] = tmp;
198                    }
199
200                }
201
202            par[jj].cost = objFunc(par[jj].position , nVar);
203
204            if(par[jj].cost<par[jj].bestCost)
205            {
206                for(int zz=0 ; zz<nVar ; zz++)
207                {
208                        par[jj].bestPosition[zz] = par[jj].position[zz];
209                }
210
211                par[jj].bestCost = par[jj].cost;
212            }
213
214            if(par[jj].bestCost<globalBest)
215            {
216                for(int zz=0 ; zz<nVar ; zz++)
217                {
218                        globalBestPosition[zz] = par[jj].bestPosition[zz];
219                }
220
221                globalBest = par[jj].bestCost;
222            }
223        }
224
225        w = w*wdamp;
226    }
227 }
228
229 // Prints the results (see line 95)
230 void PSO :: result()
231 {
232    cout<<"********** Optimization Result **********"<<endl;
233    cout<<"Optimum objective function value is: "<<globalBest<<endl;
234
235    for(int ii=0 ; ii<nVar ; ii++)
236    {
237        cout<<"Decision Variable No. ["<<ii+1<<"] is: "<<globalBestPosition[ii]<<
        endl;
238    }
239 }
240
241 // Returns the final best position parameters (see line 97)
```

```
242  double * PSO :: output()
243  {
244      return globalBestPosition;
245  }
```

Appendix G

Implementing blade element momentum method in C

The following code is an implementation of the BEM algorithm. The code is written in C language to make it as simple as possible. Note that the code implements just the basic BEM flowchart and does not calculate the gearbox or generator efficiency.

The code starts with `Init()` function which initializes the turbine characteristics. As it is commented, the geometrical and operational characteristics of the turbine are given as input data to the code. Then the code continues to simulate the turbine, according to the 8-step procedure explained in Chapter 3. We used the same naming convention as illustrated by Fig. 3.16. Therefore, the reader can easily understand what is going on.

```c
1  #include <stdio.h>
2  #include <math.h>
3
4  // Turbine geometrical parameters (Code input)
5  int bladenum;      // number of blades
6  int elnum;         // number of blade elements
7  double *r;         // radius of each element (m)
8  double *beta;      // twist at the corresponding radius (degree)
9  double *chord;     // chord at the corresponding radius (m)
10 double R;          // blade's radius (m)
11 double Area;       // turbine's swept area (m^2)
12
13 // Turbine operational conditions (Code input)
14 double Omega;      // rotational speed (rpm)
15 double u0;         // wind velocity    (m/s)
16 double uRated;     // rated velocity   (m/s)
17 double uCut_in;    // cut-in velocity  (m/s)
18 double uCut_off;   // cut-out velocity (m/s)
19
20
21 // Calculated power parameters (Code output)
22 double Cp;         // coefficient of performance
23 double Momentum;   // captured momentum
24 double Thrust;     // captured thrust
25 double Power;      // captured power (W)
26 double *PN;        // differential normal force of each element
27 double *PT;        // differential tangential force of each element
28
```

```
29  // Constant parameters
30  const double PI = acos(-1.0);
31  const double rho=1.225;  // standard air density (kg/m^3)
32
33
34  void Init()
35  {
36      // This function initiates the input parameters
37      Omega   =27.1;
38      uCut_in =4.0;
39      uRated  =14.0;
40      uCut_off=25;
41      bladenum=3;
42      elnum   =17;
43      r=new double[elnum+1]{4.5, 5.5, 6.5, 7.5, 8.5, 9.5, 10.5, 11.5, 12.5, 13.5,
            14.5, 15.5, 16.5, 17.5, 18.5, 19.5, 20.3, 20.54};
44      chord=new double[elnum+1]{1.63, 1.597, 1.54 , 1.481, 1.42 , 1.356, 1.294,
            1.229, 1.163, 1.095, 1.026, 0.955, 0.881, 0.806, 0.705, 0.545, 0.265, 0};
45      beta=new double[elnum+1]{20, 16.3, 13, 10.05, 7.45, 5.85, 4.85, 4, 3.15, 2.6,
            2.02, 1.36, 0.77, 0.33, 0.14, 0.05, 0.02, 0};
46
47      PT = new double[elnum];
48      PN = new double[elnum];
49
50      R = r[elnum];
51      Area=PI*R*R;
52  }
53  void Step1(double & a, double &aprime)
54  {
55      // Initial guesses for a and a'
56      a = 0.0;
57      aprime = 0.0;
58  }
59  void Step2(int i, double &phi, double a, double aprime,double w)
60  {
61      // Calculates phi
62      double z1 = (1 - a) * u0;
63      double z2 = (1 + aprime) * r[i] * w;
64      phi = atan2(z1, z2);
65  }
66  void Step3(int i, double &al, double phi, double theta_p)
67  {
68      // theta_p is the pitch angle assuming to be 0.0 for this simulation
69      // al is the angle of attack
70      al = (phi - beta[i]*PI/180 - theta_p);
71  }
72  void Step4(double &Cl, double &Cd, double al)
73  {
74      // Calculates cl and cd according to interpolating functions
75      double alpha=al*180/PI;
76      if(alpha<0)  alpha=0;
77      if(alpha>11) alpha=11;
78
```

```
79      if(alpha<6)
80      {
81          Cd=3.1e-05*alpha*alpha*alpha-0.0003*alpha*alpha+0.001*alpha+0.008;
82      }
83      else
84      {
85          Cd=6.323e-05*alpha*alpha*alpha*alpha*alpha-0.002496*alpha*alpha*alpha*
        alpha
86                  +0.039124*alpha*alpha*alpha-0.304439*alpha*alpha+1.1762*alpha
        -1.795;
87      }
88
89      alpha=al*180/PI;
90      if(alpha<0)  alpha=0;
91      if(alpha>21) alpha=21;
92      if(alpha<7.6)
93      {
94          Cl=-0.0008*alpha*alpha+0.115*alpha+0.51;
95      }
96      else
97      {
98          Cl=-0.000403*alpha*alpha*alpha*alpha+0.02227*alpha*alpha*alpha
99                  -0.448*alpha*alpha+3.849*alpha-10.47;
100     }
101     return;
102 }
103 void Step5(double Cl, double Cd, double phi, double &cn, double &ct)
104 {
105     // Calculates cn and ct
106     cn = Cl * cos(phi) + Cd * sin(phi);
107     ct = Cl * sin(phi) - Cd * cos(phi);
108 }
109 double Step6(int i, double cn, double ct, double phi, double & a, double & aprime)
110 {
111     // Obtains a and a'
112     double f = 0, F=0;
113     double tempa2 = 0.0;
114     double Error = 0.0;
115     double tempaprime2 = 0.0;
116
117     // Applies Prandtl's tip loss correction
118     f = (bladenum / 2.0);
119     f *= (R - r[i]);
120     f /= (r[i] * sin(phi));
121     F = (2.0 / PI) * (acos(exp(-f)));
122     double sigma = (bladenum * chord[i]) / (2.0 * PI * r[i]);
123     double k = (4 * F * pow(sin(phi), 2.0)) / (cn * sigma);
124
125     // Applies Glauert correction
126     if (a < 0.2)
127     {
128         tempa2 = 1.0 / (((4 * F * pow(sin(phi), 2)) / (sigma * cn)) + 1);
129         tempaprime2 = 1.0 / (((4 * sin(phi) * cos(phi)) / (sigma * ct)) - 1);
```

```
130        }
131     else
132     {
133         tempa2 = (0.5) * (2 + k * (0.6) - pow((pow((k * (0.6) + 2), 2) + 4 * (0.04
            * k - 1)), 0.5));
134         tempaprime2 = 1.0 / (((4 * sin(phi) * cos(phi)) / (sigma * ct)) - 1);
135     }
136     Error = fabs(tempa2 - a);
137     a = tempa2;
138     aprime = tempaprime2;
139
140     return Error;
141 }
142 void Step7(int i, double ct, double cn, double phi, double a)
143 {
144     // Calculates PT and PN for element i
145     double vrel = (u0 * (1 - a)) / (sin(phi));
146     PT[i] = 0.5 * ct * rho * chord[i] * pow(vrel, 2);
147     PN[i] = 0.5 * cn * rho * chord[i] * pow(vrel, 2);
148 }
149
150 void Step8(int i)
151 {
152     // Performs integration and calculates Momentum and Thrust
153     double A, C, D, E, dM, dT;
154     A = (PT[i + 1] - PT[i]) / (r[i + 1] - r[i]);
155     C = (PT[i] * r[i + 1] - PT[i + 1] * r[i]) / (r[i + 1] - r[i]);
156     D = (PN[i + 1] - PN[i]) / (r[i + 1] - r[i]);
157     E = (PN[i] * r[i + 1] - PN[i + 1] * r[i]) / (r[i + 1] - r[i]);
158     dM = (1.0 / 3.0) * A * (pow(r[i + 1], 3) - pow(r[i], 3)) + 0.5 * C * (pow(r[i
            + 1], 2) - pow(r[i], 2));
159     dT = 0.5 * D * (r[i + 1] * r[i + 1] - r[i] * r[i]) + E * (r[i + 1] - r[i]);
160     Momentum+=dM;
161     Thrust+=dT;
162 }
163
164 int main()
165 {
166     // Defines a and a'
167     double a,aprime;
168     double phi, al, cn, ct, Cd, Cl, Error, ErrorMax = 1.0e-5;
169     int loop;
170     // Initializes the input parameters
171     Init();
172
173     u0=uCut_in;
174     // Loop over wind velocity from uC to uF
175     while(u0<=uCut_off)
176     {
177         Momentum=0.0;
178         Thrust=0.0;
179         int i = 0;
180         double w=Omega*PI/30;
```

```
181        for (i = 0; i < elnum; i++)
182        {
183            Step1(a, aprime);
184            Error=1.0;
185            loop = 0;
186            // in some cases, the algorithm may not converge to a specific
187            // value for a and a'. Therefore, we use loop condition to detect
188            // the problem and stop the code of being locked.
189            while (Error > ErrorMax && loop < 100)
190            {
191                if (loop ==100) printf("not converged for a and a'!");
192                Step2(i,  phi, a, aprime, w);
193                Step3(i,  al, phi, 0.0);
194                Step4( Cl,  Cd, al);
195                Step5(Cl, Cd, phi,  cn,  ct);
196                Error = Step6(i, cn, ct, phi, a, aprime);
197                Step7(i, ct, cn, phi, a);
198                loop++;
199            }
200        }
201        // Calculates Momentum and Thrust for one blade
202        for (i = 0; i < elnum ; i++)
203        {
204            Step8(i);
205        }
206
207        // Calculates total Momentum, Thrust, Power, and cp
208        double Pwind=0.5*1.225*Area*u0*u0*u0;
209
210        Momentum *= bladenum;
211        Thrust *= bladenum;
212        Power=Momentum*w;
213        Cp=Power/Pwind;
214
215        // Outputs the results
216        printf("%lf  %lf  %lf  %lf  %lf  %d\n",u0, Pwind, Power,Thrust, Cp,loop);
217
218        // Increments wind velocity
219        u0+=1;
220    }
221
222    return 0;
223 }
```

Index

Printed in the United States
by Baker & Taylor Publisher Services